Camera Traps in Animal Ecology

Allan F. O'Connell • James D. Nichols
K. Ullas Karanth
Editors

Camera Traps in Animal Ecology

Methods and Analyses

Springer

Editors
Allan F. O'Connell, Ph.D.
Research Wildlife Biologist
U.S. Geological Survey
Patuxent Wildlife Research Center–
Beltsville Lab
10300 Baltimore Avenue
Beltsville, Maryland 20705
USA
aoconnell@usgs.gov

James D. Nichols, Ph.D.
Senior Scientist
U.S. Geological Survey
Patuxent Wildlife Research Center
12100 Beech Forest Road
Laurel, Maryland 20708
USA
jnichols@usgs.gov

K. Ullas Karanth, Ph.D.
Senior Conservation Scientist
Wildlife Conservation Society
Centre for Wildlife Studies
26-2, Aga Abbas Ali Road (Apt: 403)
Bangalore, Karnataka 560042
India
ukaranth@gmail.com

ISBN 978-4-431-54648-1 ISBN 978-4-431-99495-4 (eBook)
DOI 10.1007/978-4-431-99495-4
Springer Tokyo Dordrecht Heidelberg London New York

Softcover re-print of the Hardcover 1st edition 2011

Springer is part of Springer Science+Business Media (www.springer.com)

Preface

This book deals with equipment known as "camera traps," which are used to detect wild animals. The inspiration for this book was the 9th International Mammal Congress held in Sapporo, Japan, in August 2005. At that time, camera traps had become sufficiently popular as a field sampling technique to warrant a separate symposium in the Congress. The ten presentations given in Sapporo that focused on camera traps encompassed a variety of mammal species, ranging from bats to large carnivores, and included surveys and studies from various countries and habitats. The analytical approaches that were used to make inferences about the target populations were somewhat limited, however, and that sparked an interest in the production of this volume, which focuses primarily on (1) the sampling methodologies of using camera traps and (2) the analytical techniques best suited to make sound inferences from photographic evidence.

Over the last two decades, the use of various noninvasive techniques for the sampling of animal populations has increased significantly. Technological advances have allowed practitioners to sample and monitor animal populations without ever physically capturing or handling animals, and to feel confident that the equipment will operate acceptably under a variety of environmental conditions. A sensitivity to animal welfare, reduction of time, effort, and expenses in the collection of scientific data, and the basic motivation to be more efficient have increased our interest in noninvasive sampling methodologies. Noninvasive sampling methods are particularly well suited to animals that are elusive, often occur at low densities, and are difficult to capture or detect. Arguably, the most popular noninvasive sampling technique among those reviewed by Long et al. (2008) is remote photography using camera traps (Kays and Slauson 2008). In short, camera traps are a system or device with a trigger or sensor that activates a camera to take a photograph when an animal is present. The resulting image of occurrence is then used as a piece of scientific information, generating a permanent record of the event. Camera traps are relatively easy to use, and much of their popularity stems from their ability to sample animals remotely with no requirement for humans to manually operate the equipment (except for occasional equipment checks and to retrieve images or film). The cost of purchasing camera traps is becoming more reasonable all the time as technology improves and the number of manufacturers continues to increase.

More than a decade has passed since the publication of the landmark papers by Karanth (Karanth 1995; Karanth and Nichols 1998) that discussed estimation of tiger abundance and density from camera trap data. On the heels of that work, Karanth et al. (2004) explored the potential application of camera traps as scientific tools and addressed questions such as why sample, what to sample, and how to sample rare or elusive animals in an inferential framework. Karanth et al. (2006) later demonstrated how camera trap data can also be used to assess changes in vital rates for tigers over long periods of time. We viewed it as a sign of encouragement for this technique when Karanth et al. (2004), among others in the volume by Thompson (2004), received a favorable review even in a journal devoted to surveys of crypto-zoological creatures such as Bigfoot (Arment 2005)!

Given the notable increase in the scientific use of camera traps over the past several years, we believed that the time was right for a thorough review that details the various aspects of camera trapping, focusing on the estimation techniques that are currently available for analyzing the scientific information (i.e., photographs) and for making strong inferences. Despite the widespread use of camera traps, many of the techniques best suited for data analysis and statistical inference remain somewhat of an enigma to many practitioners. All too often, the concepts underlying the approaches to inference are not presented adequately or are scattered throughout the scientific literature. In this volume, we provide the conceptual frameworks underlying the inference methods themselves, and the manner in which these methods can contribute to the larger endeavors of science and management.

As we went to press, the most recent Mammal Congress (10th) in Mendoza, Argentina, had just passed into our rearview mirror, having hosted another symposium on the use of camera traps in animal ecology. Many of the contributing authors from this volume participated in the symposium, providing a thorough review of what is possible when the modern technology of camera trapping is allied with sound estimation techniques. Thus, it is our hope that this volume will serve as a solid foundation for designing studies and interpreting the data, and also as a flexible springboard for future theoretical and empirical development.

Beltsville, Maryland, USA — Allan O'Connell
Laurel, Maryland, USA — Jim Nichols
Bangalore, India — Ullas Karanth

Literature Cited

Arment, C. 2005. Sampling rare or elusive species: concepts, designs, and techniques for estimating population parameters (Book review). North American Biofortean Review 7:3–4

Karanth, K. U. 1995. Estimating tiger *Panthera tigris* populations from camera trap data using capture–recapture models. Biological Conservation 71:333–338

Karanth, K. U. and J. D. Nichols. 1998. Estimation of tiger densities in India using photographic captures and recaptures. Ecology 79:2852–2862

Karanth, K. U., J. D. Nichols, and N. S. Kumar. 2004. Photographic sampling of elusive mammals in tropical forests. Pages 229–247 *in* W. L. Thompson, editor. Sampling rare or elusive species. Island Press, Washington, DC

Karanth, K. U., J. D. Nichols, N. S. Kumar, and J. E. Hines. 2006. Assessing tiger population dynamics using photographic capture recapture sampling. Ecology 87:2925–2937

Kays, R. W. and K. M. Slauson. 2008. Remote cameras. Pages 110–140 *in* R. A. Long, P. MacKay, W. J. Zielinski, and J. C. Ray, editors. Noninvasive survey methods for carnivores. Island Press, Washington, DC

Long, R. A., P. MacKay, W. J. Zielinski, and J. Ray, editors. 2008. Noninvasive survey methods for carnivores. Island Press, Washington, DC

Thompson, W. L. 2004. Sampling rare or elusive species: concepts, designs, and techniques for estimating population parameters. Island Press, Washington, DC

Acknowledgments

Thanks first go to the folks at Springer Japan who offered symposium organizers in Sapporo an opportunity to publish technical volumes based on symposium topics. We are especially grateful to Aiko Hiraguchi, Kaoru Hashimoto, Motoko Takeda, and Dharmendra Sundardevadoss at Springer for their good nature, open-mindedness, and most of all, their patience in capably handling all the various submissions, resubmissions, and what seemed like an endless number of late-night (or morning, depending on the part of the world) e-mails. We wish also to acknowledge Dr. James Sanderson, who at the time of the Sapporo mammal meetings and for some time before, was working for Conservation International and running camera traps in remote forests around the world (and as far as we know, he still is). Jim recognized that the timing was right to organize and convene a symposium on camera trapping. He first circulated this book offer to participants in the symposium, and he deserves credit for moving this effort off the starting blocks.

The contributors to this volume represent a diverse group of individuals working in literally every corner of the globe; to their credit, they were tolerant and understanding throughout this entire process. That certainly made our jobs easier and, in the end, a rewarding experience. Thanks also to Andrew Gilbert, Arjun Gopalaswamy, Michael Haramis, Tim Jones, Roland Kays, Fred Servello, David Shindle, Ted Simons, Graham Smith, Mathias Tobler, Nimish Vyas, and Arielle Waldstein for their timely and insightful reviews.

Finally, a large measure of debt and gratitude must also go to all those field biologists and practitioners who labor tirelessly operating camera traps. These individuals often work in some of the world's harshest environments and under what can only be described as the most taxing conditions. Their work has done much to further protection and conservation of the world's animal species, some of which are dangerously close to extinction. Their efforts, dedication, and commitment have not gone unnoticed, and we hope that this volume heightens the realization that work with camera traps has important implications for the future of wildlife conservation.

AOC would like to acknowledge the support of the U.S. Geological Survey and the National Park Service's Inventory and Monitoring Program. To Andrew Gilbert and Neil Talancy, who set this journey in motion as a result of their willingness to learn about camera traps, a most gracious "thank you".

JDN would like to acknowledge the support of the U.S. Geological Survey and the U.S. Fish and Wildlife Service.

KUK would like to acknowledge the funding support for his long-term camera-trap studies received from the Wildlife Conservation Society, Panthera Foundation, National Fish and Wildlife Foundation, Liz Claiborne–Art Ortenberg Foundation, U.S. Fish and Wildlife Service, Global Enduro, UK, and the Government of India.

About the Editors

Allan O'Connell is a research wildlife biologist with the U.S. Geological Survey's Patuxent Wildlife Research Center in Maryland. His research focuses on wildlife management issues for U.S. federal resource agencies. His current work includes the design of multiple technique sampling and monitoring programs to assess biodiversity, the use of camera traps to estimate population parameters, and investigating predator ecology to improve the management of endangered species.

James Nichols is a senior scientist with the Patuxent Wildlife Research Center. He is an expert on capture–recapture sampling methods, population modeling, and adaptive management. He has authored or co-authored more than 350 scientific publications, including two books, four edited volumes, and nine monographs, on various aspects of wildlife population ecology. He is a recipient of the 2007 U.S. Presidential Rank Award for Meritorious Service and has received national recognition for his work from various universities, the U.S. Fish and Wildlife Service, U.S. Geological Survey, The Wildlife Society, American Statistical Association, and the U.S. Forest Service.

Ullas Karanth is an internationally known conservation scientist (see *www.wikipedia.org*). Based in India, he is a senior conservation scientist for the Wildlife Conservation Society, where his long-term research has focused on the ecology and conservation of tigers and their prey. He has more than 70 scientific publications to his credit. His work has been featured in the world's media including the *New York Times, Time magazine, National Geographic, BBC, CNN, Discovery*, and others. He is the recipient of the Sierra Club's prestigious international *EarthCare* Award and the World Wildlife Fund's J. Paul Getty Award for Conservation Leadership.

Contents

Chapter 1
Introduction

Allan F. O'Connell, James D. Nichols, and K. Ullas Karanth

1.1 Evolution of Camera Trapping

Cameras (and in a more general sense photography) are a staple of modern society, and from an early age we have learned that these devices can document every aspect of our lives. From a societal perspective, the use of cameras in science has a long history in fields such as astronomy and medicine; amongst varied opinions, photography has been suggested as a demystifying force in nature (Marien 2002). It should come as no surprise then that advances in photography and cameras would inevitably find their way into the conservation arena, and over time, become a preferred tool for sampling animal populations. In recent years, the use of camera traps in the study of wild animals has undeniably improved our understanding of their ecological relationships and more recently, population dynamics. There are now literally hundreds of studies and surveys being conducted that involve camera traps, from urban parks to the most remote jungles. A search of the published literature in the *Web of Science* database recently pulled up 180 citations for "camera traps" over the past 5 years. In addition, the grey literature on this topic probably includes many more hundreds of contributions, especially when one considers the global interest in documenting biodiversity. Camera traps have been used to sample species ranging from the very common [e.g., white-tailed deer *Odocoileus virginianus*, raccoons *Procyon lotor*] to species that are rare, elusive, and often enigmatic [e.g., snow leopards *Uncia uncia*].

The evolution of the camera trap (i.e., remote trigger) as a scientific tool has spanned nearly the entire Twentieth Century (see Chap. 2), but the speed of innovation has varied, depending on societal interests, cultural tendencies, and preference

A.F. O'Connell (✉)
U.S. Geological Survey, Patuxent Wildlife Research Center, Beltsville, MD 20705, USA
e-mail: aoconnell@usgs.gov

J.D. Nichols
U.S. Geological Survey, Patuxent Wildlife Research Center, Laurel, MD 20708, USA

K.U. Karanth
Wildlife Conservation Society – India Program, Centre for Wildlife Studies,
26-2, Aga Abbas Ali Road (Apt: 403), Bangalore, Karnataka 560042, India

A.F. O'Connell et al. (eds.), *Camera Traps in Animal Ecology: Methods and Analyses*,
DOI 10.1007/978-4-431-99495-4_1, © Springer 2011

based on long-held traditions. For example, an increasing interest in animal welfare has prompted an interest in non-invasive sampling techniques (Long et al. 2008). Technological advances in fields such as electronic engineering have benefitted camera traps due to automation of system components, miniaturization, and networked systems, all attributes of modern camera trap systems. These advances have enabled practitioners to use camera traps more efficiently and to sample a variety of species under vastly different environmental conditions. Finally, as camera trap systems have become more efficient and reliable, the costs of purchase, operation, and maintenance have decreased significantly, further stimulating interest.

The development and use of powerful analytical techniques to make appropriate inferences from camera trap data has experienced a similar evolution. However, despite the advantages of estimation methods and probability sampling, these approaches have been slow to become the methods of choice. In contrast, many programs that use camera traps continue to rely on judgment or convenience sampling and then use the resulting data as indices. Although such programs are relatively inexpensive and easy to implement, they typically provide biased parameter estimates. Making valid inferences about populations requires addressing two important sources of variation: detectability and spatial variability (Lancia et al. 1994; Anderson 2001; Pollock et al. 2002). Addressing spatial variability requires some consideration of probabilistic sampling, where data from the sampled sites allow for inferences to the larger area of interest (i.e., study area). The sampling design must also be efficient (in terms of precision) in estimating the parameters of interest, such as rate of change in abundance. Use of indices requires strong assumptions about detectability and the relationship between the count and the parameter of interest (e.g., abundance, density). However, without information about the validity of such assumptions, an index is of little value in making inferences to populations (Thompson et al. 1998, Chap. 6).

Factors such as economics or logistics are frequently used as grounds for not accounting for detectability or spatial variability. Indeed, these factors represent real issues that can limit the scope of most scientific endeavors. However, early consideration of study objectives in the context of either advancing science or informing management is critical for making inferences about the targeted population that are useful to the overall endeavor (Yoccoz et al. 2001; Nichols and Williams 2006). In the case of species such as jaguars *Panthera onca* or tigers *Panthera tigris*, which occur at low densities or have movement patterns that do not lend themselves to traditional sampling methods, investigators first need to consider an appropriate sampling scheme for the area of interest. Research and monitoring of animal populations require an investment in time and effort before field work begins to ensure that these programs can generate reasonable estimates of relevant parameters.

Throughout this volume we have asked that contributors provide the conceptual frameworks that underpin the techniques discussed. We believe that opportunities are lost when biologists rush to set up camera traps without adequate critical thought devoted to their sampling design, program objectives, and expected outcomes. We implore the users of camera traps to explore the "toolbox" of estimation methods for analyzing these data, and we hope that the contents of this volume offer everyone that possibility.

1.2 Book Organization and Chapter Summaries

Camera trapping involves much more than just the analytical approaches discussed previously. A variety of considerations are relevant to the successful conduct of field research or surveys with camera traps. Knowledge of biology of the study species is critical. So too is an understanding of the environmental, social and resource availability contexts in which the camera trap study is to be conducted.

This volume does not provide a broad account of the use of camera traps for photographing animals in the wild. Instead, we have focused more narrowly on the use of camera traps in the conduct of science and management. This focus has led us to organize the volume around specific quantities, information, and associated methodological and analytical approaches. Our purpose is threefold: (1) to recount the story of camera traps, what they were during the early stages of development, what they have become, and what lies ahead, (2) to provide information on how best to use camera traps in the sampling of animal populations, and (3) to provide information on the concepts and technical aspects of the analytical techniques best suited to make reliable inferences from camera trap data. To achieve these goals, we review the historical aspects of how camera trapping has evolved in modern times, starting with the basics of setting up equipment and keeping systems operable. In the early Twentieth Century, maintaining equipment was as much of an issue as documenting a target species. We devote a chapter to the various types of equipment available, because as camera traps have become increasingly popular as tools of scientific study, they are also popular with the sporting public. As a result, there is no shortage of systems or manufacturers available. We have tried to minimize the confusion one might encounter when faced with having to decide which system is best for a particular use. Following these chapters, the remainder of the volume is devoted to synthesizing how camera trap studies can be conducted to contribute to science and/or management. Without first having a clear picture of study objectives in terms of gaining reliable knowledge (Romesburg 1981), sampling with camera traps, as with most other animal sampling programs, can have a predilection to never-ending studies or surveys that provide little in the way of useful information to managers. Use of "targeted" monitoring that is a component of a larger program of science or management is presented as a means of maximizing the chances for success when conducting camera trap studies (Nichols and Williams 2006). Details are provided on the use and application of a number of analytic techniques, some old and others relatively new, that allow the practitioner to estimate important population parameters such as density and abundance. In some cases we start with the basic conceptual framework underlying a technique and then provide case studies as examples. In other chapters, due to the diverse scope of work, we review a specific topic (e.g., behavior) in terms of investigations that have been conducted with camera traps, hoping to leave the reader with a broad, but in-depth overview.

The uses of camera traps in wildlife ecology can be viewed as falling along a continuum from basic inventories, where the objective is to document the occurrence of a species (e.g., Gilbert et al. 2008), to the use of photographic images as a basis for constructing statistical models designed to assess population status (i.e., abundance, density) (Karanth and Nichols 1998) and investigate population dynamics (Karanth et al. 2006). Analogous work has been completed with camera traps in situations where individuals cannot be identified, but the importance of detectability is considered when making inferences about population status and change (Zielinski et al. 1997; O'Connell et al. 2006). This volume is focused on the portion of the continuum that includes technical topics contributing to science and management. We discuss the current state of the art and also try to peer into the future. In anticipation of what we have pulled together on the following pages, we provide below brief summaries of each chapter. If the reader has interest in only a specific topic or technique, this summary information may focus and direct that interest to a specific chapter. On the other hand, if one views this volume as a primer for scientific and management uses of camera traps, then these summaries provide a synthetic picture.

In Chap. 2, Kucera and Barrett review the history of cameras traps, dating back more than 100 years to a time when taking pictures of animals was simply referred to as "wildlife photography". They discuss the key individuals who led the transition from the large bulky cameras and cumbersome flash equipment of the early Twentieth Century (and even earlier), to the simple, remote trigger systems of the early 1900s, through to the technological advances of the mid 1900s that reduced the size of cameras, batteries, and other system components. This chapter also includes a review of the various types of studies for which these systems were used (e.g., nest predation), the various animal groups studied, and the unique sites where practitioners deployed camera traps. The authors take us through the various stages of how camera trapping transitioned from photography as an art form used by the Zoological Photographic Club in Britain to the cutting-edge science of recent years that has now set the stage for camera traps as a scientific technology of the future.

In Chap. 3, Swann, Kawanishi, and Palmer review the modern-day equipment used in camera trapping. The popularity of the technique has resulted in a variety of manufacturers and no shortage of equipment options. Even for the most experienced field personnel, the myriad of camera traps now available for purchase can lead to multiple choices for virtually any system component: sensors, triggers, images, and power supplies. In addition, the authors offer some guidance on ideas related to different system conditions, types of studies, and target species. The chapter also offers the reader a primer on the technology and principles underlying the application of remote photography in modern ecological research.

In Chap. 4, Nichols, Karanth and O'Connell narrow the discussion to camera trapping designed to inform science or management. They propose that the information gained from camera trapping is most useful when it is focused on increasing the knowledge base for science or assisting managers in making informed decisions about the resources they are charged with managing. The chapter promotes critical thinking in terms of constructing and discriminating among hypotheses, collecting data, developing models and deducing associated predictions, and using estimation methods to make strong inferences about the target resources. The discussion turns to the utility of formal decision-making processes (i.e., structured decision-making) and the four components of such processes: objectives, management actions, models and monitoring. Structured decision-making approaches should be useful in helping managers to make informed decisions in the face of different types of system uncertainty. Monitoring programs serve critical roles in structured decision processes, and a clear recognition of these roles is very useful in designing associated sampling programs, including programs based on camera traps. Whether the goal is science or management, this chapter sets the stage for use of the estimation methods discussed in most of the chapters that follow.

In Chap. 5, Bridges and Noss review how camera traps have been used to study animal behavior or ethology. The authors review various behavioral topics studied with camera traps such as nest predation, foraging, circadian rhythms, sociality and niche partitioning, reproduction, and habitat use. The chapter provides an overview of each of these topics, a summary of what has been published, and descriptions of how camera trap data in behavioral studies can be analyzed. The authors also discuss the benefits of using remote sampling systems to gain insights to behavior that contrast with insights based on more traditional approaches (e.g., radio telemetry). The reader can expect to become familiar with the scope of topics in behavioral ecology that have been studied using camera traps and to gain some understanding as to what will be possible in the future.

In Chap. 6, O'Brien reviews the concepts behind the use of camera trap data to estimate abundance, density, and relative abundance of animal populations. The main focus of this chapter is on the use of capture–recapture (CR) techniques to provide inferences about the abundance of a population for a single species. The chapter discusses the fundamental concepts underlying CR estimation, emphasizing the importance of spacing (i.e., spatial variability) of camera traps, and differences in the designs of studies providing data to be used with closed, open, and mixed time scale models. The reader is taken through the evolution of the various approaches used to estimate density and relative abundance, two topics with long and varied histories, especially with camera traps. Several different methods that are used to determine density from photographic information are reviewed, from less formal approaches (e.g., mean maximum distance moved (MMDM), nested grid analysis) to the more formal (and recent) spatially explicit CR models based on maximum likelihood and Bayesian inference methods. Given the widespread use of camera trap data for developing indices of relative or indirect abundance, this chapter closes with a brief discussion of a topic that will be of interest to many individuals. In summation, this chapter is much more than a simple compendium of methods, as O'Brien offers an informed opinion about the relative utilities of different approaches.

In Chap. 7, Karanth, Nichols, and Kumar discuss the use of "closed" CR models to estimate abundance of tiger populations. The use of closed models involves a relatively short sampling period, the assumption being that no changes occur in the target population (e.g., births, deaths, immigration, or emigration) during the study period. The authors provide details on all aspects of conducting a study of tigers using camera traps and discuss the various issues confronted when sampling a large, rare, wide-ranging species. A thorough review is provided, including recommendations for equipment and field practices, choice of trap sites, survey design and trap placement, and data analyses.

In Chap. 8, Maffei, Noss, Silver, and Kelly summarize the camera trap surveys used to estimate abundance and density of the jaguar, the largest felid found in the Americas and probably the most intensively studied anywhere. Under mounting pressure due to habitat loss and fragmentation, poaching, etc., the jaguar has been the focus of >80 individual studies, some of which are still in progress with others planned. The authors review the methodological differences among these surveys and the difficulties in developing sampling designs that properly deal with spatial variability. Despite great interest in this elusive species, the authors recommend treating these initial surveys as "preliminary", considering the rapid development of analytical techniques to address issues of spatial variability and density estimation (see Chap. 10).

In Chap. 9, Karanth, Nichols, Kumar, and Jathanna review the use "open" CR models for assessing the population dynamics of tigers. Under these models, vital rates can be estimated, as the models can accommodate changes in the sampled populations (e.g., birth, deaths, emigration, immigration) across time and space. In addition to capture probabilities, modeling data from open populations requires survival probability parameters. The authors discuss the various analytical requirements for the open model approach of Cormack–Jolly–Seber (Leberton et al. 1992) and Pollock's robust design (1982), which combine open and closed population modeling. The various issues surrounding model selection are also presented. Finally, the authors use their 9 year camera trap survey of tigers in Nagarahole Park, India, to demonstrate the utility of the robust design approach.

In Chap. 10, Royle and Gardner advance the use of spatially explicit CR models to estimate the density of a population using camera traps. They develop an extension of the classical closed CR model by modeling the observation process (i.e., encounters of animals with traps) as a function of both trap locations and the spatial distribution of animals. Key to this approach is the use of a Bayesian framework with data augmentation, a technique originally developed to deal with individual covariates in closed CR models (Royle et al. 2009). We believe that this work formally links the encounter information collected in camera trap arrays to the concepts of home range and territory. Spatially explicit CR models represent an important new methodological advance, and we expect rapid development of this approach in the future.

In Chap. 11, O'Connell and Bailey present the basics of occupancy estimation (see also Chaps. 12 and 13), a technique that simultaneously estimates the probabilities of occurrence and detection for a species. Basic approaches are briefly reviewed including inferences based on single and multiple seasons of data, single and multiple species, use of multiple methods (that include camera traps), and species interactions. The relationship of occupancy to abundance/density is also briefly discussed, along with the role of camera trapping in large-scale monitoring programs that use occupancy as a state variable. Relatively new models are reviewed for which camera traps have potential as a useful sampling technique: habitat occupancy dynamics, multiple states and multiple scales. Study design considerations and software options are also provided [e.g., GENPRES (Bailey et al. 2007)].

In Chap. 12, Kéry deals with the difficult issue of estimating size, composition, and the dynamics of entire communities (i.e., species richness). In the context of this chapter, species replace individuals. The chapter begins with a review of the options available for estimating species richness, with a focus on the basic framework of CR and occupancy approaches to community inference. Kéry notes that closed CR models designed to estimate the abundance of a population can also be used to estimate species richness. In the context of this chapter, species replace individuals. The utility of CR models stems from the fact that they accommodate imperfect detection as part of the observation process that generates the counts of species. Occupancy-based estimation approaches (Chap. 11) are described for use with single or multiple sites and with closed or open systems. Model construction and options (e.g., robust design, hierarchical frameworks), inferential capabilities, and specific instances where model assumptions may not hold are meticulously presented. Sampling design considerations are reviewed (i.e., spatial variability), and various software options [e.g., COMDYN (Hines et al. 1999)] are provided.

In Chap. 13, O'Brien, Kinnaird, and Wibisono use some of the approaches presented in Chap. 12 for estimating species richness and related parameters. They use the same maximum likelihood methods and CR models presented earlier on a camera trap survey in Bukit Barisan Selatan National Park in Indonesia. The objective was to produce estimates of species richness for medium- and large-sized, terrestrial and semi-terrestrial mammals, and the four largest species of terrestrial birds. The chapter includes a discussion of the influence of rare species on the precision of estimates, as well as the pitfalls one encounters with community-level inference. In the end, camera trapping was deemed a successful technique for estimating species richness, but require adequate sampling effort and spatial coverage.

In Chap. 14, Nichols, O'Connell, and Karanth look at what the future may hold for camera traps. Prospects and updates for equipment and technological advances are reviewed, along with development of websites and cooperative databases for the sharing of camera trap data. Recommendations are provided on how camera trap data might be used in new ways to improve estimates of population parameters discussed previously in this volume. The authors discuss the potential for supplementing camera trap data with information gained from other techniques such as DNA analyses and radio telemetry. The potential for multispecies occupancy modeling and additional inference about community dynamics is also discussed.

The counting of biological organisms has a rich history of methods development, with many key advances in field sampling and treatment of the resulting data (Elphick 2008). We believe that camera traps will ultimately be viewed as an important methodological development for counting and detecting animals, and for making inferences about populations and communities and the changes they undergo. Looking forward, we hope that this volume contributes in some small way to the evolution and development of camera trap methods and analytical options.

References

Anderson, D. R. 2001. The need to get the basics right in wildlife field studies. Wildlife Society Bulletin 29:1294–1297

Bailey, L. L., J. E. Hines, J. D. Nichols, and D. I. MacKenzie. 2007. Sampling design trade-offs in occupancy studies with imperfect detection: examples and software. Ecological Applications 17:281–290

Elphick, C. S. 2008. How you count counts: the importance of research methods in applied ecology. Journal of Applied Ecology 145:1313–1320

Gilbert, A. T., A. F. O'Connell, Jr., E. M. Annand, N. W. Talancy, J. R. Sauer, and J. D. Nichols. 2008. An inventory of terrestrial mammals at National Parks in the Northeast Temperate Network and Sagamore Hill National Historic Site. U.S. Geological Survey, Reston, Virginia. Scientific Investigations Report 2007–5245. 158 pp

Hines, J. E., T. Boulinier, J. D. Nichols, J. R. Sauer, and K. H. Pollock. 1999. COMDYN: software to study the dynamics of animal communities using a capture–recapture approach. Bird Study 46(suppl.):S209–217

Karanth, K. U. and J. D. Nichols. 1998. Estimation of tiger densities in India using photographic captures and recaptures. Ecology 79:2852–2862

Karanth, K. U., J. D. Nichols, N. S. Kumar, and J. E. Hines. 2006. Assessing tiger population dynamics using photographic capture recapture sampling. Ecology 87:2925–2937

Lancia, R. A., J. D. Nichols, and K. H. Pollock. 1994. Estimating the number of animals in wildlife populations. Pages 215–253 *in* T. Bookhout, editor. Research and management techniques for wildlife and habitats. The Wildlife Society, Bethesda, MD

Lebreton, J. D., K. P. Burnham, J. Clobert, and D. R. Anderson. 1992. Modeling survival and testing biological hypotheses using marked animals: a unified approach with case studies. Ecological Monographs 62:1–118

Long, R. A., P. MacKay, W. J. Zielinski, and J. C. Ray, editors. 2008. Noninvasive survey methods for carnivores. Island Press, Washington, DC

Marien, M. W. 2002. Photography: a cultural perspective. Harry N. Abrams, Inc., New York, NY

Nichols, J. D. and B. K. Williams. 2006. Monitoring for conservation. Trends in Ecology and Evolution 21:668–673

O'Connell, A. F. Jr., N. W. Talancy, L. L. Bailey, J. R. Sauer, R. Cook, and A. T. Gilbert. 2006. Estimating site occupancy and detection probability parameters for mammals in a coastal ecosystem. Journal of Wildlife Management 70:1625–1633

Pollock, K. H. 1982. A capture–recapture design robust to unequal probability of capture. Journal of Wildlife Management 46:757–760

Pollock, K. H., J. D. Nichols, T. R. Simon, G. L. Farnsworth, L. L. Bailey, and J. R. Sauer. 2002. Large scale wildlife monitoring studies: statistical methods for design and analysis. Envirometrics 13:105–119

Romesburg, H. C. 1981. Wildlife science: gaining reliable knowledge. Journal of Wildlife Management 45:293–313

Royle, J. A., J. D. Nichols, K. U. Karanth, and A. Gopalaswamy. 2009. A hierarchical model for estimating density in camera-trap studies. Journal of Applied Ecology 46:118–127

Thompson, W. L., G. C. White, and C. Gowan. 1998. Monitoring vertebrate populations. Academic, San Diego, CA

Yoccoz, N. G., J. D. Nichols, and T. Boulinier. 2001. Monitoring of biological diversity in space and time. Trends in Ecology and Evolution 16:446–453

Zielinski, W. J., R. L. Truex, C. V. Ogan, and K. Busse. 1997. Detection surveys for fishers and American martens in California, 1989–1994: summary and interpretations. Pages 372–392 *in* G. Proulx, H. N. Bryant, and P. M. Woodard, editors. Martes: taxonomy, ecology, techniques, and management. Provincial Museum of Alberta, Edmonton, AB, Canada

Chapter 2
A History of Camera Trapping

Thomas E. Kucera and Reginald H. Barrett

2.1 Introduction

The human desire to observe wild animals without disturbing them goes back at least to hunter-gatherers who constructed blinds. Our ability to do so was greatly enhanced with the development of photography and other, even more recent, innovations such as small, portable batteries, electric lights, and digital equipment. These technologies allow us to make undisturbed observations on a wide variety of wildlife, in a wide variety of habitats, at all hours, and under the most challenging of conditions. Our early ancestors were motivated by a desire for animal products. Today, desires for undisturbed observations of wildlife range from recreation and an aesthetic appreciation of nature to increasing our scientific understanding of animal populations and their relationship to their environment.

Modern photographic equipment, camera-triggering devices, and compact power sources allow us unprecedented, unobtrusive access into wildlife habitats using automated camera traps. Even people with no scientific training can now address simple questions such as "What animal is in my backyard at night?" Wildlife scientists are using modern remote camera equipment to answer more sophisticated questions such as "What animal species occur in a certain area?", "What are they doing?", and even "How many are there?" Detecting cryptic or rare species, delineating species distributions, documenting predation, monitoring animal behavior, and estimating population size and even vital rates are topics that are now being addressed by scientists using remote photography. Such pictures can be worth much more than words alone. This review will briefly describe the development and use of remote photographic equipment up to the refinement of techniques for quantitatively assessing the demographics of wildlife. This last topic is treated in various chapters in the current volume.

T.E. Kucera (✉)
BIR Associates, 22 Reservoir Road, San Rafael, CA 94901, USA
e-mail: tom_kucera@hotmail.com

R.H. Barrett
Department of Environmental Science, Policy, and Management, University of California, Berkeley, CA 94720, USA

A.F. O'Connell et al. (eds.), *Camera Traps in Animal Ecology: Methods and Analyses*, DOI 10.1007/978-4-431-99495-4_2, © Springer 2011

2.2 Early Developments

Photography was invented and refined in the Nineteenth Century (Newhall 1982). Heavy, bulky equipment and slow film and lenses notwithstanding, the new technique was soon applied to photographing nature. Guggisberg (1977) described one of the first successful attempts to photograph wild animals by Professor G. Fritsch, a German explorer in South Africa in 1863. In another instance, one of the earliest examples of "endangered species" photography, a captive quagga *Equus quagga* was photographed at the London Zoo in the early 1870s; by that time it had already become extinct in the wild. In 1870, Charles A. Hewins of Boston produced a photo of a white stork *Ciconia ciconia* on a nest at Strassburg. One of the earliest uses of wildlife photography for scientific purposes was during 1872–1876 on an oceanographic voyage by the English vessel *HMS Challenger*. On this expedition, C. Newbold, a corporal with the Royal Engineers, photographed rookeries of rockhopper penguins *Eudyptes chrysocome* and breeding albatrosses *Diomedia* spp.

Wildlife photography became popular in the late Nineteenth Century. According to Guggisberg (1977), by the year 1900 there were four million camera owners in Britain; the Zoological Photographic Club was founded in 1899. Technological advances resulted in smaller, more portable cameras. The "Bird-land Camera" was a type of reflex camera developed by English bird photographer Oliver Pike in the early 1900s and marketed as "Specially designed for Natural History Photography". In the United States, A. G. Wallihan (1906) published "Camera shots at Big Game," a collection of photographs of elk *Cervus elaphus*, mule deer *Odocoileus hemionus*, pronghorn *Antilocapra americana*, mountain lions *Felis concolor*, bobcats *Lynx rufus*, and other wildlife taken in the Rocky Mountains; the book's introduction was by Theodore Roosevelt.

These early wildlife photographs were taken by the photographer manually releasing a shutter. Technological developments that produced much faster shutter speeds allowed Eadweard James Muybridge in 1878 to line up a dozen cameras and have them triggered by a horse breaking strings as it galloped past. This not only demonstrated that all four feet of a horse are off the ground at certain points in a gallop, but was the beginning of a rigorous understanding of animal locomotion, and ultimately led to the development of motion pictures (Guggisberg 1977; Newhall 1982). This was also one of the first examples of an animal taking its own picture.

George Shiras in the 1890s was the first to develop a method using a trip wire and a flash system in which wild animals photographed themselves. His "flashlight" photographs won a gold medal at the 1900 Paris World Exhibition and were published in National Geographic Magazine (Guggisberg 1977; Shiras 1906, 1908, 1913. Shiras recorded numerous wildlife species with trip wires, including American mink *Mustela vison*, raccoons *Procyon lotor*, white-tailed deer *O. virginianus*, North American porcupines *Erithizon dorsatum*, muskrats *Ondatra zibethicus*, snowshoe hares *Lepus americanus*, striped skunks *Mephitis mephitis*, American beavers *Castor canadensis*, black and turkey vultures *Coragyps atratus* and *Cathartes aura*, northern bobwhite quail *Colinus virginianus*, cardinals *Cardinalis cardinalis*, Eastern gray squirrels *Sciurus carolinensis*, Virginia opossums *Didelphis virginiana*, gopher tortoises *Gopherus polyphemus*, caribou *Rangifer tarandus*, moose *Alces alces*, grizzly bears *Ursus arctos*, and elk. Shiras was successful in photographing so many wild species in part because of the variety of methods he developed to induce the animal to pull the trip wire. For example, he often used bait tied to the trip wire that attracted animals and induced them to pull on it, such as cheese for photographing raccoons and carrion for vultures. He also placed the wire across likely travel routes to photograph elk. Shiras used a particularly clever way to photograph a beaver. He tied the trip wire to a dislodged stick in the beaver's dam; at night, when the beaver repaired the dam, it took its own picture.

In the early decades of the Twentieth Century, there were several other successful attempts around the world to have animals to take their own pictures. The German sportsman and photographer Carl Georg Schillings adapted Shiras' methods to the wildlife of East Africa in 1903 and 1904. Using bait such as a live donkey, and photographing at waterholes, Schillings (1905, 1907a, b) produced spectacular photographs of many wildlife species including African lions *Panthera leo*, leopards *P. pardus*, spotted hyenas *Crocuta crocuta*, and jackals *Canis* sp., all taken by the subjects themselves. William Nesbit (1926) published the first detailed guide to outdoor photography, and stated that "flashlight trap photography," where a wild animal takes its own picture by tripping a wire, "is a most fascinating sport and is deservedly becoming more and more popular" (Nesbit 1926:62). He acknowledged the assistance of and included photos by Frank Chapman, William T. Hornaday, and George Shiras, the last of whom he described as "the father of this class of animal photography" (Nesbit 1926:303), and included brief biographies and literature citations of a "Who's who in nature photography." The book provided detailed descriptions of camera equipment, baits to attract different animals, high-speed flash apparatus, and trip wires to release the shutter. Nesbit also published a photo of the first wild tiger *P. tigris* taken with this apparatus, by F. W. Champion of the Indian Forest Service. Champion (1928, 1933) subsequently published several books describing his experiences and including many photographs of tigers and other animals such as leopards *P. pardus*, leopard cats *Felis bengalensis*, jungle cats *F. chaus*, fishing cats *F. viverrinus*, striped hyenas *H. hyaena*, sloth bears *U. ursinus*, and ratels *Mellivora capensis*. In Michigan, Harris and DuCharme (1928) used Nesbit's apparatus, and some they made themselves, to photograph beavers and other animals using trails made by beavers.

In a purely scientific context, Frank M. Chapman, Curator of Ornithology at The American Museum of Natural History in New York, worked with trip wires and bait to document the species present on the then-recently established research island of Barro Colorado in Panama. In his "census of the living" (Chapman 1927:332), using Nesbit's apparatus, he successfully photographed mountain lions, ocelots *Leopardus pardalis*, white-lipped peccaries *Tayassu pecari*, Baird's tapirs *Tapirus bairdii*, and coatimundis *Nasua* sp. in the tropical forest. This is likely the first explicit attempt to document the species present in an area with remote photography. Chapman also discussed distinguishing individual animals in the photographs;

based on one animal's markings, he concluded that he had several photographs of the same mountain lion and at least one different individual in another photograph. He also made inferences about the animals' behavior. For example, he noted that several of the cats seemed to be aware of the trip wire and attempted to step over it; the peccaries showed no such awareness. These themes of recognizing individuals and observing animal behavior have been developed greatly in more recent years.

Another early developer of the animal-triggered remote camera was Tappan Gregory, an attorney from Chicago. Gregory (1927) described taking remote photographs of a porcupine and a white-footed mouse *Peromyscus leucopus*, using a trip wire to discharge a flash. He subsequently developed more sophisticated methods with which he successfully recorded photographic images of a wide variety of North American wildlife (Gregory 1930), and worked on scientific endeavors with the U. S. Bureau of Biological Survey, Chicago Academy of Sciences, Smithsonian Institution, and the National Zoo. On scientific expeditions, using the camera traps he developed, he obtained photographs of wolves *Canis lupus* in Louisiana in 1934 and mountain lions in northern Mexico in 1937. Gregory (1939) published detailed plans of his camera traps, and discussed at length their operation, including mounting them on a tree, setting up a field darkroom, and safety issues regarding the use of magnesium flash powder. Stanley P. Young (1946) of the Bureau of Biological Survey, who lead the expedition to Mexico, used several of the mountain lion photographs in his book, and discussed the use of catnip oil to attract the animals to a treadle that, when stepped on, operated the camera.

2.3 The Modern Era

By the mid-Twentieth Century, smaller photographic equipment and the replacement of the clumsy and dangerous magnesium flash powder with flash bulbs allowed further refinement of remote wildlife photography. Several plans for remote cameras to record wildlife activity were published during this time. Gysel and Davis (1956) described an inexpensive photographic unit powered by a 6-V battery that operated when an animal pulled on bait attached to a string. In a somewhat cumbersome sequence of events involving two knife switches, a solenoid, and a modified mouse trap, a single photo was taken by a camera with a synchronized-flash unit. Designed to be housed in a wooden box, this system reportedly performed well in all seasons in Michigan. Gysel and Davis (1956) photographed eastern fox squirrels *Sciurus niger* taking seeds in a study of forest trees, a striped skunk taking a dead rabbit from a trap, and red squirrels *Tamiasciurus hudsonicus* and blue jays *Cyanocitta cristata* taking mourning dove *Zenaida macroura* eggs in a nest predation study. By placing the trip wires across den entrances, they identified the size of foxes using den sites, and determined which species used different kinds of ground dens.

Pearson (1959, 1960) designed a photographic system to monitor the activity patterns of small mammals, particularly California voles *Microtus californicus*, in runways in California. His system employed a 16-mm movie camera, operated one frame at a time so that several hundred exposures could be made without resetting the system. Pearson (1959) described two triggering systems for his cameras, neither of which used a trip wire. In one system, a treadle placed in the runway closed an electric switch when a mouse ran across it and caused a photograph to be taken. The other used a beam of deep red light that was positioned across the runway such that when interrupted by an animal, an exposure was made. He included a clock, ruler, thermometer, and hygrometer in the field of view of the camera. By using ear-tags and patterns of clipping fur, Pearson (1959) was able to recognize individual mice over time. Most photographs were of voles and western harvest mice *Reithrodontomys megalotis*, but he also identified 26 other species of mammals, birds, and lizards in his photographs. He was able to go beyond simple species identification, however, and described daily and annual activity patterns of the two mouse species as well those of brush rabbits *Sylvilagus bachmani* and shrews *Sorex* spp., and he described effects of temperature and relative humidity on the activity of shrews and western fence lizards *Sceloporus occidentalis*.

Other investigators used equipment based on that described by Pearson (1959). Using the treadle placed in runways, Osterberg (1962) studied the activity patterns of northern short-tailed shrews *Blarina brevicauda* and meadow voles *M. pennsylvanicus* in Michigan, and related them to weather, time of day, and season. Buckner (1964) used the design employing the light beam positioned across the runway to release the shutter. Working in a tamarack *Larix laricina* bog in Manitoba, he photographed nine small mammal species, and contrasted the daily activity patterns of snowshoe hare, red squirrel, and red-backed vole *Clethrionomys gapperi*. He adapted the system to operate from a 6-V car battery, increasing its portability, and suggested that the system might be of use in "…obtaining seasonal population estimates of small mammals" (Buckner 1964:79).

Dodge and Snyder (1960) presented detailed plans for a more portable remote camera system that, unlike the one described by Pearson (1959), did not require 110-V A.C. power but operated off a 6-V car battery and allowed multiple exposures without resetting the apparatus. Their design incorporated a light beam that when broken by the body of an animal activated a solenoid connected to the camera's shutter. They also used a movie camera that advanced one frame each time the shutter was activated, thus allowing a series of pictures to be taken. Abbott and Dodge (1961) used a similar apparatus in a study of forest seed predation. Abbott and Coombs (1964) described an even more portable device that used a 35-mm camera with a bulk film magazine that allowed up to 420 exposures, rather than the usual 36, and thus could be left in the field longer without changing film. The 35-mm film produced larger negatives than the 16-mm movie cameras used in the earlier designs. Powered by 6-V motorcycle batteries, this unit weighed 22 kg. Winkler and Adams (1968) developed a movie camera system to study the activity of terrestrial carnivores around bat caves. This system employed an automobile battery, four 100-W aircraft landing lamps, and a photoelectric-cell trigger.

Winkler and Adams (1968) were able to photograph 31 separate 2-sec movie sequences per roll of film, and identified raccoons and striped skunks as they entered and exited bat caves.

Although much of this earlier work focused on mammals, remote camera systems were also developed for avian research. Cowardin and Ashe (1965) described a system to count waterfowl that employed a 35-mm half-frame camera that took 72 exposures. It was controlled by a timer that took pictures every 15 min. They placed the cameras in randomly selected quadrats in different marsh habitats to estimate waterfowl use. Temple (1972) developed a time-lapse photographic system to observe the nesting behavior of peregrine falcons *Falco peregrinus*. He used an inexpensive Super-8 movie camera attached to an electronic timer. With a capacity of 3,600 frames on a roll of Super-8 film, the camera could be left in place for days without changing film. Because this system did not function at night, no flash capability was required, and thus battery requirements were minimal. The system weighed 4 kg. Diem et al. (1973) described camera systems using either a Super-8 or 35-mm camera that could withstand the rigors of a Wyoming winter. Although more expensive than the Super-8 cameras, the 35-mm cameras allowed the use of telephoto and wide-angle lenses. The cameras were attached to an intervalometer and took a picture at intervals from 5 to 15 min. They were used in studies of breeding colonies of California gulls *Larus californicus* and American white pelicans *Pelecanus erythrorhynchos*, as well as big game and livestock grazing and large-mammal movements across highways. Powered by a 6-V battery, the systems weighed between 2.2 and 5.8 kg, and thus were substantially more portable than earlier designs, and operated in temperatures as low as –35°C. Goetz (1981) developed a remote photographic system to study predation on wild turkey *Meleagris gallopavo* nests using a Polaroid camera that had an automatic flash, exposure control, and film advance and contained its own power supply in the film pack. He modified the camera to be triggered through a microswitch beneath the nest platform, and reported excellent results under all light conditions. An obvious advantage of such a system is that the exposed film is available immediately. The system as described was limited to ten pictures using flash. An inherent limitation on using Polaroid film is low temperature inhibiting the chemical developing process; it would have unlikely been useful in winter temperatures below freezing.

Echoing the work of Chapman (1927) in the Neotropics, Seydack (1984) described the operation of a 35-mm camera system to census rainforest mammals in South Africa. He connected a trip plate placed on a trail to an autowinding camera and flash; a photo was taken when an animal weighing 2 kg or more stepped on the plate. The camera was powered by a 6-V battery and had a flash capacity of 16 bulbs. He deployed six camera systems systematically along paths within 100-ha survey blocks. Seydack (1984) left the cameras out for 1 month, and then moved them to the next survey block. He repeated this procedure six times over 3 years. He detected 14 species, and made estimates of population density for bushbuck *Tragelaphus scriptus*, identifying at least 61 individuals by coat pattern and, in males, horn morphology. He could also recognize individual leopards by their patterns of spots and honey badgers *Mellivora capensis* by differences in their white lateral stripe. Seydack (1984) grouped the species he detected into: (1) those that are individually recognizable and thus for which a density estimate may be calculated; (2) those not individually recognizable but, like the African porcupine *Hystrix cristata* and large-spotted genet *Genetta trigrina*, are relatively abundant, and (3) those not individually recognizable but are either rare or difficult to detect due to a behavioral characteristic. He concluded that there is "…a great potential for the photo-recording census technique as a versatile tool of quantitative research and general wildlife censusing" (Seydack 1984:14).

Hiby and Jeffery (1987) and Nicholas et al. (1991) used remote photographic systems to record the presence of Mediterranean monk seals *Monachus monachus* at haul-out sites in caves on the Greek island of Kefallinia. Because these rare seals are particularly sensitive to human disturbance, remote photography seemed appropriate to detect seals' use of caves. They used automatic 35-mm cameras, operated by a trip wire made of fishing line, attached to the walls of suspected haul-out caves. They identified four individual Mediterranean monk seals using the caves.

Carthew and Slater (1991) described an automatic photographic system that employed a pulsed infrared beam as a triggering device. When the beam is intercepted by an animal, the infrared sensor sends a signal to a modified automatic, 35-mm camera with a dedicated flash, automatic exposure control, and a quartz data-back to record date and time on each frame. They used this system to observe animals passing along trails or the tops of logs, and to identify diurnal and nocturnal pollinators visiting flowering plants in Australia. Griffiths and Van Schaik (1993a) noted the utility of remote cameras in studying rainforest animals. They used remote photography to document the changed activity patterns and avoidance of areas used by humans by a variety of larger mammals in Sumatra (Griffiths and Van Schaik 1993b).

Mace et al. (1994) devised a remote photographic system for use in a systematic survey of grizzly bears in Montana. They adapted an automatic, 35-mm camera to be activated by a microwave motion and a passive infrared heat sensor. Using blood as an attractant at systematically deployed survey stations over 817 km^2, they photographed grizzly and black bears *U. americana* as well as 21 other species of wildlife, documented grizzly bear distribution, and ultimately were able to generate estimates of the abundance of grizzly bears in their study area.

2.4 Forest Carnivores

In the early 1990s there was an increasing awareness among wildlife managers in the United States that the conservation status of a suite of small and mid-sized carnivores, including the American marten *Martes americana*, fisher *M. pennanti*, wolverine, and lynx, was of concern. An ad hoc group of federal and state agency biologists and university researchers formed the Western Forest Carnivore Committee to gather what information existed on these species and to develop reliable, non-lethal methods to detect their presence. One issue that immediately presented itself was assessing the distribution of these shy, low density species. Because trapping them had been illegal for decades in most states, there was no recent reliable information on their occurrence throughout most of their historic range. During this period, Fowler and Golightly (1993) and Jones and Raphael (1993) developed and deployed inexpensive, 110-size cameras for field surveys of forest carnivores. Reminiscent of the system deployed by Shiras and Champion nearly a century earlier, these cameras operated when an animal pulled on bait attached by a line to the camera's shutter release. They allow only one photograph to be taken without resetting the camera, and their utility is limited by severe weather and snow. Kucera and Barrett (1993) described the use of the commercially available Trailmaster® remote camera systems for detecting wildlife. With features similar to those described by Carthew and Slater (1991), the Trailmaster® comprises an automatic, 35-mm camera triggered when a pulsed infrared beam deployed over bait or across a trail is broken (see Swann et al., Chap. 3). Kucera and Barrett (1993) and Kucera (1993) used these systems to document the contemporary distribution of rare and shy carnivores in remote areas of California. Data from these remote camera stations combined with those from sooted-track-plate surveys formed the basis for describing the first contemporary distribution of fishers (Zielinski et al. 1995) and American martens (Kucera et al. 1995) in California since the work of Grinnell et al. (1937).

Remote photographic techniques also played a large part in describing non-lethal methods to generate reliable distribution data on a variety of rare carnivores, which was developed from efforts of the Western Forest Carnivore Committee (Zielinski and Kucera 1995) These authors also discussed the strategy behind designing surveys for rare carnivores at both relatively small and larger regional levels, and provided guidelines for conducting such surveys and detailed instructions for using the equipment. This document provided general guidance for developing survey protocols for carnivore surveys throughout western North America and served as a guide for practitioners everywhere attempting to use cameras in the study of wildlife populations.

2.5 Expanding Applications

Several investigators since Goetz (1981) have employed remote photography to investigate avian nest predation. Laurance and Grant (1994) and Major and Gowing (1994) identified nest predators of birds in Australia using different designs of remote cameras built specifically for them. Laurance and Grant (1994) identified nine species, including mammals, birds, and reptiles, visiting the artificial ground nests, and concluded that white-tailed rats *Uromys caudimaculatus* were the most common predator. Major and Gowing (1994), using a somewhat different apparatus to study predation on the nests of a tree-nesting passerine, identified the most important predator as the black rat *Rattus rattus*. Leimgruber et al. (1994) studied nest predation with infrared-triggered cameras at artificial nests in forests blocks of different size in Virginia. They found 13 species preying on nests, and related predation rates more to vegetation structure than to the size of the block of forest. They also suggested that simply removing a few larger predators such as striped skunks and raccoons from a diverse predator community would have little effect on nest predation. Danielson et al. (1996) described another design for a remote camera to photograph nest predation events. They constructed a system in which an egg was placed on a microswitch; a photograph was taken when the egg was moved.

Through the 1990s, remote photography was being used in an increasing variety of studies. Sadighi et al. (1995) used the Trailmaster® system to monitor timber rattlesnakes *Crotalis horridis* in Massachusetts. They were able to recognize one individual through a scar on its head, and to count rattle segments as an indication of age on another. They used black and white film, but noted that by using color film, more individuals could probably be recognized by unique coloration and patterning. They also noted that the cameras documented the presence of a snake with much less human effort involved than did an active search effort. Browder et al. (1995) presented a design for an automatic, 35-mm camera; they used it in an investigation of the scavengers of carcasses of migratory fishes, identifying mammal, bird, and reptile scavengers. Pei (1995) used remote photography to study activity patterns of the spinous country rat *Niviventer coxingi* in Taiwan. Foster and Humphrey (1995) employed automatic camera units to document wildlife use of highway underpasses in southern Florida. They documented mountain lion, bobcat, white-tailed deer, raccoons, alligators *Alligator mississipiensis*, and black bears using the underpasses, and based on their data discussed implications for planning and designing such structures to reduce collisions with vehicles while allowing animal movement. Jacobson et al. (1997) used an infrared-triggered remote camera to census white-tailed deer at bait stations. They identified individual male deer by antler and other morphological characteristics and estimated population size over several years.

Karanth (1995) used automated camera traps to individually identify tigers in Nagarahole, India, and then estimate their numbers using photographic captures under a formal capture–recapture (CR) modeling. His work was subsequently extended to several sites across India to estimate tiger densities (Karanth and Nichols 1998; Karanth et al. 2004). Densities of tigers (O'Brien et al. 2003; Kawanishi and Sunquist 2004), jaguars *P. onca* (Silver et al. 2004; Silver 2004; Soisalo and Cavalcanti 2006), leopards (Henschel and Ray 2003) and ocelots (Trolle and Kéry 2005) have been estimated using similar methods by other workers. More recently, application of CR models to camera trap data was further extended by a 9-year study that estimated survival, recruitment, temporary emigration, transience, and rates of population change in a tiger population in Nagarahole (Karanth et al. 2006).

In their review of the primary literature, Cutler and Swan (1999) reported that the topics of published research using remote photography in wildlife ecology most frequently comprised nest predation, feeding ecology, nesting behavior, and evaluation of photographic equipment. Activity patterns, population parameters, and species detections were less common themes. Although researchers continue to investigate these topics with remote photography, the pattern may have changed. The more recent literature reveals a widening array of topics being investigated using camera traps in a truly impressive variety of habitats and locations. Fedriani et al. (2000) employed camera trapping and leg-hold trapping to assess habitat relations and relative abundance of coyotes *C. latrans*, gray foxes *Urocyon cinereoargenteus*, and bobcats in southern California. Somewhat similarly, Jacamo et al. (2004) studied niche relations among the maned wolf *Chrysocyon brachyurus*, crab-eating fox *Dusicyon thous*, and hoary fox *D. vetulus* in central Brazil using camera traps to assess habitat and activity patterns. McCullough et al. (2000) used camera traps along with radiotelemetry to investigate the ecology of the small, forest-dwelling Reeves' muntjac *Muntiacus reevesi* in Taiwan. They also produced population estimates based on CR models. By placing remote cameras in fig trees, Otani (2001) quantified the foraging frequency of Japanese macaques *Macaca fuscata* on figs and discussed the implications for seed dispersal in the forest. Beck and Terborg (2002) studied seed predation on palm *Astrocaryum murumuru* var. *macrocalyx* seeds under solitary trees versus dense groves in eastern Peru, and photographically identified several unexpected predators on the seeds. Kitamura et al. (2004) used remote photography to study seed dispersal and seed predation in forests in Thailand.

DeVault and Rhodes (2002) and DeVault et al. (2004) identified 17 species of vertebrates, including mammals, birds, and reptiles, scavenging on carcasses of small mammals in the eastern U.S. and suggested that scavenging may provide a larger component of the diet of some species than was previously thought. Main and Richardson (2002) assessed wildlife response to prescribed burning of forests in southwest Florida using camera traps distributed within forests before and after burning. Sequin et al. (2003) found that social and territorial status greatly affected the likelihood that a coyote would be captured by a remote camera. The dominant territory holders were most wary and rarely photographed; lower-status individuals and transients were detected on film much more often. Bridges et al. (2004) used remote cameras to monitor the denning behavior of black bears. Such cameras produced minimal disturbance to the animals, and provided insights into den emergence, behavior around the dens, and ages of cubs when they emerged (see Bridges and Noss, Chap. 5).

A particularly dramatic and valuable recent use of remote photography has been to document the presence of rare or presumed-extinct animals. For example, Surridge et al. (1999) documented a previously undescribed species of striped rabbit *Nesolagus timminsi* on the Southeast Asian mainland some 1,500 km north of the known range of the critically endangered Sumatran striped rabbit *N. netscheri* on the Island of Sumatra. Jeganathan et al. (2002) documented the presence of Jerdon's coursers *Rhinoptilus bitorquatus*, a critically endangered, poorly known, nocturnal, cursorial bird inhabiting scrub jungle in India, using both camera traps and track surveys . They recommend that relatively inexpensive and rapid track surveys be conducted for the bird, and that camera traps be used to confirm any suspected tracks. Holden et al. (2003) documented the presence and distribution of the endangered Asian tapir *T. indicus* in a national park in Sumatra, in an area where neither they nor park rangers ever saw the animals. Using camera traps, these investigators not only documented a surprisingly widespread distribution of the tapirs in the park, but discovered that they often occurred in pairs, and were found in a variety of habitat types in addition to primary forest. Lee et al. (2003) documented an expanded range of the Sulawesi palm civet *Macrogalidia musschenbroekii*, a little-known and endemic viverid, with the use of camera traps. Gonzalez-Esteban et al. (2004) documented the distribution of the European mink *Mustela lutreola* in northern Spain with remote photography, and recommended this method over live-trapping on the bases of cost and effort. In the Atlantic Forest of eastern Brazil, Kierulff et al. (2004) documented the distribution of the highly endangered buff-headed capuchin monkey *Cebus xanthosternos* in 13 forest fragments using camera traps baited with bananas. They also documented the presence of four other primate species, and gathered data such as the minimum number of individuals present, and number of infants. Recently, during an effort using camera traps to assess changes in the distribution of American martens over time in a study area in California's Sierra Nevada, Moriarty et al. (2009) produced photographs of a wolverine, the first documented in California since 1922. Subsequent genetic studies indicated that it was probably a dispersing male from the northern Rocky Mountains.

Mammals are not the only targets of detection using remote cameras. Lok et al. (2005) used camera traps to supplement other survey techniques to document the avifauna of Bawangling Nature Reserve, on the tropical island of Hainan in the South China Sea. Some of the bird species captured on film were classified as Vulnerable or Near Threatened, several considered very rare, and some had never before been captured on film.

The results of other remote camera surveys have been less encouraging from a conservation standpoint. Tilson et al. (2004) surveyed an area of southern China comprising eight reserves in five provinces for the presence of the south China tiger *P. t. amoyensis*. They found no evidence of tigers and little potential prey. The absence of photographic detections mirrored the absence of reported livestock depredations, and the authors conclude that it is likely that no tigers remain in this area. Numata et al. (2005) detected 18 species of mammals with camera traps with in and adjacent to a forest reserve in peninsular Malaysia, but these did not include the Asian elephant *Elephas maximus*, tiger, or sun bear *Helarctos malayanus*, and the authors concluded they are locally extinct. Among the species detected were domestic dogs used for poaching and hunting, and domestic cattle. Numata et al. (2005) did, however, confirm the presence of the Asian tapir in primary forest on the reserve; there is little published information on the current status and distribution of this species. In a forest reserve on Malaysian Borneo, Wong et al. (2005) used remote photography to monitor the physical condition, and document the starvation, of radiocollared sun bears and bearded pigs *Sus barbatus*. This occurred during a period of famine resulting from a fruit scarcity in the lowland tropical rainforest during a periodic, intermast interval.

Silveira et al. (2003) concluded that, despite relatively high initial costs, camera trapping was preferred over track surveys and direct counts in conducting rapid faunal assessments of mammals for conservation purposes. Similarly, Srbek-Araujo and Chiarello (2005) concluded that camera traps were an efficient way to inventory medium- and large-sized mammals in neotropical forests. Trolle (2003) used camera trapping and other methods to survey mammals in the Rio Japuri region of Brazil, and detected 13 of 41 mammal species using both baited and unbaited camera traps. In northern Mexico, Lorenzana-Pina et al. (2004) used camera traps to inventory medium and large mammals. They detected 18 wild mammal species, an estimated 80% of the medium- and large-sized mammals in their study area. Yasuda (2004) conducted a camera trap study of mammal diversity and abundance in central Japan, and developed guidelines for a minimum trapping effort to detect several species. Hirakawa (2005) developed a novel camera trap technique to detect bats. Knowing that insectivorous bats are attracted to any moving object of an appropriate size, he attached a pencil eraser to a line connected to a camera; when bats attacked the eraser, apparently mistaking it for insect prey, a photograph was taken. Research also confirms that remote photography is not the best tool for every job. In comparing survey methods for bobcats, Harrison (2006) found that detector dogs produced many more detections than did remote cameras, hair snares, or scent stations.

Conservation organizations now routinely incorporate the use of remote photography in their efforts to document and preserve biodiversity around the world (Henschel and Ray 2003; Sanderson and Trolle 2005). The Wildlife Conservation Society produced the first-ever photograph of the rare servaline genet *G. servalina* in Tanzania (Brink et al. 2002; Anonymous 2002). Sanderson and Trolle (2005) of Conservation International presented a photograph of the Siamese crocodile *Crocodylus siamensis* in Cambodia, previously thought to have been extirpated throughout much of its range. Staff of the World Wildlife Fund recently documented a rhinoceros on the island of Borneo, one of the last of a subspecies of the critically endangered Sumatran rhino *Dicerorhinus sumatrensis* (Anonymous 2006). The World Wildlife Fund has an online posting (http://worldwildlife.org/cameratrap/) of photographs taken at camera traps from remote places around the world.

Other novel uses of remote photography continue to be reported. In Australia, Glen and Dickman (2003a) used remote cameras to evaluate the possibility that poisoned baits set out to kill European red foxes *Vulpes vulpes* and wild dogs as part of a program to protect the spotted-tailed quoll *Dasyurus maculatus*, an endangered marsupial carnivore, would be taken by native, non-target species. As part of this research, Glen and Dickman (2003b) compared animal identifications from tracks left near baits to those from photographs taken of animals visiting the baits and found the track identifications inaccurate and unreliable, especially in unfavorable weather conditions. Following this, Claridge et al. (2004) investigated the behavior of the spotted-tailed quoll with the use of a remote, digital camera, alleviating the need to process film and getting results immediately in the field. Hegglin et al. (2004) used camera traps to document the uptake of bait laced with a rabies vaccine by red foxes in Zurich, Switzerland. With the data they gathered, they were able to recommend designs of bait stations to facilitate vaccination efficiency and reduce loss of such baits to non-target species. Using remote cameras in addition to other sampling techniques, Mazurek and Zielinski (2004) investigated the value to wildlife of legacy trees, those old trees left in an otherwise commercially harvested redwood *Sequoia sempervirens* forest in northwestern California. Using the cameras, they detected 13 species not detected by other survey methods. Rao et al. (2005) used camera traps to document the effect of hunting on the distribution and relative abundance of wildlife near a National Park in Myanmar. O'Connell et al. (2006) developed models of site occupancy to be used in large-scale monitoring programs for medium-sized and large mammals from detection data generated at an array of sampling techniques that included camera traps.

Other important topics of wildlife conservation have been studied using camera traps. Staller et al. (2005) used remote video photography to document predation on northern bobwhite *Colinus virginianus* nests. Nest predation was attributed to many more predator species than anticipated, and included nine-banded armadillos *Dasypus novemcinctus* and bobcats. This work also verified the inaccuracy of using of only nest remains to make identifications of nest predators. The use of remote photography for fixed-place monitoring, notably in studies of highways and wildlife, is common. Ng et al. (2004) documented the use of highway undercrossings by wildlife in southern California using remote photography. Goosem (2005) incorporated remote photography into a multifaceted scheme of monitoring wildlife use of crossing structures designed for a highway in Brisbane, Australia.

From the early work of Muybridge, Shiras, Nesbit, and Chapman, remote wildlife photography has developed into a modern, high-tech field, and is being used to address an increasing variety of scientific and conservation issues. Combining human curiosity and ingenuity, these remote camera techniques have allowed previously unimaginable access into the lives of many wildlife species. Developments have been driven by advances in technology such as the electronic flash, smaller batteries, and, most recently, digital and web-based photography. Yasuda and Kawakame (2002) described an "online" remote video system that streamed video images from a digital camera through a server to a computer. This provided real-time monitoring of wildlife and automatic storage of the digital images on the computer. Locke et al. (2005) described a web-based digital photographic system that could be used in remote areas. Triggered by a motion and heat sensor and with batteries that are continuously recharged with solar panels, the system can monitor wildlife at a remote site indefinitely, providing essentially real-time photographs without visits by humans to change film or batteries. Photographic results from this system can be seen at http://www.video-monitoring.com/wtek/.

A variety of commercially produced models are now available through outdoor and equipment suppliers and their internet outlets (e.g., www.cabelas.com). For example we have used RECONYX™ camera traps at all the water sources on a research station in central California to monitor wildlife on the 10 km^2 property. We have obtained nearly two million photos of terrestrial vertebrates ranging from western toads *Bufo* boreas to rattlesnakes to mountain lions to California condors *Gymnogis californianus*. These systems can be left in the field for up to 4 months at a time, during which as many as 20,000 photos are collected, documenting the presence of wildlife every second an animal is within range. We have even "captured" poachers. In another ongoing project we deploy the same camera systems on a rotating basis every square kilometer over a 300 km^2 region of the southern Sierra Nevada. The sites are baited for carnivores and checked weekly. Results are collected on site by reading compact flash cards with card readers. These major advances in technology now allow monitoring of wilderness wildlife at a very reasonable cost.

More than 100 years ago, the pioneering remote photographer Carl Georg Schillings recognized the effect of the modern world on its wild inhabitants. In passages that seem prescient, Schillings bemoaned the destruction of native fauna and flora, and observed that "Civilized man will destroy all that appears to him harmful or valueless, and will try to preserve only those animals and plants which he deems useful or ornamental" (Schillings 1905:2). He placed his photography and specimen collecting in the explicit context of increasing "…the pleasure and education of young and old" (Schillings 1905:10). We are confident that technological advances in remote photography will continue, at least in part as a spinoff from security concerns. We hope that developments in the field of remote wildlife photography continue to satisfy and pique human curiosity, increase scientific understanding, and promote the conservation of wild species and their habitats.

References

Abbott, H. G. and A. W. Coombs. 1964. A photoelectric 35-mm camera device for recording animal behavior. Journal of Mammalogy 45:327–330

Abbott, H. G. and W. E. Dodge. 1961. Photographic observations of white pine seed destruction by birds and mammals. Journal of Forestry 59:292–294

Anonymous. 2002. Shy predator comes out of the shadows. Nature 417:890–891

Anonymous. 2006. Endangered rhino stumbles into the limelight. Nature 441:920

Beck, H. and J. Terborg. 2002. Groves versus isolates: how spatial aggregation of *Astrocaryum murumuru* palms affects seed removal. Journal of Tropical Ecology 18:275–288

Bridges, A. S., J. A. Fox, C. Olfenbuttel, and M. R. Vaughn. 2004. American black bear denning behavior: observations and applications using remote photography. Wildlife Society Bulletin 32:188–193

Brink, H., J. E. Topp-Jorgensen, and A. R. Marshall. 2002. First record in 68 years of Lowe's servaline genet. Oryx 36:323–327

Browder, R. R., R. C. Browder, and G. C. Garman. 1995. An inexpensive and automatic multiple-exposure photographic system. Journal of Field Ornithology 66:137–143

Buckner, C. H. 1964. Preliminary trials of a camera recording device for the study of small mammals. Canadian Field-Naturalist 78:77–79

Carthew, S. M. and E. Slater. 1991. Monitoring animal activity with automated photography. Journal of Wildlife Management 55:689–692

Champion, F. W. 1928. With a camera in tiger-land. Chatto and Windus, London, England

Champion, F. W. 1933. The jungle in sunlight and shadow. Chatto & Windus, London, England

Chapman, F. M. 1927. Who treads our trails? National Geographic Magazine 52:330–345

Claridge, A. W., G. Mifsud, J. Dawson, and M. J. Saxon. 2004. Use of infrared digital cameras to investigate the behavior of cryptic species. Wildlife Research 31:645–650

Cowardin, L. M. and J. E. Ashe. 1965. An automatic camera device for measuring waterfowl use. Journal of Wildlife Management 29:636–640

Cutler, T. C. and D. E. Swan. 1999. Using remote photography in wildlife ecology: a review. Wildlife Society Bulletin 27:571–581

Danielson, W. R., R. M. DeGraaf, and T. K. Fuller. 1996. An inexpensive compact automatic camera system for wildlife research. Journal of Field Ornithology 67:414–421

DeVault, T. L. and O. E. Rhodes, Jr. 2002. Identification of vertebrate scavengers of small mammal carcasses in a forested landscape. Acta Theriologica 47:185–192

DeVault, T. L., I. L. Brisbin, Jr., and O. E. Rhodes, Jr. 2004. Factors influencing the acquisition of rodent carrion by vertebrate scavengers and decomposers. Canadian Journal of Zoology 82:502–509

Diem, K. L., L. A. Ward, and J. J. Cupal. 1973. Cameras as remote sensors of animal activities. Proceedings of the XIth International Congress of Game Biologists 11:503–509

Dodge, W. E. and D. P. Snyder. 1960. An automatic camera device for recording wildlife activity. Journal of Wildlife Management 24:340–342

Fedriani, J. M., T. K. Fuller, R. M. Savajot, and E. C. York. 2000. Competition and intraguild predation among three sympatric carnivores. Oecologia 125:258–270

Foster, M. L. and S. R. Humphrey. 1995. Use of highway underpasses by Florida panthers and other wildlife. Wildlife Society Bulletin 23:95–100

Fowler, C. H. and R. T. Golightly. 1993. Fisher and marten survey techniques on the Tahoe National Forest. Final Report. Agreement No. PSW-90-0034CA. Arcata, CA: Humboldt State University Foundation and U. S. Department of Agriculture, U. S. Forest Service. 119 pp

Glen, A. S. and C. R. Dickman. 2003a. Effects of bait-station design on the uptake of baits by non-target animals during control programs for foxes and wild dogs. Wildlife Research 30:147–149

Glen, A. S. and C. R. Dickman. 2003b. Monitoring bait removal in vertebrate pest removal: a comparison using track identification and remote photography. Wildlife Research 30:29–33

Goetz, R. G. 1981. A photographic system for multiple automatic exposures under field conditions. Journal of Wildlife Management 45:273–276

Gonzalez-Esteban, J., I. Villate, and I. Irizar. 2004. Assessing camera traps for surveying the European mink, *Mustela lutreola* (Linnaeus, 1761), distribution. European Journal of Wildlife Research 50:33–36

Goosem, M. 2005. Wildlife surveillance assessment Compton Road Upgrade 2005: review of contemporary remote and direct surveillance options for monitoring. Report of the Brisbane City Council. Cooperative Research Centre for Tropical Rainforest Ecology and Management. Rainforest CRC, Cairns. Unpublished report

Gregory, T. 1927. Random flashlights. Journal of Mammalogy 8:45–47

Gregory, T. 1930. Deer at night in the North Woods. Charles C. Thomas Publisher Ltd., Springfield, IL

Gregory, T. 1939. Eyes in the night. Thomas Y. Crowell Co., New York, NY

Grinnell, J., J. S. Dixson, and J. M. Linsdale. 1937. Fur-bearing mammals of California, Vol. 1. University of California Press, Berkeley

Griffiths, M. and C. P. Van Schaik. 1993a. Camera trapping: a new tool for the study of elusive rain forest animals. Tropical Biodiversity 1:131–135

Griffiths, M. and C. P. Van Schaik. 1993b. The impact of human traffic on the abundance and activity periods of Sumatran rain forest mammals. Conservation Biology 7:623–626

Guggisberg, C. A. W. 1977. Early wildlife photographers. Taplinger Publ. Co., New York, NY

Gysel, L. W. and E. M. Davis. 1956. A simple automatic photographic unit for wildlife research. Journal of Wildlife Management 20:451–453

Harris, W. P. and H. DuCharme. 1928. Notes on set camera work with beavers in Northern Michigan. Journal of Mammalogy 9:17–19

Harrison, R. L. 2006. A comparison of survey methods for detecting bobcats. Wildlife Society Bulletin 34:548–552

Hegglin, D., F. Bontadina, S. Gloor, J. Romer, U. Muller, U. Breitenmoser, and P. Deplazes. 2004. Baiting red foxes in an urban area: a camera trap study. Journal of Wildlife Management 68:1010–1017

Henschel, P. and J. Ray. 2003. Leopards in African rainforests: survey and monitoring techniques. Wildlife Conservation Society. Available at http://www.savingwildplaces.com/swp-researchmethods

Hiby, A. R. and J. S. Jeffery. 1987. Census techniques for small populations, with special reference to the Mediterranean monk seal. Symposia of the Zoological Society of London 58:193–210

Hirakawa, H. 2005. Luring bats to the camera – a new technique for bat surveys. Mammal Study 30:69–71

Holden, J., A. Yanuar, and D. J. Maryr. 2003. The Asian tapir in Kerinci Seblat National Park, Sumatra: evidence collected through photo-trapping. Oryx 37:34–40

Jacamo, A. T. A., L. Silveira, and J. A. F. Diniz-Filho. 2004. Niche separation between the maned wolf (*Chrysocyon brachyurus*), the crab-eating fox (*Dusicyon thous*), and the hoary fox (*Dusicyon vetulus*) in central Brazil. Journal of Zoology 262:99–106

Jacobson, H. A., J. C. Kroll, R. W. Browning, B. H. Koerth, and M. H. Conway. 1997. Infra-red triggered cameras for censusing white-tailed deer. Wildlife Society Bulletin 25:547–556

Jeganathan, P., R. E. Green, C. G. R. Bowden, K. Norris, D. Pain, and A. Rahmani. 2002. Use of tracking strips and automatic cameras for detecting critically endangered Jerdon's coursers *Rhinoptilus bitorquatus* in scrub jungle in Andhra Pradesh, India. Oryx 36:182–188

Jones, L. L. C. and M. Raphael. 1993. Inexpensive camera systems for detecting martens, fishers, and other animals: guidelines for use and standardization. Gen. Tech. Rep. PNW-GTR-306. Pacific Northwest Research Station. U. S. Department of Agriculture, U. S. Forest Service, Portland OR. 22 pp

Karanth, K. U. 1995. Estimating tiger *Panthera tigris* populations from camera-trap data using capture–recapture models. Biological Conservation 71:333–338

Karanth, K. U. and J. D. Nichols. 1998. Estimation of tiger densities in India using photographic captures and recaptures. Ecology 79:2852–2862

Karanth, K. U., R. C. Chundawat, J. D. Nichols, and N. S. Kumar. 2004. Estimation of tiger densities in the tropical dry forests of Panna, Central India, using photographic capture–recapture sampling. Animal Conservation 7:285–290

Karanth, K. U., J. D. Nichols, N. S. Kumar, and J. E. Hines. 2006. Assessing tiger population dynamics using photographic capture–recapture sampling. Ecology 87:2925–2937

Kawanishi, K. and M. E. Sunquist. 2004. Conservation status of tigers in a primary rainforest of Peninsular Malaysia. Biological Conservation 120:333–348

Kierulff, M. C. M., G. R. dos Santos, G. Canale, C. E. Guidoizzi, and C. Cassano. 2004. The use of camera-traps in a survey of the buff-headed capuchin monkey. Neotropical Primates 12:56–59

Kitamura, S., S. Suzuki, T. Yumoto, P. Poonswad, P. Chuailua, K. Plongmai, N. Noma, T. Maruhashi, and C. Suckasam. 2004. Dispersal of *Aglaia spectabilis*, a large-seeded tree species in a moist evergreen forest in Thailand. Journal of Tropical Ecology 20:421–427

Kucera, T. E. 1993. Seldom-seen carnivores of the Sierra Nevada. Outdoor California 54(6):1–3

Kucera, T. E. and R. H. Barrett. 1993. The Trailmaster camera system for detecting wildlife. Wildlife Society Bulletin 21:505–508

Kucera, T. E., W. J. Zielinski, and R. H. Barrett. 1995. The current distribution of American martens (*Martes americana*) in California. California Fish and Game 81:96–103

Laurance, W. F. and J. D. Grant. 1994. Photographic identification of ground-nest predators in Australian tropical rainforest. Wildlife Research 21:241–248

Leimgruber, P., W. J. McShea, and J. H. Rappole. 1994. Predation on artificial nests in large forest blocks. Journal of Wildlife Management 58:254–260

Lee, R. J., J. Riley, I. Hunowu, and E. Maneasa. 2003. The Sulawesi palm civet: expanded distribution of a little known endemic viverid. Oryx 37:378–381

Lorenzana-Pina, G. P., R. A. Castillo-Gomez, and C. A. Lopez-Gonzalez. 2004. Distribution, habitat association, and activity patterns of medium and large-sized mammals of Sonora, Mexico. Natural Areas Journal 24:354–357

Locke, S. L., M. D. Cline, D. L. Wetzel, M. T. Pittman, C. E. Brewer, and L. A. Harveson. 2005. From the field: a web-based digital camera for monitoring wildlife. Wildlife Society Bulletin 33:761–765

Lok, C. B. P., L. K. Shing, Z. Jian-Feng, and S. Wen-Ba. 2005. Notable bird records from Bawangling National Nature Reserve, Hainan Island, China. Forktail 21:33–41

Mace, R. D., S. C. Minta, T. Manley, and K. E. Aune. 1994. Estimating grizzly bear population size using camera sightings. Wildlife Society Bulletin 22:74–83

Main, M. and L. Richardson. 2002. Response of wildlife to prescribed fire in southwest Florida pine flatwoods. Wildlife Society Bulletin 30:213–221

Major, R. E. and G. Gowing. 1994. An inexpensive photographic technique for identifying nest predators on active nests of birds. Wildlife Research 21:657–666

Mazurek, M. J. and W. L. Zielinski. 2004. Individual legacy trees influence vertebrate wildlife diversity in commercial forests. Forest Ecology and Management 193:321–334

McCullough, D. R., K. C. J. Pei, and Y. Wang. 2000. Home range, activity patterns, and habitat relations of Reeves' muntjac in Taiwan. Journal of Wildlife Management 64:430–441

Moriarty, K. M., W. L. Zielinski, A. G. Gonzales, T. E. Dawson, K. M. Boatner, C. A. Wilson, F. V. Schlexer, K. L. Pilgrim, J. P. Copeland, and M. K. Schwartx. 2009. Wolverine confirmation in California after nearly a century: native or long-distance migrant? Northwest Science 83:154–162

Nesbit, W. 1926. How to hunt with the camera. E. P. Dutton & Company, New York, NY

Newhall, B. 1982. The history of photography. The Museum of Modern Art, New York, NY. 320 pp

Ng, S. J., J. W. Dole, R. M. Sauvajot, S. P. D. Riley, and T. J. Valone. 2004. Use of highway undercrossings by wildlife in southern California. Biological Conservation 115:499–507

Nicholas, K. S., A. R. Hiby, N. A. Audley, and T. Melton. 1991. The design of camera housings and automatic triggering devices for use with the monk seal register. Pages 59–62 *in* Establishment of a register of monk seal (*Monachus monachus*) within the European community. Institut Royal des Sciences Naturelles de Belgique and the Sea Mammal Research Unit, Cambridge. 29 Rue Vautier B-1040 Brussels, Belgium

Numata, S., T. Okuda, T. Sugimoto, S. Nishimura, K. Yoshida, E. S. Quah, M. Yasuda, K. Muangkhum, and N. S. M. Noor. 2005. Camera trapping: a non-invasive approach as an additional tool in the study of mammals in Pasoh Forest Reserve and adjacent fragmented areas in peninsular Malaysia. Malayan Nature Journal 57:29–45

O'Brien, T. G., M. F. Kinnaird, and H. T. Wibisono. 2003. Crouching tigers, hidden prey: Sumatran tigers and prey populations in a tropical forest landscape. Animal Conservation 6:131–139

O'Connell, A. F., Jr., N. W. Talancy, L. L. Bailey, J. R. Sauer, R. Cook, and A. T. Gilbert. 2006. Estimating site occupancy and detection probability parameters for meso- and large mammals in a coastal ecosystem. Journal of Wildlife Management 70:1625–1633

Osterberg, D. M. 1962. Activity of small mammals as recorded by a photographic device. Journal of Mammalogy 43:219–229

Otani, T. 2001. Measuring fig foraging frequency of the Yakushima macaque by using automatic cameras. Ecological Research 16:49–54

Pearson, O. P. 1959. A traffic survey of *Microtus-Reithrodontomys* runways. Journal of Mammalogy 40:169–180

Pearson, O. P. 1960. Habits of *Microtus californicus* revealed by automatic photo records. Ecological Monographs 30:231–249

Pei, K. 1995. Activity rhythm of the spinous country rat (*Niviventer coxingi*) in Taiwan. Zoological Studies 34:55–58

Rao, M., T. Myint, T. Zaw, and S. Hitun. 2005. Hunting patterns in tropical forests adjoining the Hkakaborazi National Park, north Myanmar. Oryx 39:292–300

Sadighi, K., R. M. DeGraaf, and W. R. Danielson. 1995. Experimental use of remotely-triggered cameras to monitor occurrence of timber rattlesnakes (*Crotalus horridus*). Herpetogical Review 26:189–190

Sanderson, J. G. and M. Trolle. 2005. Monitoring elusive mammals. American Scientist 93:148–155

Schillings, C. G. 1905. With flash-light and rifle: a record of hunting adventures and of studies in wild life in equatorial East Africa. Translated by H. Zick. Harper & Brothers Publishers, New York, NY

Schillings, C. G. 1907a. Mit Blitzlicht und Büchse: Neue Beobachtungen und Erlebnisse in der Wildnis inmitten der Tierwelt von Äquatorial-Ostafrika, Third edition. R. Boigtlander, Berlag in Leipzig

Schillings, C. G. 1907b. In Wildest Africa. Volumes I and II. Translated by Frederic Whyte. Hutchinson & Co., London

Sequin, E., M. M. Jaeger, P. F. Brussard, and R. H. Barrett. 2003. Wariness of coyotes to camera traps relative to social status and territory boundaries. Canadian Journal of Zoology 81:2015–2025

Seydack, A. H. W. 1984. Application of a photo-recording device in the census of larger rain-forest mammals. South African Journal of Wildlife Research 14:10–14

Shiras, G. 1906. Photographing wild game with flashlight and camera. National Geographic Magazine 17:366–423

Shiras, G. 1908. One season's game bag with a camera. National Geographic Magazine 19:387–446

Shiras, G. 1913. Wild animals that took their own pictures by day and by night. National Geographic Magazine 24:763–834

Silveira, L., A. T. A. Jacomo, and J. A. F. Diniz-Filho. 2003. Camera trap, line transect census and track surveys: a comparative evaluation. Biological Conservation 114:351–355

Silver, S. C. 2004. Assessing jaguar abundance using remotely triggered cameras – English. Available at http://www.savingwildplaces.com/swp-researchmethods

Silver, S. C., L. E. T. Ostro, L. K. Marsh, L. Maffei, A. J. Noss, M. J. Kelly, R. B. Wallace, H. Gomez, and G. Ayala. 2004. The use of camera traps for estimating jaguar *Panthera onca* abundance and density using capture/recapture analysis. Oryx 38:148–154

Soisalo, M. K. and S. M. C. Cavalcanti. 2006. Estimating the density of a jaguar population in the Brazilian Pantanal using camera-traps and capture–recapture sampling in combination with GPS radiotelemetry. Biological Conservation 129:487–496

Srbek-Araujo, C. and A. G. Chiarello. 2005. Is camera-trapping an efficient method for surveying mammals in Neotropical forests? A case study in south-eastern Brazil. Journal of Tropical Ecology 21:121–125

Staller, E. L., W. E. Palmer, J. P. Carroll, R. P. Thornton, and D. C. Sisson. 2005. Identifying predators at northern bobwhite nests. Journal of Wildlife Management 69:124–132

Surridge, A. K., R. J. Timmins, G. M. Hewitt, and D. J. Bell. 1999. Striped rabbits in Southeast Asia. Nature 400:726

Temple, S. A. 1972. A portable time-lapse camera for recording wildlife activity. Journal of Wildlife Management 36:944–947

Tilson, R., H. Defu, J. Muntifering, and P. J. Nyhus. 2004. Dramatic decline of wild South China tigers *Panthera tigris amoyensis*: field survey of priority tiger reserves. Oryx 38:40–47

Trolle, M. 2003. Mammal survey in the Rio Jauperi region, Rio Negro Basin, the Amazon, Brazil. Mammalia 67:75–83

Trolle, M. and M. Kéry. 2005. Camera-trap study of ocelot and other secretive mammals in the northern Pantanal. Mammalia 69:409–416

Wallihan, A. G. 1906. Camera shots at big game. Doubleday, Page & Co., New York. 77 pp + plates

Winkler, W. G. and D. B. Adams. 1968. An automatic movie camera for wildlife photography. Journal of Wildlife Management 32:949–952

Wong, S. T., C. Servheen, L. Ambu, and A. Norhayati. 2005. Impacts of fruit production on Malayan sun bears and bearded pigs in lowland tropical forest of Sabah, Malaysian Borneo. Journal of Tropical Ecology 21:627–639

Yasuda, M. 2004. Monitoring diversity and abundance of mammals with camera traps: a case study on Mount Tsukuba, central Japan. Mammal Study 29:37–46

Yasuda, M. and K. Kawakame. 2002. New method of monitoring remote wildlife via the Internet. Ecological Research 17:119–124

Young, S. P. 1946. The Puma, mysterious American cat. The American Wildlife Institute, Washington, DC. 358 pp

Zielinski, W. J. and T. E. Kucera, editors. 1995. American marten, fisher, lynx, and wolverine: survey methods for their detection. USDA For. Service General Technical Report PSW-GTR-157. Available at http://www.fs.fed.us/psw/publications/documents/gtr-157/

Zielinski, W. J., T. E. Kucera, and R. H. Barrett. 1995. The current distribution of fishers (*Martes pennanti*) in California. California Fish and Game 81:104–112

Chapter 3
Evaluating Types and Features of Camera Traps in Ecological Studies: A Guide for Researchers

Don E. Swann, Kae Kawanishi, and Jonathan Palmer

3.1 Introduction

Methods for capturing animals on film when researchers are not present have been used in ecological research for decades, but use of "remote photography," also called camera trapping, dramatically increased with the advent of commercial wildlife camera traps in the early 1990s (Kucera and Barrett 1993). Researchers planning to use camera traps now face an exciting but bewildering array of commercial choices, as well as many options for designing or customizing their own camera traps.

Choosing among different camera traps is complicated by the fact that they are designed for use in many different applications, under many field conditions, and for a wide range of target species. In addition to applications in hunting and wildlife viewing, research applications include studies of nest ecology, detection of rare species, estimation of population size and species richness, as well as research on habitat use and occupation of human-built structures (see review in Cutler and Swann 1999; see other chapters in this volume). These different applications may have very different requirements. For example, a trap to detect rare species in a remote area needs to be rugged, reliable, and capable of taking photographs for several weeks after it is set, while a trap used to observe feeding behavior of hatchling birds in a nest must be quiet, unobtrusive, and capable of taking a large number of images in succession.

Differences in field conditions and target species may also influence choice of camera traps. Weather conditions are a prime example: high humidity at a tropical field site creates completely different technological challenges for camera

---

D.E. Swann (✉)
Saguaro National Park, Tucson, AZ 85730, USA
e-mail: don_swann@nps.gov

K. Kawanishi
Taman Negara - Main Range Wildlife Corridor Project,
68000, Ampang, Selangor, Malaysia

J. Palmer
The Wildlife Conservation Society, Bronx, NY 10460, USA

A.F. O'Connell et al. (eds.), *Camera Traps in Animal Ecology: Methods and Analyses*,
DOI 10.1007/978-4-431-99495-4_3,

equipment than cold and snow. Ecological work in an urban environment where vandalism is an issue requires different armoring/camouflage of equipment than is needed in a wilderness setting. Similarly, traps used to study passerine birds require different trigger systems, lighting sources, and focal length, than traps used to study large mammals, reptiles, or other taxonomic groups.

Camera traps have received wide coverage in both the scientific and popular literature because they provide opportunities to gather large amounts of data where little information was previously available and because the photos generated can be used in wildlife education. Nevertheless, stories of equipment failures in the field are legion among ecologists who use camera traps. Under worst-case scenarios in remote field settings, they have been known to fail to gather enough data for studies to be successful. Many of these problems can be alleviated with extensive pre-planning, including review of the literature, consultation with experts, and practice, but we are not aware of published studies that assist researchers in choosing camera traps or making pre-planning more efficient.

The purpose of this chapter is to aid ecologists who are considering use of camera traps in deciding which types of systems and features are most appropriate for their particular study design, field conditions, and target species. We present an overview of the basic technology of camera traps, including new and developing technology, and summarize advantages and disadvantages of different trigger types, housing, software options, cameras, and features, currently available for commercial camera traps. In addition we review the literature on problems frequently encountered in the field and different ways these problems have been addressed. Our goal is not to recommend specific brands but to provide a framework that will allow researchers to more fully evaluate their needs and options in order to more efficiently and successfully use this technology.

3.2 Benefits and Problems with Camera Traps in the Field: A Review

The major ecological uses of wildlife camera traps have been to study nest ecology (Major and Gowing 1994; Liemgruber et al. 1994; Picman and Schriml 1994; Savidge and Seibert 1988; Laurance and Grant 1994) and record activity patterns of vertebrates (Carthew and Slater 1991; Griffiths and van Schaik 1993; van Schaik and Griffiths 1996; Bridges et al. 2004b; Jacomo et al. 2004; Rivero et al. 2005; see Cutler and Swann 1999 for a review). In recent years they have received increase use in documenting presence of rare species (Surridge et al. 1999; Delgado et al. 2004), rare events (Hirakawa and Sayama 2005), rare or melanistic individual animals (Martyr 1997; Azlan and Sharma 2002) species richness (see O'Brien et al., Chap. 13), and in estimating population parameters such as occupancy (O'Connell et al. 2006; Nichols et al. 2008); abundance or density (Mace et al. 1994; Karanth and Nichols 1998; O'Brien et al. 2003; Trolle and Kéry 2003; Wallace et al. 2003; Kawanishi and Sunquist 2004; Maffei et al. 2004, 2005;

Silver et al. 2004; Jackson et al. 2005; Soisalo and Cavalcanti 2006; see Karanth et al., Chap. 9), and survivorship and recruitment (Karanth et al. 2006). The great advantage of wildlife camera traps in comparison with other sampling methods such as direct observation, trapping, or tracking is that they can record very accurate data without the animal being captured or the researcher being present. In some ways these data are superior to human observations because, unlike data produced by live-trapping or observations, they can be reviewed by other researchers.

The benefits of using camera trap systems in ecological research are well-represented in the scientific literature. The problems with using them in the field have received less attention, but are well-known to experienced researchers. Primary among the issues is loss of data due to equipment failure. In response to a laudatory article on the Trailmaster® system in the Wildlife Society Bulletin (Kucera and Barrett 1993), Rice (1995) wrote about his experiences with chronic mechanical problems while using the same system in a tropical setting. Subsequent papers have also expressed similar concerns about camera traps in both tropical (Kawanishi 2002; Henschel and Ray 2003) and non-tropical settings (Khorozyan 2004; Roberts et al. 2006). Specific problems include both the failure of trigger mechanisms to activate the camera (and thus not record an animal event) or multiple photographs that contain no animals. Camera failure can be a nightmare for biologists using camera traps in remote locations because of the time it can take to realize the failure. For example, if visits to camera trap sites for exchange of films and batteries occur monthly and if prints or digital photos cannot be viewed in the field, two months of data may be lost before trouble-shooting can even begin.

Many factors influence the performance of camera traps. Poor performance is usually caused by a combination of weather, user experience, user skill, unique field conditions such as damage by animals, and poorly engineered equipment. In addition, there are great differences among types of camera traps in terms of their sensitivity, zones of detection, and performance under different environmental conditions (Swann et al. 2004). Some of these factors can be alleviated, whereas others (such as weather) are beyond a researcher's immediate control. It is therefore important to know the problems of potential equipment and select traps suitable for local application. In the following sections we review the basic technology of camera traps and the different types and features of systems currently used in ecological studies.

3.3 Types of Camera Traps

Camera traps can be categorized in various ways, but the major difference is between non-triggered and triggered systems (Cutler and Swann 1999). *Non-triggered* systems include cameras that are programmed to record images either continuously or at regular, pre-set time intervals. In contrast, *triggered* camera traps are inactive until they are triggered by an event of some kind, usually the arrival of an animal. The trigger may be mechanical, such as the animal stepping on a pressure plate that fires the camera, but more typically the trigger is an infrared light

Table 3.1 Different types of camera trap systems and where they are most and least applicable in ecological studies

System	Most applicable	Least applicable
Non-triggered	When animals are resident, occur in the open, or have high visitation rates, or when continuous data (e.g., to establish absence) is important	For rare species or infrequent events (power requirements can be large and photo analysis can be time-consuming)
Triggered – general	When camera traps must be left in the field for a long period of time, but events are infrequent	When events of interest are frequent or continuous (triggered systems may be more complicated and less reliable than non-triggered systems for recording continuous or frequent events)
Triggered – mechanical	When activity of interest requires an animal to physically step in a specific location or pull on a bait or object of interest	When activity of interest may not be recorded by a physical trigger
Triggered – infrared – general	When activity of interest does not require animal to physically step in a specific location or pull on an object; when animal is too light or fast for a mechanical trigger	When the activity of interest involves a physical activity by animals (triggered traps may be more complicated and less reliable than mechanical ones)
Triggered – infrared – active	When it is important that non-target species (e.g., those below certain height) do not trigger the trap, especially if film or digital images are limited. In general, when activity of interest is in a precise location but would not be detected by a mechanical trigger	In areas where vegetation growth is so fast that it will obscure the beam between maintenance visits. In areas where wind, rain, or snow frequently breaks the light trigger or blows vegetation or other objects across it
Triggered –infrared – passive	When a larger zone of detection of target species is desirable. In areas where wind, rain, and snow are common	In hot environments where equipment may fail to detect the differential temperature of target species

source. Some commercial camera traps can be programmed both for non-triggered time-lapse and infrared-triggered operation (Table 3.1).

The differences among triggered and non-triggered camera traps are significant, and it is important to think carefully before choosing one instead of the other. In general, non-triggered camera traps are most appropriate when the event of interest occurs frequently (such as feeding behavior), or when a continuous record is required. Triggered camera traps are more appropriate when the event of interest is infrequent or discontinuous, such as when it is important to record the presence of a species or individual at a site. Also, non-triggered camera traps tend to require more power, which makes them less useful for remote situations. Bird studies often use non-triggered systems, while nearly all large mammal studies use triggered traps, although this difference may be partially cultural among biologists.

3.3.1 Non-triggered Camera Systems

Non-triggered (time-lapse or continuous recording) camera traps vary greatly, but usually consist of a camera unit (a still camera with a time lapse timer, or a digital video camera), a power source, and a wire or wireless connection between them. Recent advances have made this technology much smaller and less expensive than in the past, and commercial systems are available for less than US $200. However, high-end systems with remotely-controlled cameras, satellite connections, and solar panels for continuous power are also receiving use, particularly for monitoring animals that occur as groups in the open, such as sea birds, seals, and grazing animals.

Non-triggered camera traps are most commonly used for animal behavior and nest ecology (Cutler and Swann 1999). Continuous recording can also be advantageous over a trigger when it is important to know that an animal is absent from a location during a given time interval, as sometimes a rare animal may appear, but fail to trigger the camera trap. Non-triggered cameras may also fail less frequently in the field than triggered camera traps because there are fewer parts. Prior to the widespread use of commercial triggered camera traps, time-lapse cameras were used in studies of use of specific locations, such as wildlife water catchments (e.g., Bleich et al. 1997).

The great disadvantage of non-triggered camera traps is that many of them require larger amounts of power for continuous operation (however, some time-lapse systems require less power). This is often not a limiting problem in studies of nesting birds, where a long cord can connect to a direct power source, but makes use of these traps in remote field locations difficult unless solar power is used. A further disadvantage, for many studies, is the large amount of time required to review images to find target animals or events for analysis.

Use of non-triggered camera traps have become very popular as web cams operating in real time for educational use. Internet sites where viewers can view wild animals ranging from grizzly bears to bats to barn owls have proliferated in recent years (e.g., http://www.animalcameras.com). From a research point of view, the wide availability of this technology and the ease with which it can be distributed has created the potential for large-scale studies of nest ecology using citizen scientists who operate nest-box cameras in their backyards (Proudfoot 1996; Hudson and Bird 2006; Huebner and Hurteau 2007).

3.3.2 Triggered Camera Traps

Mechanical camera traps use either pressure pads (Griffiths and van Schaik 1993; Mudappa 1998; Fedriani et al. 2000; York et al. 2001; Moruzzi et al. 2002) or bait lines in which a line connects the trigger of a camera and the bait, decoy, or egg

(Picman and Schriml 1994; Cresswell et al. 2003; Glen and Dickman 2003; González-Esteban et al. 2004). A modification of the bait line works in a similar way as the pressure pad does whereby a micro-switch triggers a camera when bait is removed from a platform and completes the circuit (DeVault et al. 2004). Mechanically triggered camera traps have become less common in recent years due to improvements in light-triggered technology, but are appropriate for studies where an animal might closely inspect a small area or move an object of interest; they are often used in nest predation studies.

Light-triggered camera traps use a light beam (typically, an infrared beam) as a trigger and can be either "active" or "passive." *Active* infrared-triggered traps emit a continuous beam from a transmitter to a receiver that is like an invisible string; when the beam is broken, a message is sent to the camera to record an image. *Passive* camera traps, which are far more common than active systems, consist of two separate sensors placed side by side that read a background temperature signature; when an animal passes in front of the sensors, both the movement detected by the two sensors and temperature change due to the animal's different temperature signature trigger the camera trap to record an image (Swann et al. 2004). Active traps (e.g., the Trailmaster® 1500 unit) consist of a separate receiver and transmitter, as well as a camera and a connection (usually a cord) between the camera and the receiver, while passive camera traps usually bundle all of these components into a single box.

In general, active infrared camera traps work best for target animals of an identifiable height and allow for precise linear detection of the animal. However, active infrared traps tend to have a much higher rate of false triggers than passive traps, as events can be triggered whenever objects that are not animals (including not only blowing or growing vegetation, but even rain or snow) move through the infrared beam (Kawanishi 2002; Henschel and Ray 2003). Passive infrared traps have a wider "zone" of detection than active units, which allows detection of a greater range in animal sizes. However, detection zone sizes vary among passive units (Swann et al. 2004). Passive units are usually not as prone to being triggered by non-animals, but false triggers do occur, particularly when the zone of detection is very large. Most commercial passive infrared units come in a single unit and are easier to set up in the field than active traps, and as a result, the number of commercial passive camera traps has proliferated in recent years.

The greatest disadvantage of triggered camera traps in comparison to non-triggered traps, in addition to false triggers, is that sensors sometimes fail due to environmental conditions or operator error. However, their great advantage for wildlife studies, particularly studies of rare animals, is that they use little power and can be easily utilized for long periods in remote settings.

3.4 Camera Trap Features and Trade-offs Among Them

Most commercial triggered and non-triggered camera traps have special features that can greatly influence the number and quality of photographic images produced. As with all technologies, trade-offs exist among the different features that are available.

For example, because power is required for camera traps but direct current is not usually available in the field, commercial traps include many battery options. Batteries may be large or small; large batteries provide power for much longer periods, but are heavier to carry in the field. Rechargeable batteries are more expensive to purchase but will save money in the long run; however, rechargeable batteries do not last as long as alkaline batteries of the same weight.

Of course, one of the most important trade-offs is between the number of features available and their cost. Because camera traps range in price from less than $100 to thousands of dollars, it is fair to say that there is one that is available for nearly any project budget. Below is a summary of different features of camera traps and the trade-offs associated with them.

3.4.1 System Components

Non-triggered camera traps generally have at least two components (the camera and a power source), but triggered traps contain several components that either operate independently, or are bundled within a single unit. For most traps these components are the infrared sensor, the camera, and connective cords, but other components may include supplemental lighting or power sources. When the components are set up separately, the advantage is flexibility that may lead to greater photograph quality. For example, if the camera is separate from the sensor it may be set a greater distance from, or at an angle to, the sensor, and background lighting may be more effectively utilized. Increasingly, non-triggered systems in particular can be customized to allow researchers to remotely control video cameras by tilting the camera and zooming or panning to animal subjects (e.g., robotic camera systems from See More Wildlife Systems; http://www.seemorewildlife.com).

The major disadvantage of multiple parts is that if any of the parts fail, the entire system may fail. Several studies have documented that single unit camera traps are less likely to fail in the field than multiple-unit systems. Cord failure resulting from chewing or pulling by animals is often cited as a major problem (Sequin et al. 2003), but other examples include animal damage, misalignment, or battery failure of the sensor or camera (Main and Richardson 2002; Bridges et al. 2004a). In general, single units that contain all the components are also easier to transport, set up, and armor, and are usually recommended for research with multiple camera units in more remote field settings (Kawanishi 2002; Henschel and Ray 2003). However, some commercial systems with multiple components now provide wireless connections as an option (e.g., Faunatech, available from http://www.faunatech.com).

3.4.2 Housing

Housing for camera traps varies by color, weight, size, shape, weather-proofing, and durability (Table 3.2). Different color options, such as camouflage, may be desirable if reduced visibility of the trap to wildlife or humans is desired.

Table 3.2 Housing and external hardware options available for camera traps

Feature	Purpose
Camouflage color	Lowers visibility of traps to animals and people
Water proofing	Essential in rainy or humid environments to prevent equipment malfunction
Tree cables and locks	Reduces risk of vandalism and theft
Compact unit size	Smaller camera traps designed for nest boxes where little room is available
Noiseless housing	Reduces noise produced by unit, such as by camera shutter click, so that animals are not alarmed
Sound recording	Allows recording of sound of animal with images
Armoring of housing	Reduces risk of damage by animals, vandalism and theft
Armoring of connective cables	Reduces risk of damage by animals
Extra-length connective cables (for non-triggered systems)	Allows >100 m connectivity of cameras with other hardware for viewing, power, or transmission

The most durable traps are made of sturdy metal and are larger and heavier, but if the units need to be carried for long distances it may be desirable to use traps that are smaller and are made of a lighter material such as plastic.

Weather-proof housing is essential in areas of extreme weather, and it is important to carefully check manufacturer's specifications for different field situations. Tropical field conditions require units that are water-proof, with components that can all be kept dry during periods of high heat and humidity – which is often every day. A pack of fresh silica gel kept inside the unit can determine the air-tightness of the unit. If the color changes from blue to pink, the unit is not airtight. Camera traps in tropical areas must be equipped with rubber seals and any holes, such as for sensors and screws, must be sealed to prevent moisture from seeping in. Moisture damages computer components, causes sensors to malfunction, oxidizes metal parts, and causes film to stick inside the camera. In addition, ants and termites nest in housing that is not air-tight; termites eat interior parts, and ants can be hazardous to operators. Lastly, moisture from human perspiration can easily build up in non-waterproofed camera traps when being transported in a backpack. The extreme humidity is the biggest challenge to electronic equipment in tropical environments (Kawanishi 2002). Some camera traps are now manufactured specifically to work in these conditions.

Units that are advertised as effective to 100°F will be unsuitable in many desert areas. In general, due to the inherent qualities of infrared sensors, most passive infrared camera traps do not perform as well in extreme heat (Swann et al. 2004).

Vandal-proof housing and locks may not be necessary in some areas, but are absolutely critical in others. However, housing to protect camera traps from animals is very important in most field situations, although the animal problems may vary greatly depending on the species involved. Cables and cords that are chewed by small animals, such as rodents, may be reinforced by the manufacturer, or reinforced in the field with tinfoil or covered with a toxic substance. Many large mammals such as bears and elephants will destroy cameras when they encounter them, and armored systems have been developed for popular commercial units such as the Camtrakker® (Grassman et al. 2005) and Trailmaster® (Karanth and Nichols 2002).

3.4.3 Software and Programming

Most commercial camera traps now contain software that can assist in system operation and increase functionality, many of which are very helpful for ecological studies (Table 3.3). Typical software features include options set by the user to record additional data (such as date and time of triggered events), to only record images at pre-determined times (such as at night) so as to save power or film or to pause the camera during events so that multiple images of the same individual are not recorded. The latter two options are useful if film is being used and needs to be conserved, but is less important for digital cameras. Recording of additional data such as time and date, or even temperature, may be important for many studies. Many camera traps also come with power options that reduce battery use and increase battery longevity, which is essential in remote field situations. Increasingly, camera trap features can be programmed directly from a laptop or personal computer.

Probably the most important issue that researchers planning a field project should recognize is that field biologists and their technicians vary in their technological ability and interest, and that more complicated camera traps are not suitable for all projects. As with system components, the greater flexibility that comes with more options may be offset by the greater chance than an important feature may be improperly set or may fail under extreme field conditions. Many camera traps are equipped with default software options that are relatively easy to use, with additional features available for users who desire to use them.

Table 3.3 Software options available for camera traps. The most important trade-off with software features is the ability for researchers and field technicians to use them effectively

Feature	Purpose
Event data recording	To record time and date of event as data that can be downloaded
Event recording options	To record events 24 h per day, or only at specific intervals to record events in order to reduce battery and film use
Image recording options	Camera delay; to prevent capture of too many images, which may use up film and power
Speed options	To reduce time between triggering of animal event and recording of photographic images
Power options	Sleep mode, used to reduce battery use and increase battery longevity
Sensitivity settings	Allow sensitivity to be set to detect target species; more sensitive settings will detect smaller, lighter species such as birds
Password protection	Software theft devices, to prevent theft of data
Audio devices for target animals	Features such as "whistle stop," to cause target species to stop in front of camera for better photograph
Data loggers	To record environmental variables, such as temperature, while camera is operating

3.4.4 Power

Electrical power for camera traps is often a limiting factor for their use in ecological studies, although great strides have been made in recent years to diminish power needs. Power can be provided by alternating current (AC), as from a cord plugged into a standard electrical outlet, or by direct current (DC) in the form of batteries or solar energy (Table 3.4). Most commercial triggered traps run on batteries, while most non-triggered systems run on AC. When it is possible to study animals near sources of AC, this power source has great advantages, but electrical outlets seldom exist in remote field settings.

Batteries used in camera traps may be alkaline, lithium-ion, or rechargeable, and commercial systems often provide the option to use any of these. Alkaline or lithium batteries are preferred by most researchers because they are reliable, give uniform power, and require no maintenance, as they are disposed of after use. Lithium-ion batteries are also reliable and have a longer life than alkaline, but are more expensive.

Where portability and longevity are important, careful choice of batteries is essential. Manufacturers often provide a range of performance indicators for their alkaline batteries, but the most reliable comparison can be achieved by researching (on the internet) the average watt-hours delivered by the battery. For example, an array of three D-cell pro batteries has approximately the same watt-hour rating as an external lithium battery pack. Typically, batteries for professional photography have the highest watt-hour rating. External battery packs are an option in many commercial camera traps, and may be constructed or armored to provide protection against animals and the elements.

Table 3.4 Power options for camera traps and their advantages and disadvantages

Feature	Advantages and disadvantages
Alternating current (AC)	Direct current is inexpensive and easy to use, but not realistic for remote field studies
Solar direct current (DC)	Solar power provides continuous power so that fewer field visits are required but is expensive and heavy to transport. May be ideal for camera traps set for long periods in one location
Alkaline batteries	Very dependable and provide uniform power expectancy. More expensive over time than rechargeable batteries and have a shorter battery life than lithium-ion
Lithium-ion batteries	Very dependable and provide uniform power expectancy. Longer life than alkaline and as reliable, but are more expensive. More expensive than rechargeable over time
Rechargeable batteries	Initial cost high but economical over time. Battery life is shorter than fresh alkaline and declines over time. More labor needed for batter management. May not work as well in humid areas if camera traps are not water-proofed
Fuel cells	Volume and power-size ratios competitive with lithium-ion batteries. More "environmentally friendly." Technology is still immature and not yet available in commercial camera traps

Rechargeable batteries, which have a higher initial cost but can be economical in the long run, do not tend to last as long in the field as alkaline or lithium batteries. In part because they need to be recharged at an electrical source and require greater maintenance on the part of the researcher, they tend to be less reliable. Solar-powered fuel cells have great potential for remote systems, and are available for some commercial applications; however, this technology remains too expensive for use in most ecological studies.

3.4.5 Camera Types

All camera traps, triggered or non-triggered, can utilize any type of camera once the trap begins to record images. Different traps use still and video cameras, as well as film and digital cameras, and the current options can be bewildering in their scope. Currently, film cameras used in many commercial units provide better quality images than digital cameras due to faster shutter speeds and the inherent trade-offs between power, speed, and image quality. However, digital technology is rapidly improving and digital cameras are rapidly replacing film cameras in commercial camera traps.

The obvious advantages of digital over film are that many more images can be captured on digital than on film cameras. Images can be stored, easily downloaded, and viewed on the computer and then selectively printed and distributed. Most digital camera traps are currently able to work for 30 continuous days, taking >20 pictures per day, using three D-cell, 15 watt-hour batteries. Shutter speeds of digital cameras are continuing to increase.

Use of video cameras, standard in most non-triggered camera systems, is an option for some triggered camera traps as well, but requires a larger power source. The great advantage of digital is that it provides continuous footage, which is particularly useful for capturing animal behavior (see Bridges and Noss, Chap. 5) or rare events, or where animals are in the open and can be viewed easily for long periods. The disadvantage of video is that has greater power needs, so that it is most commonly used where AC power sources are nearby or solar is available.

3.4.6 Camera and Lighting Options

Nearly all commercial camera traps include as standard features automatic focus and automatic flash as well as image time and date stamps (Table 3.5). Both film and digital cameras vary in their features as well as cost. Higher resolution digital photos provide better quality, but larger images take up more disk space.

During the night or in dark environments, lighting can make a significant difference in photo quality. Conventional commercial camera traps include a film or digital

Table 3.5 Camera features and lighting options for camera traps

Feature	Purpose, advantages and disadvantages
Time and date recording	Records time and date directly on image
Auto-focus	Auto-focus to allow images to be focused when animals are at different distances from camera
High resolution camera	Higher resolution for better quality images; lower resolution allows more images to be stored
In-field camera display	View pictures in the field with camera display
Miniature camera size	Can be inserted into nest boxes and other small spaces
Auto-flash	Auto-flash allows night images (but consumes more power)
Flash focus	Allows larger flash range
Infrared flash	Decreases risk of target animals being alarmed by flash (but image quality is not as good as with regular flash)
Supplemental lights	Better image quality than regular flash (but drains power and may alter animal behavior)
Strobe flash	Better image quality (but drains power and may alter animal behavior)
Thermal imagery	Allows lighting of animals at night with no disturbance (but poor image quality)

instamatic camera that is modified to work with an infrared sensor; during low-light situations additional lighting is provided by an instant flash attached to the camera. This is sufficient for most field studies, but may be inadequate where animals are dark-colored and the background is very deep, or if photographs with greater clarity are needed. Supplemental lighting (Table 3.5) can be added in a number of ways, either as an additional commercial option, or by researcher invention. One method is to install "slave" flashes that respond instantly to the flash of the camera, and another is to increase flash capacity directly. Improved camera flashes, such as strobe flashes, can also increase image quality. Both supplemental lighting and strobe flashes require greater power than regular flashes. Some video-type cameras also deploy thermal imaging, which does not disturb animals but is a more expensive technology.

A few studies have looked at the negative response of animals to camera traps, and researchers have long been concerned that this might bias the results of ecological studies (Wegge et al. 2004), especially of behavioral studies (Major 1991; Major and Gowing 1994; Laurance and Grant 1994; Liemgruber et al. 1994; Sequin et al. 2003; Hegglin et al. 2004; Ball et al. 2005). However, one nest predation study (Thompson and Burhans 2003) found little evidence that camera presence affected predation. Besides olfactory and acoustic cues, camera flash may scare animals so that they avoid (or even destroy) camera traps. The major alternative light source is infrared, which is not usually detectable by mammals or birds. Infrared light can be provided as a flash source (Claridge et al. 2004), or can be operated continuously during pre-programmed hours of operation. Continuous infrared is often a standard feature in commercial nest-box cameras. The disadvantage of infrared light is that it does not provide as high quality images as other light sources.

3.5 Discussion

3.5.1 Working with Camera Traps in the Field

Choosing a trap to work with in the field is a difficult task because of the growing number of commercial camera traps and the complexity of features. We hope that this chapter can provide a framework for making decisions about different features of camera traps as the technology continues to improve.

The ubiquitous use of commercial camera traps in the field during the past decade has produced experts who have learned a great deal from their mistakes. Swann et al. (2004) review specific commercial infrared-triggered camera traps and discuss common field errors based on technical aspects of different traps. They conclude that researchers should choose camera traps based on the size of the target species and target area, and offer suggestions related to aspects of setting camera traps such as height and distance from target, anchoring, vegetation removal, and other factors that can be controlled by researchers. Because many problems in the field are related to inexperience of researchers, their strongest recommendation is that new researchers carefully read the instructions provided with new equipment and to practice using new equipment extensively before beginning fieldwork. As with any new skill or technology, there is no substitute for asking questions or working in the field with experienced practitioners before embarking on a research project.

3.5.2 Emerging Technology

Recent improvements in camera trap technology are many, and include the development of single-unit infrared traps and improvements in digital technology that allow many images to be stored between equipment checks, as well as non-triggered robotic cameras operated at very remote sites to view bird and sea mammal colonies. Most importantly, as with most developing technologies, nearly all traps have become smaller and less expensive, and as competition in this market increases we believe that there will likely be a decrease in the number of commercial traps available in the future.

An immediate future improvement in camera traps is likely to be the improved image quality of digital cameras. Much of the recent success of digital cameras lies in the use of CMOS (complementary metal–oxide–semiconductor) chips. Both CMOS and CCD (charge-coupled device) chips were invented during the 1960s, but it was only in the 1990s, with improvements in lithographic techniques, did CMOS become a viable technology for digital cameras. Compared to CCD, CMOS chips consume less power and have lower fabrication costs. After the release of the first mass production multi-megapixel CMOS sensor camera (the Canon© EOS-D30), wider adoption of CMOS by camera trap producers such as Bushnell®

and Cuddeback® occurred in approximately 5 years. It appears to be only a matter of time before production modules of camera traps attain the <50 ms shutter lag achieved by many leading digital single lens reflex cameras using CMOS technology.

Less immediately, researchers should expect some improvements in power options. No matter how much technology reduces the power consumption of camera traps, there will always be a demand to increase the power available to remote, hard to reach camera traps. Fuel cells provide the most promising solution, with the potential increased longevity and ease of refueling, not to mention the hope for environmental benefits. Canon's© 2005 release of a digital camera with a fuel cell with the same volume energy density as current lithium-ion batteries, suggests we are still a number of years from fuel cell technology being able to widely benefit the camera trapping community.

Other contributions to this volume describe how camera traps are allowing ecologists to answer questions that have been previously difficult to answer. This trend will continue as camera trap power and performance issues are resolved and they become even less expensive and easier to deploy in the field. For example, multiple cameras set at a single site can improve the ability to identify individuals, which will improve precision of population estimates. Other researchers are currently developing plans for remote, continuous monitoring of a landscape using multiple camera traps. One promising trend is to couple infrared camera trap technology with wireless networks of camera traps in remote settings that can communicate with each other and with researchers back at the office via satellite (C. Bray, pers. commun., Fig. 3.1). Just as it took many years after the development of commercial camera traps for many of the current ecological applications to be fully developed, it is probably only fair to state that we are probably only beginning to realize the potential uses of these new technologies for addressing complex ecological questions.

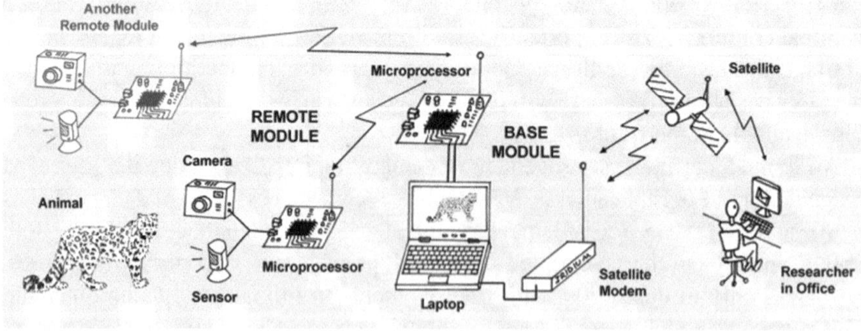

Fig. 3.1 Prototype satellite linked, wireless camera trap network (C. Bray, pers. commun.; http://scoff.ee.unsw.edu.au/posters/posters2006/Satellite%20Linked,%20Wireless%20Camera%20Trap%20Network.pdf)

References

Azlan, M. J. and D. S. K. Sharma. 2002. First record of melanistic tapirs in Peninsular Malaysia. Journal of Wildlife and Parks 20:123–124

Ball, S. J., D. Ramsey, G. Nugent, B. Warburton, and M. Efford. 2005. A method for estimating wildlife detection probabilities in relation to home-range use: insights from a field study on the common brushtail possum (*Trichosurus vulpecula*). Wildlife Research 32:217–227

Bleich, V. C., R. T. Bowyer, and J. D. Wehausen. 1997. Sexual segregation in mountain sheep: resources or predation? Wildlife Monographs 134:1–50

Bridges, A. S., J. A. Fox, C. Olfenbuttel, M. R. Vaughan, and M. B. Vaughan. 2004a. American black bear denning behavior: observations and applications using remote photography. Wildlife Society Bulletin 32:188–193

Bridges, A. S., M. R. Vaughan, and S. Klenzendorf. 2004b. Seasonal variation in American black bear *Ursus americanus* activity patterns: quantification via remote photography. Wildlife Biology 10:277–284

Carthew, S. M. and E. Slater. 1991. Monitoring animal activity with automated photography. Journal of Wildlife Management 55:689–692

Claridge, A. W., G. Mifsud, J. Dawson, and M. J. Saxon. 2004. Use of infrared digital cameras to investigate the behaviour of cryptic species. Wildlife Research 31:645–650

Cresswell, W., J. Lind, U. Kaby, J. L. Quinn, and S. Jakobsson. 2003. Does an opportunistic predator preferentially attack non-vigilant prey? Animal Behaviour 66:643–648

Cutler, T. L. and D. E. Swann. 1999. Using remote photography in wildlife ecology: a review. Wildlife Society Bulletin 27:571–581

Delgado, E., L. Villalba, J. Sanderson, C. Napolitano, M. Berna, and J. Esquivel. 2004. Capture of an Andean cat in Bolivia. Cat News 40:2

DeVault, T. L., I. L. Brisbin, and O. E. Rhodes, Jr. 2004. Factors influencing the acquisition of rodent carrion by vertebrate scavengers and decomposers. Canadian Journal of Zoology 82:502–509

Fedriani, J. M., T. K. Fuller, R. Sauvajot, and E. York. 2000. Competition and intraguild predation among three sympatric carnivores. Oecologia 125:258–270

Glen, A. S. and C. R. Dickman. 2003. Monitoring bait removal in vertebrate pest control: a comparison using track identification and remote photography. Wildlife Research 30:29–33

González-Esteban, J., I. Villate, and I. Irizar. 2004. Assessing camera traps for surveying the European mink, *Mustela lutreola* (Linnaeus, 1761), distribution. European Journal of Wildlife Research 50:33–36

Grassman, L. I. Jr., M. E. Tewes, and N. J. Silvy. 2005. From the field: armoring the Camtrakker camera-trap in a tropical Asian forest. Wildlife Society Bulletin 33:349–352

Griffiths, M. G. and C. P. van Schaik. 1993. Camera trapping a new tool for the study of elusive rain forest animals. Tropical Biodiversity 1:131–135

Hegglin, D., F. Bontadina, S. Gloor, J. Romer, U. Mueller, U. Breitenmoser, P. Deplazes, S. Glor, and U. Müller. 2004. Baiting red foxes in an urban area: a camera trap study. Journal of Wildlife Management 68:1010–1017

Henschel, P. and J. Ray. 2003. Leopards in African rainforests: survey and monitoring techniques. Wildlife Conservation Society Global Carnivore Program, Washington, DC

Hirakawa, H. and K. Sayama. 2005. Photographic evidence of predation by martens (*Martes melampus*) on vespine wasp nests. Bulletin of Forestry and Forest Products Research Institute 4:207–210

Hudson, M. A. R. and D. M. Bird. 2006. An affordable computerized camera technique for monitoring bird nests. Wildlife Society Bulletin 34:1455–1457

Huebner, D. P. and S. R. Hurteau. 2007. An economical wireless cavity-nest viewer. Journal of Field Ornithology 78:87–92

Jackson, R. M., J. D. Roe, R. Wangchuk, and D. O. Hunter. 2005. Camera-trapping of snow leopards. Cat News 42:19–21

Jacomo, A. T. D. A., L. Silveira, and J. A. F. Diniz. 2004. Niche separation between the maned wolf (*Chrysocyon brachyurus*), the crab-eating fox (*Dusicyon thous*) and the hoary fox (*Dusicyon vetulus*) in central Brazil. Journal of Zoology 262:99–106

Karanth, K. U. and J. D. Nichols. 1998. Estimation of tiger densities in India using photographic captures and recaptures. Ecology 79:2852–2862

Karanth, K. U. and J. D. Nichols, editors. 2002. Monitoring tigers and their prey: a manual for researchers, managers and conservationists in tropical Asia. Centre for Wildlife Studies, Bangalore

Karanth, U., J. D. Nichols, N. S. Kumar, and J. E. Hines. 2006. Assessing tiger population dynamics using photographic capture-recapture sampling. Ecology 87:2925–2937

Kawanishi, K. 2002. Population status of tigers (*Panthera tigris*) in a primary rainforest of Peninsular Malaysia. Dissertation, University of Florida, Gainesville, FL

Kawanishi, K. and M. E. Sunquist. 2004. Conservation status of tigers in a primary rainforest of Peninsular Malaysia. Biological Conservation 120:329–344

Khorozyan, I. 2004. Strengthening local capacities for biodiversity conservation in Armenia. Final report submitted to The Whitley Awards, UK

Kucera, T. E. and R. H. Barrett. 1993. The Trailmaster camera system for detecting wildlife. Wildlife Society Bulletin 21:505–508

Laurance, W. F. and J. D. Grant. 1994. Photographic identification of ground-nest predators in Australian tropical rainforest. Wildlife Research 21:241–248

Liemgruber, P., W. J. McShea, and J. H. Rappole. 1994. Predation on artificial nests in large forest blocks. Journal of Wildlife Management 58:254–260

Mace, R. D., S. C. Minta, T. L. Manley, and K. A. Aune. 1994. Estimating grizzly bear population size using camera sightings. Wildlife Society Bulletin 22:74–83

Maffei, L., E. Cuellar, and A. Noss. 2004. One thousand jaguars (*Panthera onca*) in Bolivia's Chaco? Camera trapping in the Kaa-Iya National Park. Journal of Zoology 262:295–304

Maffei, L., A. J. Noss, E. Cuéllar, and D. I. Rumiz. 2005. Ocelot (*Felis pardalis*) population densities, activity, and ranging behaviour in the dry forests of eastern Bolivia: data from camera trapping. Journal of Tropical Ecology 21:349–353

Main, M. B. and L. W. Richardson. 2002. Response of wildlife to prescribed fire in southwest Florida pine flatwoods. Wildlife Society Bulletin 30:213–221

Major, R. E. 1991. Identification of nest predators by photography, dummy eggs and adhesive tape. Auk 108:190–195

Major, R. E. and G. Gowing. 1994. An inexpensive photographic technique for identifying nest predators at active nests of birds. Wildlife Research 21:657–666

Martyr, D. 1997. Important findings by FFI team in Kerinci Seblat, Sumatra, Indonesia. Oryx 31:80

Moruzzi, T. L., T. K. Fuller, R. M. DeGraaf, R. T. Brooks, and W. Li. 2002. Assessing remotely triggered cameras for surveying carnivore distribution. Wildlife Society Bulletin 30:380–386

Mudappa, D. 1998. Use of camera-traps to survey small carnivores in the tropical rainforest of Kalakad-Mundanthurai Tiger Reserve, India. Small Carnivore Conservation 18:9–11

Nichols, J. D., L. L. Bailey, A. F. O'Connell, Jr., N. W. Talancy, E. H. C. Grant, A. T. Gilbert, E. Annand, T. Husband, and J. Hines. 2008. Occupancy estimation using multiple detection devices. Journal of Applied Ecology 45:1321–1329

O'Brien, T. G., M. F. Kinnaird, and H. T. Wibisono. 2003. Crouching tigers, hidden prey: Sumatran tiger and prey populations in a tropical forest landscape. Animal Conservation 6:131–139

O'Connell, A. F. Jr., N. Talancy, L. L. Bailey, J. Sauer, R. Cook, and A. T. Gilbert. 2006. Estimating site occupancy and detection probability parameters for mammals in a coastal ecosystem. Journal of Wildlife Management 70:1625–1633

Picman, J. and L. M. Schriml. 1994. A camera study of temporal patterns of nest predation in different habitats. Wilson Bulletin 106:456–465

Proudfoot, G. A. 1996. Miniature video-board camera used to inspect natural and artificial nest cavities. Wildlife Society Bulletin 24:528–530

Rice, C. G. 1995. Trailmaster camera system: the dark side. Wildlife Society Bulletin 23: 110–111

Rivero, K., D. I. Rumiz, and A. B. Taber. 2005. Differential habitat use by two sympatric brocket deer species (*Mazama americana* and *M. gouazoubira*) in a seasonal Chiquitano forest of Bolivia. Mammalia 69:169–183

Roberts, C. W., B. L. Pierce, A. W. Braden, R. R. Lopez, N. J. Silvey, P. A. Frank, and D. Ransom, Jr. 2006. Comparison of camera and road survey estimates for white-tailed deer. Journal of Wildlife Management 70:263–267

Savidge, J. A. and T. F. Seibert. 1988. A camera device for monitoring avian nest predation at artificial nests. Journal Wildlife Management 52:291–294

Sequin, E. S., M. M. Jaeger, P. F. Brussard, and R. H. Barrett. 2003. Wariness of coyotes to camera traps relative to social status and territory boundaries. Canadian Journal of Zoology 81:2015–2025

Silver, S. C., L. E. T. Ostro Linde, L. Marsh, L. Maffei, A. J. Noss, M. J. Kelly, R. B. Wallace, H. Gómez, and G. Ayala. 2004. The use of camera traps for estimating jaguar *Panthera onca* abundance and density using capture/recapture analysis. Oryx 38:148–154

Soisalo, K. M. and S. M. C. Cavalcanti. 2006. Estimating the density of a jaguar population in the Brazilian Pantanal using camera-traps and capture-recapture sampling in combination with GPS radio-telemetry. Biological Conservation 129:487–496

Surridge, A. K., R. J. Timmins, G. M. Hewitt, and D. J. Bell. 1999. Striped rabbits in Southeast Asia. Nature 400:726

Swann, D. E., C. C. Hass, D. C. Dalton, and S. A. Wolf. 2004. Infrared-triggered cameras for detecting wildlife: an evaluation and review. Wildlife Society Bulletin 32:357–365

Thompson, F. R. III and D. E. Burhans. 2003. Predation of songbird nests differs by predator and between field and forest habitats. Journal of Wildlife Management 67:408–416

Trolle, M. and M. Kéry. 2003. Estimation of ocelot density in the Pantanal using capture–recapture analysis of camera-trapping data. Journal of Mammalogy 84:607–614

van Schaik, C. P. and M. Griffiths. 1996. Activity periods of Indonesian rain forest mammals. Biotropica 28:105–112

Wallace, R. B., H. Gomez, G. Ayala, and F. Espinoza. 2003. Camera trapping for jaguar (*Panthera onca*) in the Tuichi Valley, Bolivia. Mastozoologia Neotropica 10:133–139

Wegge, P., C. P. Pokheral, and S. R. Jnawali. 2004. Effects of trapping effort and trap shyness on estimates of tiger abundance from camera trap studies. Animal Conservation 7:251–256

York, E. C., T. L. Moruzzi, T. K. Fuller, J. F. Organ, R. M. Sauvajot, and R. M. DeGraaf. 2001. Description and evaluation of a remote camera and triggering system to monitor carnivores. Wildlife Society Bulletin 29:1228–1237

Chapter 4
Science, Conservation, and Camera Traps

James D. Nichols, K. Ullas Karanth, and Allan F. O'Connell

4.1 Introduction

Biologists commonly perceive camera traps as a new tool that enables them to enter the hitherto secret world of wild animals. Camera traps are being used in a wide range of studies dealing with animal ecology, behavior, and conservation. Our intention in this volume is not to simply present the various uses of camera traps, but to focus on their use in the conduct of science and conservation. In this chapter, we provide an overview of these two broad classes of endeavor and sketch the manner in which camera traps are likely to be able to contribute to them. Our main point here is that neither photographs of individual animals, nor detection history data, nor parameter estimates generated from detection histories are the ultimate objective of a camera trap study directed at either science or management. Instead, the ultimate objectives are best viewed as either gaining an understanding of how ecological systems work (science) or trying to make wise decisions that move systems from less desirable to more desirable states (conservation, management). Therefore, we briefly describe here basic approaches to science and management, emphasizing the role of field data and associated analyses in these processes. We provide examples of ways in which camera trap data can inform science and management.

J.D. Nichols (✉)
U.S. Geological Survey, Patuxent Wildlife Research Center, Laurel, MD 20708, USA
e-mail: jnichols@usgs.gov

K.U. Karanth
Wildlife Conservation Society – India Program, Centre for Wildlife Studies,
26-2, Aga Abbas Ali Road (Apt: 403), Bangalore, Karnataka 560042, India

A.F. O'Connell
U.S. Geological Survey, Patuxent Wildlife Research Center,
Beltsville, MD 20705, USA

A.F. O'Connell et al. (eds.), *Camera Traps in Animal Ecology: Methods and Analyses*,
DOI 10.1007/978-4-431-99495-4_4, © Springer 2011

4.2 Science

4.2.1 Approaches to Science

Rather than attempt a review of the various approaches to the conduct of science, we will focus on what are widely viewed as key steps in the process. Most discussions of science begin with hypotheses, "plausible stories" about how a studied system "works." Science can then be viewed as a process of sifting through these hypotheses and trying to identify one or two that provide fair approximations to reality, in the sense that they permit explanation of past events and prediction of future ones. Some approaches to science focus on one favorite hypothesis at a time, with a sequence of pair-wise comparisons leading to an eventual winner, whereas other approaches consider multiple plausible hypotheses simultaneously, with support accumulating for the best one or two among them (e.g., Platt 1964; Hilborn and Mangel 1997; Nichols 2001; Burnham and Anderson 2002; Williams et al. 2002; Stephens et al. 2005). Here we present brief sketches of these two approaches to science, emphasizing the critical step involving confrontation of predictions based on the different hypotheses with relevant data.

Single-hypothesis approaches to science typically compare the hypothesis of interest against a competing hypothesis, one at a time. In many cases the competing hypothesis is an omnibus hypothesis that is intended to represent anything other than the hypothesis of primary interest. For example, a hypothesis about a positive relationship between prey densities and tiger *Panthera tigris* numbers might be tested against a hypothesis that simply states the relationship does not exist. Such a test is based on a confrontation between the different predictions emerging from these two hypotheses and relevant data. The predictions typically come from quantitative models, which are simply mathematical representations of the hypotheses, or at least of the key relationships in those hypotheses (Levins 1966; Hilborn and Mangel 1997; Nichols 2001; Williams et al. 2002). For example, we could hypothesize a linear model relating tiger abundance to prey density, with a positive slope (perhaps of a specified value) corresponding to our focal hypothesis and any other slope (e.g., ≤ 0) corresponding to the competing hypothesis. Neither the hypotheses, nor the associated models used to produce specific predictions, are viewed as attempts to describe reality. Rather, we recognize that we are incapable of perceiving reality in all its complexity, so our hypotheses and their associated models are viewed as simplified approximations of natural processes.

We distinguish between the two hypotheses by comparing their respective predictions against data collected in the field. The hypothesis test then requires that we assess whether the data correspond more closely to one hypothesis than the other. If the data are found to be equally likely to have been generated by either hypothesis, then we conclude that they provide little support for the focal hypothesis. This conclusion may lead us to test the focal hypothesis again, or to modify it or develop a new hypothesis. On the other hand, if the data correspond substantially more closely to the focal hypothesis, then we conclude that the competing hypothesis can be rejected and that the data support the focal hypothesis. Such a conclusion might lead to additional testing, with the idea that we develop more and more faith in the hypothesis as it withstands more tests.

A multiple-hypothesis approach to science (Chamberlin 1897) begins with a set of plausible hypotheses about the system of interest, and the process of science is used to discriminate among them and to select one (or perhaps two) that provides the best approximation to reality. As under the single hypothesis approach, models (typically quantitative) are developed for each hypothesis as a means of generating specific predictions. The study system is manipulated or simply observed, and the resulting data are compared against the model-based predictions of the different hypotheses. Faith is increased in models that predict well (small distance between prediction and observation) and decreased for models that predict poorly (e.g., Hilborn and Mangel 1997; Nichols 2001; Williams et al. 2002). Define $\Pr(\text{data}_t | m_i)$ as the probability of observing the data at time t, given that model i is a good approximation to truth (the true process that generated the data). Further define model-specific weights ($p_{i,t}$) that reflect our confidence in model i at time t, relative to that for the other models in the model set (the models representing the hypotheses under consideration). At each time step when we obtain new data, we have the potential to change our faith in the different models. We view such changes in faith (i.e., in the $p_{i,t}$) as learning, expressed formally using Bayes' theorem (e.g., Williams et al. 2002) as

$$p_{i,t+1} = \frac{\Pr(data_t \mid m_i) p_{i,t}}{\sum_i \Pr(data_t \mid m_i) p_{i,t}} \tag{4.1}$$

According to expression (4.1), our relative faith in the considered hypotheses, and thus in the members of the model set, evolves through time based on new information (data_t) and on the faith accumulated up until that time, $p_{i,t}$. If the considered hypotheses include one that is a good approximation to reality, then we would expect its model weight to become large ($p_{i,t} \rightarrow 1$) and the weights for the other models to become small. If model weights fluctuate and do not accumulate for one or two models, then we may conclude that our set of competing hypotheses does not include a good approximation to reality, leading us to consider the development of new hypotheses.

Under both single- and multiple-hypothesis approaches to science, the confrontation of model-based predictions with observed data is the critical step. Hypothesis discrimination is accomplished using a sequence of paired comparisons under the single-hypothesis approach, whereas the multiple-hypothesis approach instead considers all hypotheses simultaneously at each comparison. But in both cases, the comparisons themselves ask how closely the data correspond to model-based predictions. The relevance of this sketch of approaches to the conduct of science to this volume is to provide a focus for the use of camera traps.

4.2.2 Science and Camera Traps

The primary utility of data and estimates resulting from camera trap studies will thus be to provide the observations that are used to discriminate among competing hypotheses. For example, investigations of animal behavior using camera trap data might focus on a priori hypotheses about activity patterns of animals (e.g., nocturnal vs. diurnal). Temporal distributions of photographs of active animals (e.g., Dillon and Kelly 2007) would then be compared against predictions of competing models about activity patterns.

Camera trap data are increasingly being used to address questions about spatial and temporal dynamics of animal populations. Capture-recapture (CR) approaches (e.g., Otis et al. 1978; Williams et al. 2002; Amstrup et al. 2005) can be used to estimate abundance and density for animals that can be individually identified from photographs (e.g., Karanth 1995; Karanth and Nichols 1998). Karanth et al. (2004b) used empirical information on the proportions of prey populations typically taken by tigers and on the kill rates of individual tigers to develop a simple model predicting tiger density as a function of prey density. They obtained camera trap estimates of tiger densities and distance-sampling estimates of prey densities from 11 sites throughout India. These data represented a wide range of variation in tiger and prey density and were consistent with the mechanistic model (Karanth et al. 2004b). The field effort involved in this macroecological investigation was substantial but resulted in corroboration of an important hypothesis about the determinants of tiger density. Questions about temporal dynamics can be addressed by sampling the same areas at multiple points in time. For example, Karanth et al. (2006) used camera traps to sample tigers at Nagarahole Park periodically from 1991 to 2000. This work resulted in estimates of annual survival probability and of population growth rate that were used to draw inferences about population viability and stability.

When animals are not individually identifiable, camera trap data have been used to develop indices to relative abundance (e.g., Carbone et al. 2001). The difficulty in using such indices to discriminate among competing hypotheses is the inability to know whether observed variation in camera trap count statistics is attributable to animal abundance/density or detection probabilities or both (Jenelle et al. 2002). New approaches to abundance estimation based on occupancy modeling (Royle and Nichols 2003) or repeated counts (Royle 2004) permit valid inference about abundance in some cases where animals are not individually identifiable.

Camera trap surveys at relatively large geographic scales can be used to estimate occurrence of a species across the landscape using occupancy models (Nichols and Karanth 2002; MacKenzie et al. 2006) that do not require individual identification. For example, MacKenzie et al. (2005) used data of Kawanishi on gaur (*Bos frontalis*) at different study areas in Malaysia to test hypotheses about area-specific variation in probabilities of occupancy. O'Connell et al. (2006) used occupancy modeling in conjunction with camera trapping (and two other sampling methods) to select from a variety of hypotheses about the influence of habitat on occupancy of several species of medium- and large-sized mammals on Cape Cod, Massachusetts, USA.

Information from this study was subsequently used to draw inferences about occupancy at two different geographic scales for the same species throughout New England (Nichols et al. 2008). Indeed, most camera trap studies that have used occupancy modeling have focused on spatial variation and habitat characteristics associated with such variation. However, if the same sites are surveyed with camera traps over time (e.g., every year), then the resulting data can be used to draw inferences about occupancy dynamics and the rate parameters (probabilities of local extinction and colonization) that govern these dynamics.

Camera trap studies can also be used to test hypotheses of community ecology (Tobler et al. 2008; O'Brien et al., Chap. 13). Species list data can be obtained for species photographed in different locations, and resulting data can be used to draw inferences about species richness using CR (e.g., Burnham and Overton 1979; Nichols and Conroy 1996; Boulinier et al. 1998) or occupancy (Dorazio and Royle 2005; MacKenzie et al. 2006; Royle and Dorazio 2008) approaches to modeling. Both modeling approaches lead directly to approaches for drawing inference about variation over time and space as well (Nichols et al. 1998a,b; Williams et al. 2002; Dorazio and Royle 2005; Royle and Dorazio 2008).

4.3 Management/Conservation

4.3.1 Structured Decision-Making: Introduction

Management and conservation can be viewed as decision processes by which humans attempt to move systems to desirable states (or maintain them there) and keep them away from undesirable states. At defined decision points through time, the manager decides what action to take, and the aim is to select the action that is most likely to achieve stated objectives. As an interesting comment on educational systems worldwide, we note that although management and conservation are processes that entail decisions and actions, in many universities it is possible to obtain undergraduate and graduate degrees in wildlife management and conservation biology without ever taking a course in decision-making. We believe that serious thought about decision-making has the potential to greatly improve our effectiveness in wildlife management and conservation.

"Structured decision-making" is a general descriptor for a process that involves breaking decisions into component parts and initially focusing on each part separately. Such decomposition appears to reduce confusion and lead to a useful clarity of thought. The components are then integrated to develop the full decision process, leading to inferences about which actions are recommended. Informed decision-making in general, and structured decision-making in particular, require four basic components: objectives, management actions, models, and monitoring (Williams et al. 2002).

4.3.2 Structured Decision-Making: Components

Objectives are simply clear statements of what managers hope to achieve with their systems. Conservation objectives might focus on minimizing a probability of extinction or maintaining a population above some specified threshold level. If the management problem includes multiple objectives and corresponding trade-offs, then it is sometimes possible to develop a common currency by which different consequences for the different state variables (variables that characterize system status; e.g., population size) can be evaluated. It is also possible to deal with multiple objectives via use of constraints. For example, one might seek to maximize timber harvest on an area subject to the constraint that population size of a focal species is expected to remain above some specified level. In order to use tools such as optimization for management decisions, objectives must be translated into formal statements known as objective functions. It is important to note that objectives reflect human values and should thus be constructed with input from all relevant stakeholders. Such stakeholders include, but are not restricted to, scientists and managers.

Management actions are specified during the development of the decision process, and the management decision then entails selection of one of these actions at each decision point. We prefer that the actions be discrete (as opposed to continuous variables) and relatively small in number, as this approach offers advantages for learning and optimization. For example, North American waterfowl harvest management considers four discrete "packages" of hunting regulations ranging from very restrictive (expected to produce small harvest rates) to liberal (high harvest rates) (Nichols et al. 1995). Even when actions involve a continuous variable such as proportion of an area on which timber harvest is permitted, it is possible to simply discretize (e.g., proportion area subject to harvest may be 0, 0.05, and 0.10) in order to produce a small set of potential actions. As with objectives, potential management actions should reflect human values and should be developed with input from all stakeholders. For example, predator control is a potentially effective management action that managers may choose not to consider because of political or societal values.

Models provide a basis for predicting the consequences of management actions and are thus an essential component of any sort of informed management (Kendall 2001; Nichols 2001; Williams et al. 2002). Such models might live only in the minds of experienced managers, but there are substantial advantages to making them more explicit and mathematical. These advantages include transparency and the ability to compute optimal solutions. Construction of mathematical models for use in decision processes must focus on changes in system state variables induced by implementation of different management actions. Variables that are not under the control of managers can enter models either implicitly as environmental variation or, if they are important influences and readily measured, explicitly. Frequently, there is substantial uncertainty about how actions translate into system responses. Such uncertainty can be accommodated by the inclusion of multiple models in the decision process. Just as described for the multiple-hypothesis approach to science, multi-model management requires model weights or metrics reflecting relative degrees of faith in the different members of the model set (Williams 1996; Williams et al. 2002). Unlike objectives and management actions, the development of management models is primarily the task of scientists and managers. However, if different stakeholder groups have very different ideas about system response to management actions, then it is important that these disparate views be included in the model set for reasons of transparency and fair play.

The final component required for making informed decisions is system monitoring. Indeed, camera trapping may form the basis for monitoring programs designed to inform certain management programs. For example, management directed at tigers might require estimates of tiger density or abundance across different times and/or locations, depending on the scale of the management problem. Estimates of state and other variables obtained from monitoring programs are used for three primary purposes in informed management (Yoccoz et al. 2001; Nichols and Williams 2006). The first purpose involves the state-dependence of management decisions. That is, the decision of what management action to impose at a particular decision point is likely to depend on whether population size, for example, is greater or smaller than we would like, or instead near a desired value. The second use of monitoring data is to assess the degree to which management objectives are being met. The third use of monitoring data is to inform our knowledge of system dynamics, as encoded in our system models. Data from monitoring programs are typically used to estimate the rate parameters (e.g., rates of survival, reproduction, local patch extinction) that populate the models. In addition, for management processes that are repeated through time, adaptive management (e.g., Walters 1986; Williams et al. 2002, 2007) can be used as a means of simultaneously managing for objectives and learning about system responses to management actions. A key step in adaptive management (see below) is the comparison of system state, as identified by the monitoring program, against the predictions of the different models describing system behavior. This step permits learning in exactly the same way as described under multiple-hypothesis science, with new information about how well the different models predict, leading to updated weights reflecting their relative predictive abilities (4.1).

4.3.3 Sources of Uncertainty

Structured decision-making focuses on these four components (objectives, actions, models, monitoring) separately, hopefully leading to clear thinking about the nature of the decision process. However, even when the decision maker has fully developed these components, decision-making can still be difficult. A primary source of this difficulty is uncertainty. Some decision makers respond to uncertainty with paralysis and calls for "more information," whereas the wise decision maker simply attempts to deal with uncertainty as part of the decision process (Walters 1986;

Williams et al. 2002, 2007; Nichols and Williams 2006). We recognize four categories of uncertainty in the management of animal populations and communities (e.g., Williams et al. 2002). First, ecological or structural uncertainty refers to the common situation in which system dynamics and, in particular, responses to management actions, are not completely known. As noted above, we can deal with this uncertainty via use of multiple models that incorporate our various hypotheses about system responses to management actions. Second, environmental variation represents an important source of uncertainty in all natural systems. Third, managers typically face the issue of partial controllability, in which management actions are applied only indirectly, and immediate effects of actions are characterized by uncertainty. Finally, partial observability refers to the manager's inability to directly observe the state of nature. Instead, system state must be estimated, with resulting estimates characterized by uncertainty. Indeed, the estimation methods that are used with camera trapping data and emphasized in this volume are those that permit estimation of variances that characterize partial observability.

4.3.4 Adaptive Resource Management

Adaptive resource management is a form of structured decision-making that is useful for so-called sequential decision processes, processes for which management decisions are made periodically through time (Walters 1986; Williams et al. 2002, 2007). For example, harvest management and certain kinds of habitat manipulation (e.g., prescribed burning) are frequently characterized as sequential decision processes, with decisions (e.g., about harvest quotas or whether or not to burn) made each year or at specified time periods. At each such decision point, the manager must decide which management action to take based on objectives, available actions, models of system response (with their respective weights which govern the relative influence of different models in the decision) and the current estimated system state obtained from the monitoring program. The actual process of selecting an action may be informal or may involve formal optimization methods (Williams 1996; Williams et al. 2002). In either case, the action is taken and the system responds in some manner. Monitoring then identifies the new system state, and comparison with model-based predictions leads to an updating of model weights using (4.1). At the next decision point (e.g., next year), armed with the same objectives, set of potential actions, and model set, but with new model weights and estimate(s) of system state, the manager makes the next decision and the process proceeds in an iterative manner. Just as with the multiple-hypothesis approach to science, learning is reflected in the evolution of model weights over time, with weights becoming larger for the model(s) that predict system responses to management well and smaller for those that predict poorly.

This iterative process of adaptive management can proceed indefinitely, but there is also an opportunity to revisit the components of the process. For example, after the process has been underway for some period of time, stakeholders may choose to reconsider objectives and/or available management actions. If none of the models in the model set seems to predict well (this might lead to model weights that fluctuate and do not accumulate for a single model), then additions to the model set should be considered by scientists and managers. It is also possible that the monitoring program is not as informative as it could be, in which case it might be altered as well. This periodic revisiting of the components of adaptive management is referred to as double-loop learning (Johnson 2006; Williams et al. 2007). The only restriction on such reconsideration of process components is that it should occur at time periods that are long relative to the iterative phase of the adaptive management process described above.

4.3.5 Management, Conservation, and Camera Trapping

We are not aware of any management program as described above that is currently being informed by a monitoring program based on camera trapping. Certainly a number of camera trap studies have considered scientific hypotheses that should be relevant to management. For example the relationship between predator and prey density (Karanth et al. 2004b; Kawanishi and Sunquist 2004) is obviously relevant to the use of law enforcement and protection of prey as management actions of potential use in managing predator populations. Studies of population dynamics in areas that do (Karanth et al. 2006) and do not emphasize law enforcement and protection are relevant to management as well. Similarly, studies directed at modeling occupancy (as estimated from camera trap data) as functions of habitat covariates (e.g., O'Connell et al. 2006) are relevant to potential habitat management.

We believe that monitoring programs based on camera traps have the potential to be very useful in future structured decision-making directed at animals that are most easily sampled by camera traps. For single-population management in which abundance is the state variable of interest, camera trapping may be the most efficient means of monitoring, at least for animals that can be individually identified from photographs. In cases where occupancy or proportion of patches occupied is the state variable of primary management interest, camera trapping will be the most efficient sampling approach for a number of secretive medium- to large-sized species (Karanth et al. 2004a). Animals that are more easily sampled by camera traps than by any other means (e.g., large cats) include several charismatic species at which substantial conservation funds have been directed. It seems reasonable to expect greater accountability for use of such funding in the future, and this expectation should lead to increasingly focused management efforts guided by structured decision processes.

4.4 Discussion

This chapter has focused on the processes of science and management and may seem to some readers to be misplaced in a volume dealing with camera trapping as a sampling methodology. Our rationale lies in our belief that neither the data nor the parameter estimates resulting from camera trapping have great inherent utility. Instead, they attain value only to the extent that they inform the larger processes of science or management in which they are embedded. This view of estimation and monitoring, not as stand-alone activities but as parts of a larger process, leads to the view that aspects of sampling design should also be inherited from the larger process. Thus, there is no omnibus design for camera trap studies that should be universally applicable. Instead, design should be dependent on, and tailored to, the uses to which the resulting data and estimates are to be put.

This view does not seem to be widely held by investigators involved in camera trap sampling. All too frequently, it appears that estimates of density or abundance or occupancy based on camera trapping are viewed as ends in themselves, rather than as information useful in informing a larger process. To some extent, this view can be defended with the observation that estimation based on camera trapping is a relatively new methodology that must be properly tested and evaluated. Our claim is that basic approaches to estimation have now been provisionally worked out, and the bulk of this volume will elaborate these approaches. Given this stage of relative maturity, it is time for scientists and managers to shift primary focus from the methods themselves to their application in the conduct of science and management. So while the remainder of this volume emphasizes *how* to use camera trapping data to estimate and monitor quantities of interest, we ask the reader to consider the *why* of estimation and monitoring, with an eye towards specific uses of camera trap data to inform science and/or management.

References

Amstrup, S. C., T. L. McDonald, and B. F. J. Manly. 2005. Handbook of capture-recapture analysis. Princeton University Press, Princeton, NJ

Boulinier, T., J. D. Nichols, J. R. Sauer, J. E. Hines, and K. H. Pollock. 1998. Estimating species richness: the importance of heterogeneity in species detectability. Ecology 79:1018–1028

Burnham, K. P. and D. R. Anderson. 2002. Model selection and multi-model inference: a practical information-theoretic approach. Springer, New York

Burnham, K. P. and W. S. Overton. 1979. Robust estimation of population size when capture probabilities vary among animals. Ecology 60:927–936

Carbone, C., S. Christie, T. Coulson, N. Franklin, J. Ginsberg, M. Griffiths, J. Holden, K. Kawanishi, M. Kinnaird, R. Laidlaw, A. Lynam, D. W. Macdonald, D. Martyr, C. McDougal, L. Nath, T. Obrien, J. Seidensticker, D. Smith, M. Sunquist, R. Tilson, and W. N. W. Shahruddin. 2001. The use of photographic rates to estimate densities of tigers and other cryptic mammals. Animal Conservation 4:75–79

Chamberlin, T. C. 1897. The method of multiple working hypotheses. Journal of Geology 5:837–848

Dillon, A. and M. J. Kelly. 2007. Ocelot *Leopardis pardalis* in Belize: the impact of trap spacing and distance moved on density estimates. Oryx 41:1–9

Dorazio, R. M. and J. A. Royle. 2005. Estimating size and composition of biological communities by modeling the occurrence of species. Journal of the American Statistical Association 100:389–398

Hilborn, R. and M. Mangel. 1997. The ecological detective. Confronting models with data. Princeton University Press, Princeton, NJ

Jenelle, C. S., M. C. Runge, and D. I. MacKenzie. 2002. The use of photographic rates to estimate densities of tigers and other cryptic animals: a comment on misleading conclusions. Animal Conservation 5:119–120

Johnson, F. A. 2006. Adaptive harvest management and double-loop learning. Transactions of the North American Wildlife and Natural Resources Conference 71:197–213

Karanth, K. U. 1995. Estimating tiger *Panthera tigris* populations from camera-trap data using capture-recapture models. Biological Conservation 71:333–338

Karanth, K. U. and J. D. Nichols. 1998. Estimation of tiger densities in India using photographic captures and recaptures. Ecology 79:2852–2862

Karanth, K. U., J. D. Nichols, N. S. Kumar, W. A. Link, and J. E. Hines. 2004a. Tigers and their prey: predicting carnivore densities from prey abundance. Proceedings of the National Academy of Sciences 101:4854–4858

Karanth, K. U., J. D. Nichols, and N. S. Kumar. 2004b. Photographic sampling of elusive mammals in tropical forests. Pages 229–247 *in* W. L. Thompson, editor. Sampling rare or elusive populations. Island Press, Washington, DC

Karanth, K. U., J. D. Nichols, N. S. Kumar, and J. E. Hines. 2006. Assessing tiger population dynamics using photographic capture-recapture sampling. Ecology 87:2925–2937

Kawanishi, K. and M. E. Sunquist. 2004. Conservation status of tigers in a primary rainforest of Peninsular Malaysia. Biological Conservation 120:329–344

Kendall, W. L. 2001. Using models to facilitate complex decisions. Pages 147–170 *in* T. M. Shenk and A. B. Franklin, editors. Modeling in natural resource management: development, interpretation and application. Island Press, Washington, DC

Levins, R. 1966. The strategy of model building in population biology. American Scientist 54:421–431

MacKenzie, D. I., J. D. Nichols, N. Sutton, K. Kawanishi, and L. L. Bailey. 2005. Suggestions for dealing with detection probability in population studies of rare species. Ecology 86:1101–1113

MacKenzie, D. I., J. D. Nichols, J. A. Royle, K. H. Pollock, L. A. Bailey, and J. E. Hines. 2006. Occupancy modeling and estimation. Academic, San Diego, CA

Nichols, J. D. 2001. Using models in the conduct of science and management of natural resources. Pages 11–34 *in* T. M. Shenk and A. B. Franklin, editors. Modeling in natural resource management: development, interpretation and application. Island Press, Washington, DC

Nichols, J. D. and M. J. Conroy. 1996. Estimation of species richness. Pages 227–234 *in* D. Wilson et al., editors. Measuring and monitoring biological diversity: standard methods for mammals. Smithsonian Institution Press, Washington, DC

Nichols, J. D. and K. U. Karanth. 2002. Statistical concepts: assessing spatial distributions. Pages 29–38 *in* K. U. Karanth and J. D. Nichols, editors. Monitoring tigers and their prey. A manual for wildlife managers, researchers, and conservationists. Centre for Wildlife Studies, Bangalore

Nichols, J. D. and B. K. Williams. 2006. Monitoring for conservation. Trends in Ecology and Evolution 21:668–673

Nichols, J. D., F. A. Johnson, and B. K. Williams. 1995. Managing North American waterfowl in the face of uncertainty. Annual Review of Ecology and Systematics 26:177–199

Nichols, J. D., T. Boulinier, J. E. Hines, K. H. Pollock, and J. R. Sauer. 1998a. Estimating rates of local extinction, colonization and turnover in animal communities. Ecological Applications 8:1213–1225

Nichols, J. D., T. Boulinier, J. E. Hines, K. H. Pollock, and J. R. Sauer. 1998b. Inference methods for spatial variation in species richness and community composition when not all species are detected. Conservation Biology 12:1390–1398

Nichols, J. D., L. L. Bailey, A. F. O'Connell, Jr., N. W. Talancy, E. H. C. Grant, A. T. Gilbert, E. Annand, T. Husband, and J. E. Hines. 2008. Occupancy estimation using multiple detection devices. Journal of Applied Ecology 45:1321–1329

O'Connell, A. F., N. W. Talancy, L. L. Bailey, J. R. Sauer, R. Cook, and A. T. Gilbert. 2006. Estimating site occupancy and detection probability parameters for mammals in a coastal ecosystem. Journal of Wildlife Management 70:1625–1633

Otis, D. L., K. P. Burnham, G. C. White, and D. R. Anderson. 1978. Statistical inference from capture data on closed animal populations. Wildlife Monographs 62:1–135

Platt, J. R. 1964. Strong inference. Science 146:347–353

Royle, J. A. 2004. *N*-mixture models for estimating population size from spatially replicated counts. Biometrics 60:108–115

Royle, J. A. and R. M. Dorazio. 2008. Hierarchical modeling and inference in ecology. Academic, New York

Royle, J. A. and J. D. Nichols. 2003. Estimating abundance from repeated presence absence data or point counts. Ecology 84:777–790

Stephens, P. A., S. W. Buskirk, G. D. Hayward, and C. M. del Rio. 2005. Information theory and hypothesis testing: a call for pluralism. Journal of Applied Ecology 42:4–12

Tobler, M. W., S. E. Carrillo-Percastegui, R. L. Pitman, R. Mares, and G. Powell. 2008. An evaluation of camera-traps for inventorying large and medium-sized terrestrial rainforest mammals. Animal Conservation 11:169–178

Walters, C. J. 1986. Adaptive management of renewable resources. MacMillan, New York

Williams, B. K. 1996. Adaptive optimization and the harvest of biological populations. Mathematical Biosciences 136:1–20

Williams, B. K., J. D. Nichols, and M. J. Conroy. 2002. Analysis and management of animal populations. Academic, San Diego, CA

Williams, B. K., R. C. Szaro, and C. D. Shapiro. 2007. Adaptive management: the US Department of the Interior technical guide. Adaptive Management Working Group, US Department of the Interior, Washington, DC

Yoccoz, N. G., J. D. Nichols, and T. Boulinier. 2001. Monitoring of biological diversity in space and time. Trends in Ecology and Evolution 16:446–453

Chapter 5
Behavior and Activity Patterns

Andrew S. Bridges and Andrew J. Noss

5.1 Introduction

The study of animal behavior has long been a topic of interest to evolutionary biologists, conservation biologists, and wildlife managers. The study of activity patterns is a subset of the general study of behavior. Jürgen Aschoff, famous for his research of circadian rhythms, stated in 1954 "an animal is active when it moves parts of its body or moves itself." Despite the popularity of this research topic, recording and quantifying behavior and activity patterns in wild, free-ranging animals presents challenges and, to this end, a variety of techniques have been employed with varying degrees of success. Remote camera systems (i.e., camera traps) are the newest tool for researchers examining both animal behavior and activity patterns. Although not a panacea, camera traps have allowed researchers to overcome some of the challenges that have hampered previous efforts. In this chapter, we review applications of camera traps to the study of animal behavior and activity.

5.2 Traditional Techniques of Studying Animal Behavior and Activity

Direct observation of a study animal by a biologist physically present in the field is a time-tested tool for evaluating wildlife behavior and activity. Prior to the arrival of radio telemetry in the 1960s, direct observation was the predominant ethological technique, and despite many liabilities, is still commonly used today. Direct observation of

A.S. Bridges (✉)
Institute for Wildlife Studies, 2515 Camino del Rio South, Suite 334,
San Diego, CA 92108, USA
e-mail: bridges@iws.org

A.J. Noss
Director, Proyecto Gestión Integrada de Territorios Indígenas, Wildlife Conservation Society – Ecuador, Av. Eloy Alfaro N37-224 y Coremo, Quito, Ecuador

A.F. O'Connell et al. (eds.), *Camera Traps in Animal Ecology: Methods and Analyses*,
DOI 10.1007/978-4-431-99495-4_5, © Springer 2011

animal activity can be conducted for extended periods of time and used to evaluate reactions of the subject to environmental stimuli that otherwise would be difficult to document were a researcher not physically present with the animal. However, human presence can alter natural behavior and activity patterns of the target animals. Observational studies can also be limited by sample sizes due to logistical constraints, and spurious conclusions may be drawn if many hours of observation are conducted on only a small number of individuals.

Attaching radio transmitters to individual animals has proven to be a valuable tool for researchers examining activity patterns and behavior. Telemetry allows animals to be tracked across the landscape while generating information about their activity patterns and behavior. Consecutive triangulations in different locations and modulation in signal strength have been used as indicators that an animal was moving across a landscape. Because all animals are handled when transmitters are attached, sex, age, reproductive status and other individual-specific parameters are often known and can be included as covariates in behavioral analyses. Additionally, unlike direct observational methods, researchers employing telemetry can gather data remotely with little disturbance to the animals. They also can monitor more individuals and gather data at times and in places not conducive to direct observation. However, inferences based on radio telemetry have inherent limitations due to biases associated with trapping and handling, and because misinterpreting signal strength fluctuations or triangulation error can lead to data errors. Thus, although both observational and telemetry-based techniques have played an important role in behavioral research, new methods were needed to alleviate potential biases and allow for new avenues of investigation and insight.

5.3 Advantages and Implications for Using Camera Traps

Remote camera systems are one of the newest tools in the ethologist's arsenal. They combine many of the advantages of the traditional techniques described above while offering a number of improvements. Although camera housing, sounds, and sometimes flashes associated with some camera systems could potentially modify behavior, disturbance is likely to be less than would be expected if the researcher were present directly observing behavior (Alexy et al. 2003; Bridges et al. 2004a; Griffiths and Van Schaik 1993). Data gathered using cameras for population estimation sometimes allow researchers to concurrently quantify activity patterns of their target species (Bridges et al. 2004b; Dillon and Kelly 2006; Maffei et al. 2004). Also, if non-target species can be photographed during sampling, researchers also may be able to examine activity patterns of these species as well (Noss et al. 2003, 2004). Because it allows researchers to simultaneously study activity patterns of multiple species using a single study design, researchers have recently started examining temporal activity partitioning and associated implications for niche overlap in sympatric species (Fedriani et al. 2000; de Almeida Jacomo et al. 2004; Rivero et al. 2005; Wacher and Attum 2005).

In some cases, the use of camera traps requires identification of specific locations where animals are presumed to occur, thereby limiting the type of behavior that can be observed. For example, photographing nest sites can document predatory behavior, whereas camera traps located near carcasses or fruiting trees can observe foraging behavior. However, camera trap photographs may not provide specific information about behavior, as the animals' actions sometimes cannot be determined from a single photograph. Behavior is often inferred from the presence of the animal in a particular location. For example, animals photographed at artificial nests are often considered potential predators of natural nests. Unless the predator is photographed with the broken eggshell in its mouth, however, predation can only be confirmed by a posteriori examination of the nest (Hernández et al. 1997). Patterns of habitat use are often based on comparative photographic encounter rates, although photos cannot distinguish animals that are occupying a particular habitat from those that are simply passing through.

Triggering mechanisms vary based on the camera systems' intended applications. Many studies use commercially available camera systems that incorporate a motion and/or heat sensor to trigger the shutter when an animal is present. Some studies use actively triggered systems where breaking a beam of light triggers the camera (see Swann et al., Chap. 3). Nest predation studies sometimes utilize a trigger system attached to an egg in the nest, ensuring that only displacement or removal of the egg, presumably by a predator, will trigger the camera. Other studies use a weight sensitive treadle to activate the system.

5.4 Case Studies

Use of camera traps to study animal behavior or activity patterns can be broadly divided into the following categories: (1) circadian rhythms, (2) nest predation, (3) foraging, (4) niche partitioning and social systems, (5) habitat use, and (6) refugia and reproduction.

5.4.1 Circadian Rhythms

Because an entire animal population, not just those individuals that can be physically captured, is potentially exposed to photography, researchers can come closer to population level assessment of activity than using other currently available techniques (Bridges et al. 2004b). In research involving American black bears *Ursus americanus*, Bridges et al. (2004b) recorded activity patterns at bait sites in Virginia and found bear activity was generally vespertine but exhibited seasonal differences with bears becoming more nocturnal in the fall, possibly in response to hunting seasons and associated pursuit by bear-hounds. Examining mule deer *Odocoileus hemionus* activity, Hernández et al. (2005) used camera systems positioned at feed boxes and found that this species was most active from evening through early morning. Hicks et al. (1998) compared activity patterns of deer mice *Peromyscus* spp. in North Carolina and found that camera traps were not biased due to their ability to photograph individuals throughout the entire sample period. 'In contrast' Sherman traps generated biased data because once the traps were occupied, they were no longer available for capture to indicate activity of the target species. In Malaysia, Azlan and Sharma (2006) found that tigers *Panthera tigris*, leopard cats *Prionailurus bengalensis* and clouded leopards *Neofelis nebulosa* were heavily nocturnal, whereas leopards *Panthera pardus* and golden cats *Catopuma temminckii* were less active at night. These tigers may have been most active at dawn and dusk to maximize encounter rates with both diurnal and nocturnal prey (Laidlaw and Noordin 1998). Di Bitetti et al. (2006) examined felid activity on trails in Argentina and found that ocelots (*Leopardus pardalis*) were primarily nocturnal with no significant difference between males and females and more activity during dark-sky periods (around the new moon). Similarly, Dillon and Kelly (2006) used camera traps to conclude that both sexes of ocelots in Belize were primarily nocturnal. Examining usage timing at highway underpasses, Foster and Humphrey (1995) found Florida panthers crossed exclusively at night, raccoons *Procyon lotor* and bobcats *Lynx rufus* crossed throughout the evening into morning, and white-tailed deer *Odocoileus virginianus* and wading birds were most frequently photographed during morning hours. Among the earliest studies using cameras, Pearson (1959) determined that meadow mice *Microtus californicus* were active all day and night, harvest mice *Reithrodontomys megalotis* were nocturnal, brush rabbits *Sylvilagus bachmani* were most active in the morning, and shrews *Sorex ornaturs* were nocturnal and more active in the winter.

Several large-scale studies have measured activity patterns and circadian rhythms for multiple taxa simultaneously. In the dry forests of Bolivia, research teams set cameras primarily on trails and roads (but also on salt licks, ponds, and river banks) and described the activity patterns for a variety of mammal species including: jaguars (Maffei et al. 2004), jaguarundi *Puma yaguarondi* (Maffei et al. 2007a), Geoffroy's cats *Oncifelis geoffroyi* (Cuéllar et al. 2006), ocelots (Gómez et al. 2005; Maffei et al. 2005), crab-eating raccoons *Procyon cancrivorous* (Arispe et al. 2008; Gómez et al. 2005), crab-eating fox *Cerdocyon thous* and pampas fox *Pseudalopex gymnocercus* (Maffei et al. 2007b), lowland tapirs *Tapirus terrestris* (Gómez et al. 2005; Noss et al. 2003; Wallace et al. 2002), giant armadillos *Priodontes maximus* (Noss et al. 2004), three-banded armadillo *Tolypeutes matacus*, agouti *Dasyprocta punctata/variegata*, paca *Cuniculus paca*, puma, gray brocket deer *Mazama gouazoupira* red brocket deer *Mazama americana*, collared peccary *Pecari tajacu*), white-lipped peccary *Tayassu pecari*, and Brazilian rabbits *Sylvilagus brasiliensis* (Gómez et al. 2005; Maffei et al. 2002). In Indonesian rain forests where camera traps were set along natural and human trails, Van Schaik and Griffiths (1996) describe activity patterns for wildlife in Sumatra and Java, reporting findings for 13 carnivores, nine ungulates, three other mammal species, three birds, and three reptiles.

5.4.2 Nest Predation

The largest set of behavioral studies using camera traps has focused on predation of bird nests, permitting the identification of egg and fledgling predators at natural nests (Major and Gowing 1994; Smith 2004) or potential predators at artificial nests (J.W. Cain et al. 2003; Hernández et al. 1997; Leimgruber et al. 1994; Savidge and Seibert 1988; Sawin et al. 2003; Sieving and Willson 1999; Van der Werf 2001). Some studies also assessed the relative importance of these predators in relation to nest failure (Farnsworth and Simons 2000; Major et al. 1999; Meckstroth and Miles 2005; Picman and Schriml 1994).

Liebezeit and Luke (2002) identified nest predation behavior using surveillance cameras (video and film) and found that raptors preyed on nestlings more frequently than on eggs, whereas small mammals depredated nestlings and eggs in proportion to their availability. Buler and Hamilton (2000) compared predation activity at camera-monitored vs. camera-free nests, and concluded that predation of artificial shrub nests did not accurately mimic that of natural shrub nests, thereby emphasizing the need for research on natural nests. Using artificial nests and camera traps, Laurance and Grant (1994) and Maier and DeGraaf (2000) confirmed a number of species as predators and other species as potential predators. Cooper and Ginnett (2000) compared predation levels at artificial ground nests at varying distances from deer feeders. Picman and Schriml (1994) found that mammalian predators were nocturnal, whereas avian predators were diurnal. In one of the few non-avian nest predation studies, Hunt and Ogden (1991) identified predators and timing of predation at American alligator *Alligator mississippiensis* nests and found that American black bears were the principal egg predators with common raccoons and rice rats *Oryzomys palustris* also taking eggs.

5.4.3 Foraging

Foraging areas are places that provide food, water, or nutrients for animals and can provide insight into animal behavior and activity patterns. Photographs allow for estimates of the timing, duration and frequency of visits by particular species (Claridge et al. 2004), determine group size and sociality (Altendorf et al. 2001; Hernández et al. 2005; López González and Lorenzana Piña 2002; Miura et al. 1997; Otani 2001), and document behavior–vigilance vs. active foraging (Altendorf et al. 2001; Hernández et al. 2005; Otani 2001; Page et al. 2001).

Perovic (2002) used live pigs (*Sus scrofa*) as bait to test the effectiveness of electric fencing for modifying jaguar behavior and to deter predation on domestic pigs. Examining foraging at flowering, fruiting, and seeding trees, Beck and Terborgh (2002) monitored seed removal from palms (*Astrocaryum murumuru*) inside and outside of peccary exclosures and found that the most frequent visitors were the green acouchi *Myoprocta pratti* and spiny rat *Proechimys* spp.

Similarly, Kitamura et al. (2004) investigated seed removal of mahogany *Aglaia spectabilis* and used camera traps to photograph the consumption of fallen fruits on the forest floor. Carthew and Slater (1991) developed a technique for monitoring and identifying pollinators at flowering plants and proposed a number of applications for their newly devised cameras systems. In Malaysia, Miura et al. (1997) found varying activity patterns in mammals, birds and reptiles visiting fruiting trees. Mazurek and Zielinski (2004) evaluated the importance of older trees spared during harvest for subsequent habitat management of California redwoods *Sequioa sempervirens* and found no disproportionate use of these trees by mammals.

Devault and Rhodes (2002) and Devault et al. (2004) used cameras to identify vertebrate scavengers and timing of scavenging events at rodent carcasses in upland pine and bottomland hardwood forests. Cameras deployed near a cow *Bos taurus* carcass allowed López González and Lorenzana Piña (2002) to describe activity patterns of jaguars and turkey vultures *Cathartes aura.* Pierce et al. (1998) examined the onset of feeding behavior in mountain lions *Puma concolor* in California by placing cameras at kill sites. They found that most feeding occurred after sunset and females with new kittens fed significantly earlier than males or females in other reproductive classes.

Edelman et al. (2005) studied squirrel behavior and activity patterns by placing baits and camera traps at occupied Mt. Graham red squirrel *Tamiasciurus hudsonicus grahamensis* middens, unoccupied middens, and at random sites to identify previously marked Abert's squirrels *Sciurus aberti* and red squirrels visiting the sites. They also used visits by other mammals and birds to document species richness. Claridge et al. (2004) examined spotted-tailed quoll *Dasyurus maculates* visitation timing at communal latrines in Australia and found that visitation took place at all hours instead of the nocturnal activity pattern that had been expected.

At salt licks in the Brazilian Pantanal, Pfeifer (2006) used camera traps to document geophagy for 14 taxa including agouti, tapir, peccaries, brocket deer, cracids, columbids. Atwood and Weeks Jr. (2003) compared the mean number of visits per camera day by white-tailed deer using natural, pure salt, and commercial mineral salt licks and found more visits by females than males in all seasons. Griffiths and Van Schaik (1993) used cameras on trails in conjunction with systems set at wallows and salt licks to determine that human traffic influenced animal behavior with species responding in different ways to this disturbance. These authors (van Schaik and Griffiths 1996) later found that the size of mammals and the substrate determined if behavior patterns were predisposed to being nocturnal, diurnal, or cathemeral (equally active throughout the day and night). Morgart et al. (2005) studied water usage by placing camera traps at surface water sources to examine frequency and seasonality of use by pronghorns *Antilocapra americana*. In a unique study to examine the effects of prey vigilance, Cresswell et al. (2002) used automatic cameras to study sparrowhawk *Accipiter nisus* attacks on vigilant vs. non-vigilant prey, documenting the direction of the attack in relation to the prevailing wind. Weckel et al. (2006) compared spatial (type of trail and distance from man-made roads) and temporal (activity patterns) distribution of jaguars with that of their principal prey species to describe jaguar feeding ecology. They suggested that jaguar prey use was related to its abundance; however, jaguars may rely on foraging strategies other than chance encounters for exploiting prey.

5.4.4 Niche Partitioning and Social Systems

Several research teams have used camera traps to collect information on behavior and activity patterns, using the data to build models for evaluation of niche partitioning, overlap, and competition among sympatric species. Most of the research in this area has been conducted on carnivores with at least one report on ungulates. Fedriani et al. (2000) quantified activity time using camera traps at scent stations as a variable in examining competition between sympatric gray foxes, coyotes, and bobcats in California. Wacher and Attum (2005) used camera traps to study niche partitioning in red foxes *Vulpes vulpes* and Rueppell's sand foxes *Vulpes rueppelli* and found similar activity patterns that suggested overlap. In another study involving medium-sized canids, de Almeida Jacomo et al. (2004) used camera traps set on trails to study niche overlap in maned wolves *Chrysocyon brachyurus*, crab-eating foxes, and hoary foxes *Lycalopex vetulus* in Brazil, and found similar nocturnal foraging times but differential diurnal foraging times. Séquin et al. (2003) used camera traps to describe social structure and characterize behavior of controlling alpha, resident beta, and non-territorial transient coyotes *Canis latrans*. Camera trap records may also permit analyses of spatial and temporal separation or avoidance behavior among multiple individuals of one or more species. For example, sympatric jaguars and pumas in Chaco and Chiquitano dry forests were studied using data from camera trap surveys by measuring both temporal and spatial separation among individuals (Noss and Venticinque 2006). Rivero et al. (2005) used cameras on trails to determine that sympatric red and gray brocket deer were active at different times within a 24-h period.

5.4.5 Habitat and Corridor Usage

Camera traps have become popular tools for assessing habitat occupancy and use. This is generally accomplished by comparing relative abundance of study species across habitats based on photographic capture frequencies. Augustine (2004) studied impala *Aepyceros melampus* in glades vs. bushland habitats and used the number of herds photographed per 24-hr period to evaluate seasonal presence and found that impala selected nutrient-rich glades more frequently than acacia bushland during dry and wet seasons. Raillard and Svoboda (2000) examined habitat use by muskoxen and characterized the behavior of each animal as either foraging or resting.

A unique study by Hilty and Merenlender (2004) examined usage of corridors. They monitored vineyards near and far from core and riparian habitat corridors and found that the activity of native mammalian predators was higher in vineyards near core habitats, whereas the activity of non-native predators was higher in vineyards further from core habitats. Camera traps are frequently used to evaluate the effectiveness of a range of artificial features such as culverts, overpasses, underpasses that are intended to facilitate wildlife movements across the landscape and reduce highway mortality (Cain et al. 2003; Dodd et al. 2004; Ng et al. 2004). Many of these studies use camera traps in conjunction with other detection methods, including direct observation, tracks, and radio-telemetry (Foster and Humphrey 1995; Mata et al. 2005).

5.4.6 Refugia and Reproduction

Camera traps have also been deployed at entry or exit points of animal dens or burrows to document activity patterns. These systems allow researchers to document activity while minimizing disturbance to animals entering or exiting these sites (Bridges et al. 2004a). In addition to describing nest predation, camera traps can generate information on other aspects of reproductive activity. In the first published study to use cameras to examine bear denning behavior, Bridges et al. (2004a) used camera traps outside of American black bear dens to quantify activity patterns of bears moving in and out of den structures in Virginia. They found that bears were surprisingly active during denning and frequently exited and re-entered dens prior to final emergence. They were also able to calculate cub age at den emergence. At gopher tortoise *Gopherus polyphemis* burrows, Alexy et al. (2003) used an actively triggered camera system to monitor movements and activity of tortoises and commensals in Florida. Although nocturnal activity was also recorded, gopher tortoises were found to be strongly diurnal. Doody and Georges (2000) used camera traps in a study of hog-nosed turtle *Carettochelys insculpta* nests to determine nest initiation timing, emergence time of hatchlings, and to link mothers to specific nests.

5.4.7 Statistical Analyses

Behavioral and activity patterns lend themselves particularly well to graphical representation, and thus many papers do not report statistics, instead relying on inference drawn from histograms and proportions. Although this type of information may paint a valid picture from a biological perspective, the use of formal statistical methods allow for more robust inference. Traditional statistical methods such as analysis of variance (Augustine 2004; Campbell et al. 2006), chi-square goodness of fit tests (Foresman and Pearson 1999; Griffiths and Van Schaik 1993; Liebezeit and Luke 2002; de Almeida Jacomo et al. 2004; Campbell et al. 2006), likelihood ratio chi-square tests (Rivero et al. 2005), Kruskal–Wallace test (Pierce et al. 1998), Pearson's r (Zegers et al. 2000), F-tests (Zegers et al. 2000), and Mann–Whitney U tests (Griffiths and Van Schaik 1993) have been employed to examine differences in behavior and activity. Advances in statistical analyses of habitat selection offer new opportunities for analysis of activity pattern data. These techniques can often be applied if the analyst treats activity time as a use-vs.-availability question. For example, Bridges et al. (2004b) employed compositional analysis (Aebischer et al. 1993) in their analyses of American black bear activity patterns and Di Bitetti et al. (2006) used circular statistics (e.g., Batschelet 1981) in their analyses of ocelot circadian rhythms.

5.5 Future Applications for Camera Traps in Behavior Studies

Camera traps provide a tool that offers many advantages compared to traditional techniques for examining animal behavior. However, because some aspects of behavior may be better studied using other techniques, researchers may prefer to use multiple techniques simultaneously to provide a more thorough analysis. Replacing film-based systems with digital systems has resulted in camera traps that are relatively quiet and less likely to alarm or disturb animals. Digital systems also increase image storage capacity which reduces the frequency of researcher visitation and site disturbance. Additionally, many digital systems can also now record videos and can provide new research opportunities. Video systems inserted into previously inaccessible areas, such as nest boxes and bear dens, will provide insight into previously unattainable aspects of behavior and activity. Cameras mounted to individual animals, such as CritterCams®, will likely expose previously unknown behaviors (see Nichols et al., Chap. 14). Technological advances and continued innovation ensure camera traps will almost certainly play an increasingly important role in the study of wildlife behavior and activity.

References

Aebischer, N. J., P. A. Robertson, and R. E. Kenward. 1993. Compositional analysis of habitat use from animal radiotracking data. Ecology 74:1313–1325

Alexy, K. J., K. J. Brunjes, J. W. Gassett, and K. V. Miller. 2003. Continuous remote monitoring of gopher tortoise burrow use. Wildlife Society Bulletin 31:1240–1243

Altendorf, K. B., J. W. Laundré, C. A. López González, and J. S. Brown. 2001. Assessing effects of predation risk on foraging behavior of mule deer. Journal of Mammalogy 82:430–439

Arispe, R., C. Venegas, and D. Rumiz. 2008. Abundancia y patrones de actividad del mapache (*Procyon cancrivorous*) en un bosque Chiquitano de Bolívia. Mastozoología Neotropical 15(2):323–333

Atwood, T. C. and P. H. Weeks, Jr. 2003. Sex-specific patterns of mineral lick preference in white-tailed deer. Northeastern Naturalist. 10:409–414

Augustine, D. J. 2004. Influence of cattle management on habitat selection by impala on central Kenyan rangeland. Journal of Wildlife Management 68:916–923

Azlan, J. M. and D. S. K. Sharma. 2006. The diversity and activity patterns of wild felids in a secondary forest in Peninsular Malaysia. Oryx 40:36–41

Batschelet, E. 1981. Circular statistics in biology. Academic Press, New York, NY, USA

Beck, H. and J. Terborgh. 2002. Groves versus isolates: how spatial aggregation of *Astrocaryum murumuru* palms affects seed removal. Journal of Tropical Ecology 18:275–288

Bridges, A. S., J. A. Fox, C. Olfenbuttel, and M. R. Vaughan. 2004a. American black bear denning behavior: observations and applications using remote photography. Wildlife Society Bulletin 32:188–193

Bridges, A. S., M. R. Vaughan, and S. Klenzendorf. 2004b. Seasonal variation in American black bear *Ursus americanus* activity patterns: quantification via remote photography. Wildlife Biology 10:277–284

Buler, J. J. and R. B. Hamilton. 2000. Predation of natural and artificial nests in a southern pine forest. Auk 117:739–747

Cain, A. T., V. R. Tuovila, D. G. Hewitt, and M. E. Tewes. 2003. Effects of a highway and mitigation projects on bobcats in Southern Texas. Biological Conservation 114:189–197

Cain, J. W. III, M. L. Morrison, and H. L. Bombay. 2003. Predator activity and nest success of willow flycatchers and yellow warblers. Journal of Wildlife Management 67:600–610

Campbell, T. A., C. A. Langdon, B. R. Laseter, W. M. Ford, J. W. Edwards, and K. V. Miller. 2006. Movements of female white-tailed deer to bait sites in West Virginia, USA. Wildlife Research 33:14

Carthew, S. M. and E. Slater. 1991. Monitoring animal activity with remote photography. Journal of Wildlife Management 55:689–692

Claridge, A. W., G. Mifsud, J. Dawson, and M. J. Saxon. 2004. Use of infrared digital cameras to investigate the behaviour of cryptic species. Wildlife Research 31:645–650

Cooper, S. M. and T. F. Ginnett. 2000. Potential effects of supplemental feeding of deer on nest predation. Wildlife Society Bulletin 28:660–666

Cresswell, W., J. Lind, U. Kaby, J. L. Quinn, and S. Jakobsson. 2002. Does an opportunistic predator preferentially attack nonvigilant prey? Animal Behavior 66:643–648

Cuéllar, E., L. Maffei, R. Arispe, and A. J. Noss. 2006. Geoffroy's cats at the northern limit of their range: activity patterns and density estimates from camera trapping in Bolivian dry forests. Studies on Neotropical Fauna and Environment 41:169–177

de Almeida Jacomo, A. T., L. Silveira, and J. A. F. Diniz-Filho. 2004. Niche separation between the maned wolf (*Chrysocyon brachyurus*), the crab-eating fox (*Dusicyon thous*) and the hoary fox (*Dusicyon vetulus*) in central Brazil. Journal of Zoology 262:99–106

Devault, T. L. and O. E. Rhodes, Jr. 2002. Identification of vertebrate scavengers of small mammal carcasses in a forested landscape. Acta Theriologica 47:185–192

Devault, T. L., I. L. Brisbin, Jr. and O. E. Rhodes, Jr. 2004. Factors influencing the acquisition of rodent carrion by vertebrate scavengers and decomposers. Canadian Journal of Zoology 82:502–509

Di Bitetti, M. S., A. Paviolo, and C. De Angelo. 2006. Density, habitat use and activity patterns of ocelots (*Leopardus pardalis*) in the Atlantic forest of Misiones, Argentina. Journal of Zoology 270:153–163

Dillon, A. and M. J. Kelly. 2006. Ocelot *Leopardus pardalis* in Belize: the impact of trap spacing and distance moved on density estimates. Oryx 41:469–477

Dodd, C. K., W. J. Barichivich, and L. L. Smith. 2004. Effectiveness of a barrier wall and culverts in reducing wildlife mortality on a heavily traveled highway in Florida. Biological Conservation 118:619–631

Doody, J. S. and A. Georges. 2000. A novel technique for gathering turtle nesting and emergence phenology data. Herpetological Review 31:220–222

Edelman A. J., J. L. Koprowski, and J. R. Edelman. 2005. Kleptoparasitic behavior and species richness at Mt. Graham red squirrel middens. Pages 395–398 *in* G. J. Gottfried, B. S. Gebow, L. G. Eskew, and C. B. Edminster, editors. Connecting mountain islands and desert seas: biodiversity and management of the Madrean Archipelago II. Proceedings Rocky Mountain Research Station. RMRS-P-36. US Department of Agriculture, Forest Service, Fort Collins, CO

Farnsworth G. L. and T. R. Simons. 2000. Observations of wood thrush nest predators in a large contiguous forest. Wilson Bulletin. 112:82–87

Fedriani, J. M., T. K. Fuller, R. M. Sauvajot, and E. C. York. 2000. Competition and intraguild predation among three sympatric carnivores. Oecologia 125:258–270

Foresman, K. R. and D. E. Pearson. 1999. Activity patterns of American martens, *Martes americana*, snowshoe hares, *Lepus americanus*, and red squirrels, *Tamiasciurus hudsonicus*, in west-central Montana. Canadian Field-Naturalist 113:386–389

Foster, M. L. and S. R. Humphrey. 1995. Use of highway underpasses by Florida panthers and other wildlife. Wildlife Society Bulletin 23:95–100

Gómez, H., R. B. Wallace, G. Ayala, and R. Tejada. 2005. Dry season activity patterns for some Amazonian mammals. Studies of Neotropical Fauna and the Environment 40:91–95

Griffiths, M. and C. P. Van Schaik. 1993. The impact of human traffic on the abundance and activity patterns of Sumatran rain forest wildlife. Conservation Biology 7:623–626

Hernández, F., D. Rollins, and R. Cantu. 1997. Evaluating evidence to identify ground-nest predators in west Texas. Wildlife Society Bulletin 25:826–831

Hernández, L., J. W. Laundré, and M. Gurung. 2005. Use of camera traps to measure predation risk in a puma-mule deer system. Wildlife Society Bulletin 33:353–358

Hicks, N. G., M. A. Menzel, and J. Laerm. 1998. Bias in the determination of temporal activity patterns of syntopic *Peromyscus* in the southern Appalachians. Journal of Mammalogy 79:1016–1020

Hilty, J. A. and A. M. Merenlender. 2004. Use of riparian corridors and vineyards by mammalian predators in northern California. Conservation Biology 18:126–135

Hunt, R. H. and J. J. Ogden. 1991. Selected aspects of the nesting ecology of American alligators in the Okefenokee Swamp. Journal of Herpetology 25:448–453

Kitamura, S., S. Suzuki, T. Yumoto, P. Poonswad, P. Chuailua, K. Plongmai, N. Noma, T. Maruhashi, and C. Suckasam. 2004. Dispersal of *Aglaia spectabilis*, a large-seeded tree species in a moist evergreen forest in Thailand. Journal of Tropical Ecology 20:421–427

Laidlaw, R. and W. S. W. Noordin. 1998. Activity patterns of the Indochinese tiger (*Panthera tigris corbetti*) and prey species in peninsular Malaysia. Journal of Wildlife and Parks 16:85–96

Laurance, W. F. and J. D. Grant. 1994. Photo identification of ground-nest predators in Australian tropical rainforest. Wildlife Research 21:241–248

Leimgruber, P., W. J. McShea, and J. H. Rappole. 1994. Predation on artificial nests in large forest blocks. Journal of Wildlife Management 58:254–260

Liebezeit, J. R. and G. T. Luke. 2002. Nest predators, nest-site selection, and nesting success of the Dusky Flycatcher in a managed ponderosa pine forest. Condor 104:507–517

López González, C. A. and G. Lorenzana Piña. 2002. Carrion use by jaguars (*Panthera onca*) in Sonora, Mexico. Mammalia 66:603–605

Maffei, L., E. Cuéllar, and A. J. Noss. 2002. Uso de trampas-cámara para la evaluación de mamíferos en el ecotono Chaco-Chiquitanía. Revista Boliviana de Ecología y Conservación Ambiental 11:55–65

Maffei, L., E. Cuéllar, and A. J. Noss. 2004. One thousand jaguars (*Panthera onca*) in Bolivia's Chaco? Camera trapping in the Kaa-Iya National Park. Journal of Zoology, London 262:295–304

Maffei, L., A. J. Noss, E. Cuéllar, and D. Rumiz. 2005. Ocelot (*Felis pardalis*) population densities, activity, and ranging behavior in the dry forests of eastern Bolivia: data from camera trapping. Journal of Tropical Ecology 21:349–353

Maffei, L., A. Noss, and C. Fiorello. 2007. The jaguarundi (*Felis (Herpailurus) yaguaroundi*) in the Kaa-Iya del Gran Chaco National Park, Bolivia. Mastozoología Neotropical 14:263–266

Maffei, L., R. Paredes, A. Segundo, and A. J. Noss. 2007. Home range and activity of two sympatric fox species in the Bolivian dry Chaco. Canid News 10.4 [online] (http://www.canids.org/canidnews/10/Sympatric_foxes_in_Bolivia.pdf)

Maier, T. J. and R. M. DeGraaf. 2000. Predation on Japanese quail vs. house sparrow eggs in artificial nests: small eggs reveal small predators. Condor 102:325–332

Major, R. E. and G. Gowing. 1994. An inexpensive photographic technique for identifying nest predators at active nests of birds. Wildlife Research 21:657–666

Major, R. E., F. J. Christie, G. Gowing, and T. J. Ivison. 1999. Elevated rates of predation on artificial nests in linear strips of habitat. Journal of Field Ornithology 70:351–364

Mata, C., I. Hervás, J. Herranz, F. Suárez, and J. E. Malo. 2005. Complementary use by vertebrates of crossing structures along a fenced Spanish motorway. Biological Conservation 124:397–405

Mazurek, M. J. and W. J. Zielinski. 2004. Individual legacy trees influence vertebrate wildlife diversity in commercial forests. Forest Ecology and Management 193:321–334

Meckstroth, A. M. and K. A. Miles. 2005. Predator removal and nesting waterbird success at San Francisco Bay, California. Waterbirds 28:250–255

Miura, S., M. Yasuda, and L. C. Ratnam. 1997. Who steals the fruits? Monitoring frugivory of mammals in a tropical rain forest. Malayan Nature Journal 50:183–193

Morgart, J. R., J. J. Hervert, P. R. Krausman, J. L. Bright, and R. S. Henry. 2005. Sonoran pronghorn use of anthropogenic and natural water sources. Wildlife Society Bulletin 33:51–60

Ng, S. J., J. W. Dole, R. M. Sauvajot, S. P. D. Riley, and T. J. Valone. 2004. Use of highway undercrossings by wildlife in southern California. Biological Conservation 115:499–507

Noss, A. J. and E. Venticinque. 2006. Jaguares *Panthera onca* y pumas *Puma concolor* simpátricos: patrones de separación temporal y espacial. II Congreso de Mastozoología en Bolivia, La Paz, Bolivia

Noss, A. J., R. L. Cuéllar, J. Barrientos, L. Maffei, E. Cuéllar, R. Arispe, D. Rúmiz, and K. Rivero. 2003. A camera trapping and radio telemetry study of *Tapirus terrestris* in Bolivian dry forests. Tapir Conservation 12:24–32

Noss, A. J., R. Peña, and D. I. Rumiz. 2004. Camera trapping *Priodontes maximus* in the dry forests of Santa Cruz, Bolivia. Endangered Species Update 21:43–52

Otani, T. 2001. Measuring fig foraging frequency of the Yakushima macaque by using automatic cameras. Ecological Research 16:49–54

Page, L. K., R. K. Swihart, and K. R. Kazacos. 2001. Seed preferences and foraging by granivores at raccoon latrines in the transmission dynamics of the raccoon roundworm (*Baylisascaris procyonis*). Canadian Journal of Zoology 79:616–622

Pearson, O. P. 1959. A traffic survey of *Microtus-Rheithrodontomys* runways. Journal of Mammalogy 40:168–180

Perovic, P. G. 2002. Conservación del jaguar en el noroeste de Argentina. Pages 465–475 *in* R. A. Medellín, C. Equihua, C. L. B. Chetkiewicz, P. G. Crawshaw, Jr., A. Rabinowitz, K. H. Redford, J. G. Robinson, E. W. Sanderson, and A. B. Taber, editors. El jaguar en el nuevo milenio. Fondo de Cultura Económica, UNAM, Wildlife Conservation Society, Mexico

Pfeifer, C. I. 2006. Relações entre barreiros e a fauna de vertebrados no nordeste do Pantanal, Brasil. Master's thesis. Universidade Federal do Rio Grande do Sul, Porto Alegre, Brazil

Picman, J. and L. M. Schriml. 1994. A camera study of temporal patterns of nest predation in different habitats. Wilson Bulletin 106:456–465

Pierce, B. M., V. C. Bleich, C.-L. B. Chetkiewicz, and J. D. Wehausen. 1998. Timing of feeding bouts of mountain lions. Journal of Mammalogy 79:222–226

Raillard, M. and J. Svoboda. 2000. High grazing impact, selectivity, and local density of muskoxen in central Ellesmere Island, Canadian High Arctic. Arctic, Antarctic and Alpine Research 32:278–285

Rivero, K., D. I. Rumiz, and A. B. Taber. 2005. Differential habitat use by two sympatric brocker deer species (*Mazama americana* and *M. gouazoubira*) in a seasonal Chiquitano forest of Bolivia. Mammalia 69:169–183

Savidge, J. A. and T. F. Seibert. 1988. An infrared trigger and camera to identify predators at artificial nests. Journal of Wildlife Management 52:291–294

Sawin, R. S., M. W. Lutman, G. M. Linz, and W. J. Bleier. 2003. Predators on red-winged blackbird nests in eastern North Dakota. Journal of Field Ornithology 74:288–292

Séquin, E. S., M. M. Jaeger, P. F. Brussard, and R. H. Barrett. 2003 Wariness of coyotes to camera traps relative to social status and territory boundaries. Canadian Journal of Zoology 81:2015–2025

Sieving, K. E. and M. F. Willson. 1999. A temporal shift in Steller's jay predation on bird eggs. Canadian Journal of Zoology 77:1829–1834

Smith, M. L. 2004. Edge effects on nest predators in two forested landscapes. Canadian Journal of Zoology 82:1943–1953

Van der Werf, E. A. 2001. Rodent control decreases predation on artificial nests in o'ahu 'elepaio habitat. Journal of Field Ornithology 72:448–457

Van Schaik, C. P. and M. Griffiths. 1996. Activity patterns of Indonesian rain forest mammals. Biotropica 28:105–112

Wacher, T. and O. Attum. 2005. Preliminary investigation into the presence and distribution of small carnivores in the Empty Quarter of Saudi Arabia through the use of a camera trap. Mammalia 69:81–84

Wallace, R., G. Ayala, and H. Gómez. 2002. Lowland tapir activity patterns and capture frequencies in lowland moist tropical forest. Tapir Conservation 11:14

Weckel, M., W. Giuliano, and S. Silver. 2006. Jaguar (*Panthera onca*) feeding ecology: distribution of predator and prey through time and space. Journal of Zoology 270:25–30

Zegers, D. A., S. May, and L. J. Goodrich. 2000. Identification of nest predators at farm/forest edge and forest interior sites. Journal of Field Ornithology 71: 207–216

Chapter 6
Abundance, Density and Relative Abundance: A Conceptual Framework

Timothy G. O'Brien

6.1 Introduction

In the early 1990s, biologists began experimenting with camera traps to estimate the abundance of tigers *Panthera tigra* in the Nagarahole National Park (Karanth 1995), marking the first time that camera traps were used to sample a wildlife population in a statistically rigorous manner. Since that time, camera traps have been employed for a wide variety of uses in behavioral and ecological studies. Camera trap studies can result in capture histories of species whose members are individually recognizable via distinct natural traits or artificial markings (e.g. radio collars, tags) as well as capture histories of species that are not reliably identified as individuals. In either case, dependent on study objectives, each type of data may be used to estimate population size, species richness, site occupancy or relative abundance indices. In addition, well-designed camera trap studies usually include data on covariates at the sites where the cameras are set. Ideally, covariates are chosen based on their purported influence on abundance or other parameters of interest, including detectability (White 2005). The challenge to biologists is to use these data to the greatest extent possible, to make unbiased inferences about the state of the target wildlife population under investigation.

In this chapter, I review some design and analytical issues associated with the use of camera traps to estimate abundance, density, and relative abundance, to make comparisons among estimates and to monitor populations over time. I focus on the estimation of detection probability, abundance in both open and closed populations, and density, as well as on occupancy-abundance models, abundance indices and sources of bias. Almost all of the topics I cover here are explored in much more detail elsewhere, for example in books such as *The Estimation of Animal Abundance and Related Parameters* (Seber 1982), *Techniques for Wildlife Investigations* (Skalski and Robson 1992), *The Analysis and Management of Animal Populations* (Williams et al. 2002), and *Occupancy Estimation and Modeling* (MacKenzie et al. 2006).

T.G. O'Brien (✉)
Wildlife Conservation Society, 2300 Southern Blvd, Bronx, NY 10460, USA
Mpala Research Centre, P.O. Box 555, Nanyuki, Kenya
e-mail: tobrien@wcs.org

A.F. O'Connell et al. (eds.), *Camera Traps in Animal Ecology: Methods and Analyses*,
DOI 10.1007/978-4-431-99495-4_6, © Springer 2011

6.2 Estimation of Abundance

The estimation of abundance of wildlife populations lies at the heart of most ecological research and monitoring. Ecologists seeking to understand population processes require models that accurately track populations. Management, whether for game species or non-game species, often seeks to maintain or increase population size of target species. Management for decreasing population size falls into the more specialized realm of pest control or invasive species management. Monitoring efforts attempt to understand the effect of management interventions or covariates, such as climate change, on the population. When studying most animals, it is rarely possible to conduct a complete count or census. Sampling methods usually yield a count of some portion of the population or an index that relates to actual abundance via the probability of detection. Camera trap methods become particularly useful in studies of animals that roam over areas too large to sample completely, and when the species is elusive and difficult to observe.

Population size estimation from samples has a long history dating from LaPlace's (1786) development of ratio estimators for application in demographic analysis to Peterson's (1896) basic formulation and Lincoln's (1930) application to estimation of waterfowl populations. Subsequent development has been substantial, relaxing constraints on sources of variation in detection probability and on population closure. Closed populations models have been reviewed in Otis et al. (1978), Seber (1982), Williams et al. (2002), and Chao and Huggins (2005a,b).

Many of the published camera trap studies have the objective of estimating abundance at a site, and the study is designed to estimate the size of a single population in a spatially and temporally closed system (Karanth 1995; Karanth and Nichols 2000; Silver et al. 2004), based on surveys over a subset of the total area of interest. Only one study, to date, has used camera traps to estimate vital rates over time (Karanth et al. 2006). Often unstated, is why the estimate is needed or how the estimate will be used. Specifying the objectives for the estimation exercise, or why the abundance estimate is required, is an important first step (Yoccoz et al. 2001; Pollock et al. 2002, see Nichols et al., Chap. 4).

Estimation of abundance over large study areas, a typical objective of camera trap studies, requires that an investigator pay special attention to spatial variation in abundance. When attempting to make inference about a large area, it is rarely possible to sample the entire area of interest, and investigators should take care to select locations for sampling arrays that are representative of the area for which inferences are made. If the investigator wishes to make inferences beyond the effective sampling area (e.g., extend the inference from a sampling area to an entire park), then rules of stratification or random sampling should apply in determining the location of a sampling array (Cochran 1997; Thompson 1992). Often, sampling areas are chosen because they appear to be typical of the larger area of interest, or because they are easy to access. Representativeness, however, is not easy to assess subjectively and easily accessible areas often are not typical of areas that are not easily accessed. In the case of non-random choice of sampling areas, it may be that not all areas have a chance of being sampled, and inferences are then restricted to the portion of the population that does have a chance of being sampled.

Once a sampling area is assigned for a camera trap study, the investigator faces the problem of sampling the target species. If all of the animals in the sample area could be counted, we would have a census of the sample area. Detectability, however, is rarely complete, so we must consider the more common situation in which the detection probability, denoted by p or β, is less than 1. For the purpose of abundance estimation, detection probability can be defined as the likelihood that an individual will be detected (photographed or captured) if it is present in a sample unit during the time of the sample. Detection is a source of variability in abundance estimation because not all animals will be detected with absolute certainty during a sampling effort, individuals may vary in their detectability and detection may vary over time and space. The likelihood of detecting an individual during a sample occasion provides the key to converting the sample count statistic into an estimate of abundance or density. Detection probabilities therefore are an important component of any abundance estimation exercise or monitoring program.

The incorporation of an estimate of the detection probability distinguishes a population estimate from a population index. An abundance index is a statistic that is assumed to be correlated with the population size in some way. If the assumption holds, then an average change in the index should reflect an average change in population numbers. Abundance indices require that the index have a monotonic relationship to true abundance; however, the nature of the functional relationship between the index and abundance is crucial for meaningful inference. An index based on a monotonic linear relationship ($I=bN$) will not behave the same as an index based on a nonlinear monotonic relationship (e.g. $I=e^{bN}$). Without knowledge of the functional relationship, all the abundance index reveals is the direction of change in N. Calibration of an index with independent estimates of abundance helps improve the interpretation of the index but the calibration needs to be checked over time and space to ensure that it holds. Because of the strong assumptions needed to use an index, they should be avoided unless no reasonable alternative is possible. I return to the topic of indices at the end of this chapter.

In studies designed to estimate abundance using camera traps, camera trap placement in the sampling area affects the ability to detect individual animals and therefore the detection probability. For abundance estimation, cameras should be placed in locations that maximize the chance of detecting the target species. If camera trapping points are spaced too widely, there may be areas large enough to contain a target species' home range that are not subject to coverage by a camera. Animals living in these gaps in coverage have essentially no chance of being detected even though they reside within the sampling area. Closer spacing of cameras closes the gaps and increases the probability of detecting individual animals by increasing the number of cameras in a home range. Animals with more camera trap points in their home ranges may have higher detection probabilities than animals with fewer trap points, creating heterogeneity in detection probabilities. There are tradeoffs in camera trap deployment because close trap spacing often reduces the sample area coverage, resulting in fewer individuals being available for sampling whereas wide trap spacing may create holes, or reduce an individual's exposure to traps. Appropriate placement and spacing of camera sampling points depends on a thorough understanding of the movement patterns and habitat use of the target species.

If we consider a properly collected sample count C of animals on some study area, then C is related to the population size N by the detection probability. C can be considered a random variable that likely varies each time a sample is collected. E(C) is the expected value or average of the count over a very large number of replicated samples of the population:

$$\mathrm{E}(C) = Np \text{ or } N\beta. \tag{6.1}$$

If detection probability, p, can be estimated by some means, then abundance can be estimated as

$$\hat{N} = \frac{C}{\hat{p}}. \tag{6.2}$$

This is the most basic form of a general population estimator used in both distance and capture–recapture (CR) sampling and is often called the canonical estimator (Williams et al. 2002). The estimator can also be written as

$$\hat{N} = \sum_{i=1}^{C} 1/\hat{p} \tag{6.3}$$

to emphasize that each animal counted shares a common detection probability and that the abundance estimate relies on the detection of individuals as well as the count of individuals. This equation can then be generalized to include heterogeneity in detection probability among the C individuals in the sample by substituting $\hat{p}_i$ for $\hat{p}$ in expression (6.3).

The estimate of the variance of population size can be generally written as

$$\mathrm{var}(\hat{N}) \approx [\mathrm{var}(C)/\mathrm{E}(C)^2 + \mathrm{var}(\hat{p})/\hat{p}^2]\hat{N}^2. \tag{6.4}$$

When a population estimate is required over an area that is too large to survey completely, a representative proportion α of the area is sampled, and the estimate for the entire area incorporates the probability of sampling a particular area, α:

$$\hat{N} = C/\hat{N} = C/(\hat{p}\,\hat{\alpha}) = \sum_{i=1}^{C} 1/(\hat{p}_1\hat{\alpha}). \tag{6.5}$$

The sampling fraction is frequently known, but it is written above for the general case in which the sampling fraction is estimated. In the case of sampling that involves selection of potential sample units with probability α, the variance of the count statistic contains a spatial component that declines as the sampling fraction α increases. A survey of the complete area does not have this spatial component.

Bailey et al. (2004a, b) point out that frequently part of the population may become temporarily unavailable for sampling through temporary emigration from the study area, burrowing or hiding at a point, or through camera failures (or thefts) that create holes in the sample coverage. In this case, $\hat{p}_i = \hat{p}_{ai}\ \hat{p}_{di}$, where $\hat{p}_{ai}$ estimates the probability that the *i*th individual is available for detection and $\hat{p}_{di}$ estimates the probability that *i*th individual was detected given that it was available for detection. Under the assumption that a population is closed to additions and subtractions during the sampling period, $\hat{p}_{ai}$ is often assumed to equal 1.

Estimating the population size when $\hat{p}_{ai} < 1$ typically requires the use of open-population models (Williams et al. 2002). If temporary emigration is a random process, estimates of total population size should be unbiased (Kendall 1999). Temporary emigration during sampling occasion *j* may depend on whether the individual was in or out of the study area during occasion $j-1$. This is called Markovian emigration, and it requires two probabilities, γ'_j = the probability that an animal absent from the sampling area in time $j-1$ stays away from the study area in time *j*, and γ''_j = the probability that an animal present in the sampling area in time $j-1$, moves out of the area in time *j*. Kendall et al. (1997) present models for estimating both types of emigration when data are collected using Pollock's (1982) robust design (see below).

The interpretation of $\hat{p}$ is the probability that an individual is caught at least once during a sampling period of *K* trap occasions. When $K=2$,

$$\hat{p} = 1-(1-p_1)(1-p_2)$$

or 1 minus the probability of *not* being encountered during either trapping occasion. When $K>2$ trap occasions, this generalizes to

$$\hat{p} = 1-\prod_{i=1}^{K}(1-p_i). \tag{6.6}$$

Equation (6.6) is the general form of the detection probability corresponding to the primary sampling period. As *K* increases, the likelihood of capturing an individual at least once ($\hat{p}$) increases for a given p_i (Figs. 6.1 and 6.2). This is especially important for camera trap estimation of abundance because often the species of interest has a very low likelihood of being captured on a given occasion (low p_i). The ability to increase the number of sampling occasions for little additional cost by leaving the cameras in position for extra days may increase estimates of $\hat{p}$. If behavioral changes in detectability occur after the first capture (trap-response), and if recapture probability is not related to capture probability, then all of the information for estimating *p* and *N* is contained in the number of initial detections occurring in each sampling period.

6.2.1 Closed Capture–Recapture Models

Closed CR estimators are used to estimate the population size of a target species composed of marked or individually recognizable individuals, in a geographically and temporally closed system that precludes immigration, emigration, births and deaths. The Lincoln–Petersen estimator (Peterson 1896; Lincoln 1930) is the

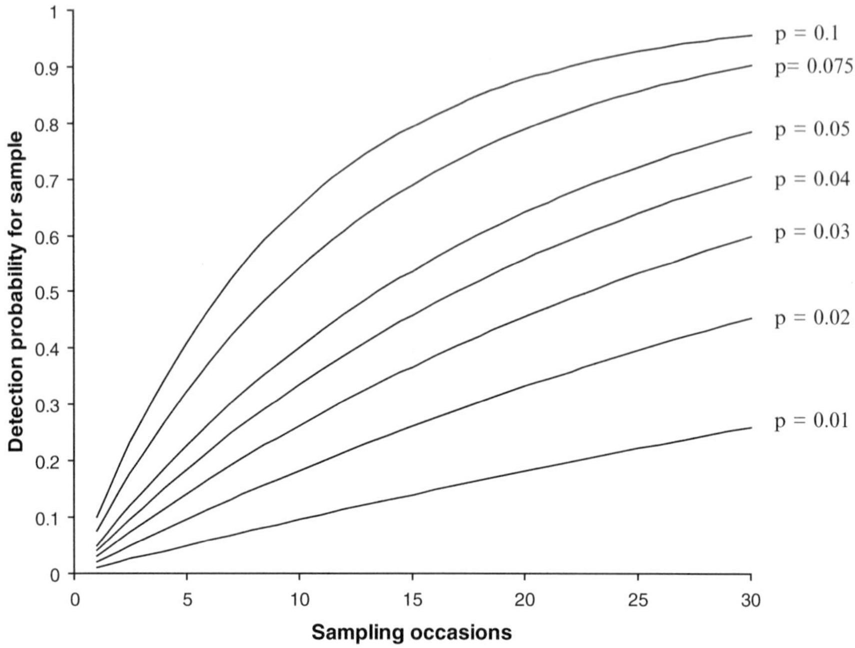

Fig. 6.1 The cumulative likelihood of capturing an individual with a Pr(detection) per sampling occasion $=p$ over $K=1\dots30$ sampling occasions. As p and K increase, the likelihood of detection approaches 1

simplest case of this family of estimators, involving just two sampling occasions ($K=2$) with a relatively short interval between occasions. Consider a population of N individuals. On the first trapping occasion, n_1 animals are photographed and identified. On the second trapping occasion, n_2 animals are photographed and identified, of which m animals were captured on the first trapping occasion. The proportion of marked animals in the population after the first capture occasion is n_1/N. If all animals have equal capture probabilities, an assumption of the estimator discussed below, then the proportion of marked animals in the population should be the same as the proportion of marked animals in the second sample. If true,

$$n_1 / N = m / n_2$$

which leads to the Lincoln–Petersen estimator,

$$\hat{N} = n_1 n_2 / m. \tag{6.7}$$

If we consider n_1 as a count statistic, C, the estimate of the capture probability associated with the count statistic is $\hat{p}_1 = m/n_2$, and we can derive $\hat{N}$ using the canonical estimator (6.2).

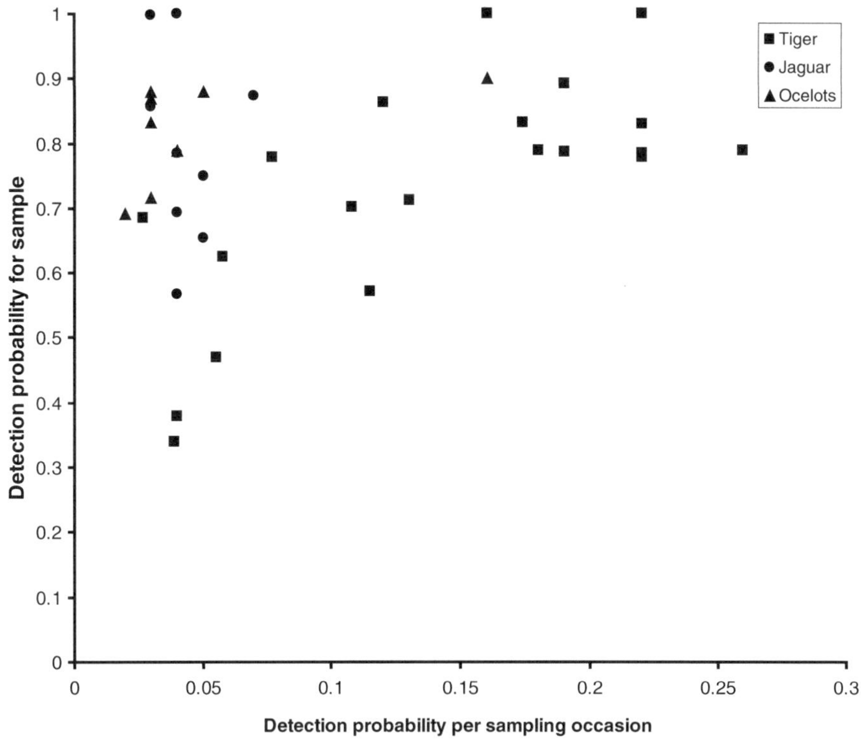

Fig. 6.2 Relationship between detection probability per sampling occasion and the probability of detecting an individual at least once during the study. Data are from published population estimates for tigers, jaguars (*P. onca*) and ocelots (*Leopardus pardalis*). Estimates of detection probabilities calculated using CAPTURE software (p^*) illustrate that camera trap studies typically result in large detection probabilities ($\hat{p}^* = 0.788$, SD = 0.164, $n = 41$ estimates)

When the number of capture occasions exceeds $K=2$, it is possible to model variation in capture probability that arises from time, behavioral response to capture, and individual heterogeneity. These K-sample models are described by Otis et al. (1978) and others as the M_0 (constant p), M_t (temporal variation in p), M_b (p varies between first capture and subsequent captures), and M_h models (p varies by individual). More complex combinations of these models are possible, but have not been used widely in the camera trap literature. M_h models are commonly used in camera trap studies because associated estimators are robust and because it seems realistic that individuals would have different probabilities of capture. Because camera traps are considered passive data collection devices, they are not expected to impart a trap response following capture (but see Wegge et al. 2004). In any event, trap response can be modeled (M_b, M_{bh}) and does not present a problem in abundance estimation.

Use of closed population models requires that three assumptions be met:

(1) Population is closed to birth, death, immigration and emigration processes. The interpretation of population closure is slightly different for the Lincoln–Petersen model vs. *K*-sample closed models. Deaths that occur during sampling (*d*) are known, and these animals can be removed from the CR data. $\hat{N}$ then estimates the population size after sampling, and $\hat{N}+d$ estimates the pre-sampling population. Losses that occur between sampling (emigration and death) are tolerated when they are equally likely for marked and unmarked individuals, but the interpretation of $\hat{N}$ is restricted to population size at the first sampling period. When there are additions to the population between sampling (births and immigration), $\hat{N}$ estimates the population size at the second sampling period. When both losses and gains occur between sampling periods, $\hat{N}$ is positively biased, and open models should be used instead. When temporary immigration and emigration occur randomly, the detection probability estimate includes the probability of an individual being present in the study area during sampling as well as the probability of capture given that it is present (Kendall 1999). In this situation, the Lincoln–Petersen closed population estimate, $\hat{N}$, and *K*-sample estimates represent a super-population composed of animals that are capable of moving into and out of the sample area. Tests of the population closure assumption are available in the software programs CAPTURE (Otis et al. 1978) and MARK (White and Burnham 1999).

(2) No tag loss during the study. Loss of identifiable individuals after the first trapping period causes detection probability to be underestimated in subsequent trap occasions. Seber (1982) describes double tag studies to estimate tag loss. This is not too much of a problem when distinctive marks are used to identify individuals, but it may be a problem when photos contain target animals but the distinguishing marks cannot be observed. Again, if this inability to detect distinguishing marks is a result of a poor photograph, and poor photographs are a random occurrence (i.e. not more likely with one animal than another), then the absence of detection is simply incorporated into the complement of detection probability. Only if some individual animals are consistently poorly photographed compared to other individuals in the sample can bias arise due to heterogeneity of detection.

(3) Sources of variation in detection probability are properly identified and modeled. Ensuring that all sources of variation in detection are included in an analysis requires that one consider the most general (i.e. complex) model. There is a tradeoff in modeling between bias and precision because the most complex models fit better, but provide less precise estimates than models with fewer parameters. One alternative is to choose the most parsimonious model (Otis et al. 1978; Burnham and Anderson 2002) using the goodness of fit tests and discriminant analysis for model selection in CAPTURE (Otis et al. 1978) or the Akaike's Information Criterion (AIC) to evaluate models in MARK (Cooch and White 2006). Another alternative is to incorporate uncertainty in model selection using the model-averaging approach to estimation (Buckland et al. 1997; Stanley and Burnham 1998; Burnham and Anderson 2002). Under model averaging, the estimate $\hat{N}$ is a weighted average of several model estimates. The weights represent the likelihood that a particular model is the "best" of a set of candidate models.

6.2.2 Open Capture–Recapture Models

When the population is open to the demographic processes of birth, death, immigration, or emigration that cause permanent changes in the population (as opposed to temporary movement in and out of a study area), open-population models may be used to incorporate gains and losses over time. Open models may be single age models in which capture and survival probabilities are assumed to be the same for all individuals in the sampled population or multiple-age models in which survival and capture probabilities are age- or age-class-specific. Camera trap studies of marked cats have normally assumed that kittens were not reliably sampled and all other animals were adults. These assumptions indicate the use of a single-age model, known as the Jolly–Seber model, that includes capture probability, survival, recruitment and abundance as parameters (Jolly 1965; Seber 1965), although multi-age models would be possible if enough data were available and individuals could be classified to age groups. Variations of the Jolly–Seber model include partially open models (Darroch 1959), which consider only losses or only gains, and reduced-parameter models in which some of the detection and/or survival parameters are assumed to be constant over time (Brownie et al. 1986). Recruitment can be obtained as a derived quantity (Jolly 1965; Seber 1965) or modeled using Pradel's temporal symmetry approach (Pradel 1996) or a "superpopulation" approach in which the superpopulation is the total number of individuals that were in the study area at any time between the first and last sampling occasion (Crosbie and Manly 1985; Schwarz and Arnason 1996).

The Jolly–Seber open model is based on two kinds of parameters: p_i is the probability that a marked individual in the population at period i is photographed during period i, and φ_i is the probability that a marked individual in the population at period i survives until period $i+1$ and remains in the study area, a condition that precludes permanent emigration. φ_i combines the probability that an individual survives with the probability that it does not permanently emigrate, and as such is not a true survival rate. An additional parameter, χ_i, is the probability that an animal alive and in the population at period i is not caught or observed again in any subsequent period. χ_i's are calculated by

$$\chi_i = (1-\varphi_i) + \varphi_i(1-p_{i+1})\chi_{i+1} \tag{6.8}$$

For a study involving K periods, $\chi_K = 1-\chi_i$ describes the fates of animals that are not recaptured after the ith period: either they fail to survive in the study area (die or move away) or they survive in the area and simply are not captured again. Two additional parameters of the Jolly–Seber model, representing the probability of release for marked and unmarked animals captured in the ith period, are assumed to be 1 in camera trap studies since camera trap "capture" does not involve a physical capture or risk of removal. Williams et al. (2002) provide a description of the structure of the probability model.

The Jolly–Seber model considers the population size at each time interval (N_i) and its components (M_i marked animals and U_i unmarked animals) and the number of new recruits (B_i) as unknown variables to be estimated. Abundance estimation at each time interval corresponds to the canonical estimator (6.2)

$$\hat{N}_i = n_1 / \hat{p}_i = \hat{M}_i n_i / m_i. \tag{6.9}$$

where $\hat{M}_i$ is equal to the estimated number of marked animals in the population at the time. $\hat{M}_i$ can be estimated by $m_i + R_i z_i / r_i$, where m_i is the number of marked animals caught in the ith period, R_i is the number of animals released in the ith period, r_i is the number of animals released at i, that are subsequently recaptured, and z_i is the number of animals caught before the ith period, not caught in the ith period, but caught at some time after i. Estimation of recruitment is based on the assumption that the population size at time $i+1$ is equal to the proportion of the population that survives from time i plus the number of additions to the population between time i and $i+1$ (B_i). B_i is estimated for the sampling periods 2 to $K-2$ by

$$\hat{B}_i = \hat{N}_{i+1} - \hat{\varphi}_i \hat{N}_i \tag{6.10}$$

The Jolly–Seber model and its variants all assume:

(1) All animals in the population at time i have the same probability of detection p_i;
(2) All animals present in the population immediately following sample period i have the same survival probability φ_i, until sampling period $i+1$;
(3) Marks or tags are not lost;
(4) Sampling periods are instantaneous (or very short);
(5) All emigration is permanent;
(6) Detection and survival of an individual are independent of the fates of other individuals.

Assumptions 1 and 2 affect the ability to make population and survival estimates. Varying detection and survival probabilities affect the underlying distribution properties of the model, leading to potentially severe bias in some estimates. Stratification of the population into groups (e.g., sexes) expected to exhibit different rate parameters is a wise way to try to achieve similar rates for individuals within groups. Individual covariates can also be used, if such covariates are thought to influence survival and/or capture probabilities. Williams et al. (2002) discuss a number of alternatives to deal with violations of assumptions including stratification and multi-state models.

Assumption 4 is important to keep the likelihoods of mortality and emigration near 0 during the sampling period. In fact, "short" period may be interpreted as a period of time short enough relative to the time interval for which survival is estimated.

Assumption 5 is used to clarify the interpretation of detection probability. Consider the case of an individual with a capture history 1001. The 1's represent a capture with detection probabilities p_i, and the 0's represent their complement $(1-p_i)$. Let γ_i be the probability of temporary emigration. When temporary emigration occurs, a 0 that is bracketed by 1's has two possible explanations: either the individual is present and not detected ($1-p_i$) or the individual has left the area (γ_i). In this situation, the interpretation of $\hat{p}_i$ changes to $(1-\gamma_i)p_i$, and the resulting estimate of abundance now estimates the superpopulation size, including animals temporarily outside the sampled area. Kendall et al. (1997) present a model that incorporates temporary emigration using a robust design approach.

Assumption 6 becomes important when estimating abundance of animals that live in groups. For individuals within groups, survival or detection may be dependent, in the sense that if one animal is caught, the likelihood of capture for other individuals in the group may also increase. Violation of this assumption results in biased variance estimates for detection and survival parameters but usually does not lead to biased point estimates.

Jolly–Seber models can be implemented under MARK (Cooch and White 2006), POPAN (Schwarz and Arnason 1996) or JOLLY (Pollock et al. 1990). They can be developed with covariates associated with detection and survival (MARK, POPAN), and models can be evaluated using goodness of fit tests (Pollock et al. 1985; Burnham et al. 1987), between model tests and AIC model selection (Burnham and Anderson 2002).

6.2.3 *Mixed Time Scale Model*

Pollock's (1982) robust design combines the strengths of closed and open population estimation while avoiding some of the limitations of each. Specifically, closed models cannot be used to estimate quantities associated with population change, and open models result in biased estimates when there is heterogeneity in detection probabilities and, sometimes, when there is temporary emigration. Under the robust design (Fig. 6.3), sampling is divided into K primary periods and l secondary sampling periods in each primary period. Secondary periods occur in a relatively short time frame and are used to develop closed CR model estimates of abundance for the

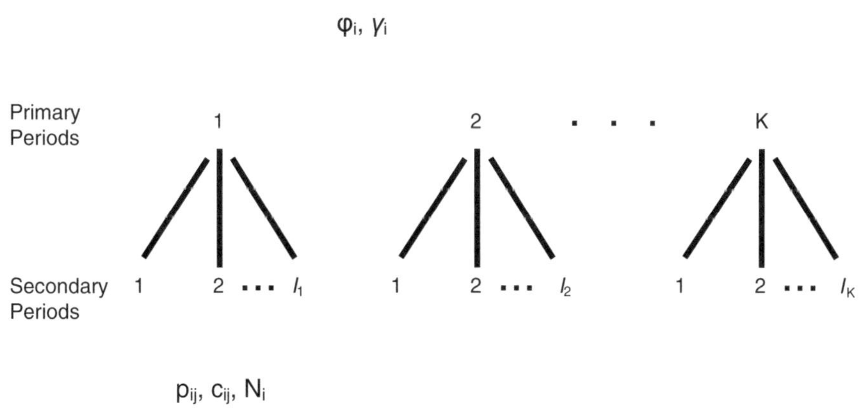

Fig. 6.3 Schematic illustration of Pollock's robust design. A long-term study is divided into K primary sampling periods, each composed of l secondary sampling periods. Primary trapping periods represent seasons or years for which information on demographic processes such as survival (φ) and recruitment (γ) is desired. The secondary periods with each primary period are closed trapping periods within which estimates of abundance (N), capture probabilities (p) and recapture probabilities (c) are made

primary trapping period. Data are then combined across primary periods, and each animal is recorded as captured in the *i*th primary period if it was observed at least once during the *l* secondary periods. Survival can then be estimated using the Jolly–Seber model, and recruitment is estimated using both closed and open model estimates.

Under Pollock's (1982) original formulation, the data from the secondary periods within a primary period are modeled independently from the data of the other primary periods (an *ad hoc* approach). Williams et al. (2002) observed that this opens the possibility that different closed models of abundance may be selected under the automated model selection algorithms within a single analysis. In the absence of a priori reasons for expecting such differences, they recommend that a common single closed population model be used for each of the *K* abundance estimates in the analysis. Use of a single model may result in more sensible estimates of recruitment when the closed and open estimates are combined. The model selection results from closed population estimates can aid in choosing the general model, since it is unlikely that covariates affecting detection probability will change substantially throughout a study. A priori expectations may also play a role in the selection of a common model. Karanth et al. (2006) used the M_h model in preference to the model selection choice (M_0) based on the robustness of the heterogeneity estimator and the belief that tigers are likely to vary in their individual capture probabilities.

Kendall et al. (1995) provide a likelihood-based framework for robust design models. The key feature of the likelihood-based models is the link between detection probabilities for the primary and secondary trapping periods. These models also were restricted in that the closed models did not include heterogeneity models for which there were maximum likelihood estimators. Pledger (2000) extended the work of Norris and Pollock (1996) to model heterogeneity using a finite mixture model that assumes a population is composed of sub-groups of animals that each share common detection probabilities. This approach allows maximum likelihood solutions for heterogeneity models. Kendall and Nichols (1995) and Kendall et al. (1997) also provide models for the case of temporary emigration in the population.

The assumptions underlying the robust design are the same as those for separate closed and open population models. Initial work on the robust design required that the population be closed during the secondary samples (although temporary emigration may be incorporated), that no tag loss occurs, that there is no heterogeneity in capture probabilities during the secondary periods, that survival between primary periods is the same for all animals, and that the fates of animals are independent. Robust design models can now accommodate open modeling of secondary period data as well (e.g., Schwarz and Stobo 1997).

6.3 Estimation of Density

Given an estimate of abundance based on an array of camera traps, estimating density is relatively straightforward if we have a method of determining the area sampled A. Given estimates of *N* and *A*, density (*D*) and its variance are

$$\hat{D} = \hat{N} / \hat{A}, \tag{6.11}$$

$$\mathrm{var}(\hat{D}) \approx \hat{D}^2[\mathrm{var}(\hat{A}) / \hat{A}^2 + \mathrm{var}(\hat{N}) / (\hat{N}^2)]. \tag{6.12}$$

Estimation of A is of primary importance, and the variance of $\hat{A}$ can be estimated using the bootstrap method (Efron and Tibshirani 1986) or the delta method (Seber 1982), depending on how the area is calculated.

At its most simplistic, A may be described by a concave polygon defined by connecting the outermost trap locations (A_{tp}; Mohr 1947). This assumes that animals do not move from outside the bounded area to inside the area or vice versa. Unless the study is conducted on a small island or a physical barrier is erected in the study area to limit movement of animals, this assumption is unlikely to be true. More often, a boundary area of width W (A_w) is added to the area defined by the polygon A_{tp} to reflect the area beyond the limit of the traps that potentially is contributing animals to the abundance estimate (Otis et al. 1978). The sampled area, also known as the effective area, is then $A(W)=A_{tp}+A_w$. Calculation of the buffer strip width (W) is critical to the estimation of density and is problematic because there is no agreed upon method of estimating W. Solutions to this problem all involve *ad hoc* methods that date back to early attempts to estimate abundance and home ranges based on trapping grids (see Hayne 1949). Dice (1938) first drew attention to this problem in small mammal studies and recommended using one-half the diameter of an average home range. Other solutions have included use of inter-trap distances (Blair 1940; Burt 1943), mean movements among traps, maximum movements among traps (Holdenried 1940, Hayne 1949), nested grids (Otis et al. 1978), and assessment lines (Smith et al. 1971).

Otis et al. (1978) recommended a nested-grid analysis to estimate the contribution of animals centered outside a trapping grid to the abundance estimate. In the simplest case, consider a camera grid defined by the four corner cameras as X km by X km or X^2 km^2. The effective area is defined as

$$A(W) = X^2 + 4XW + \pi W^2. \tag{6.13}$$

Note that $4X$ is simply the perimeter of the grid. To estimate density we use the estimate $\hat{N}$ and fix X, W is unknown. If two grids of different sizes are used, the original grid and a grid composed of interior cameras, W can be estimated and consequently $\hat{D}$. Otis et al. (1978) caution that the nested grid approach has large data requirements. At a minimum they recommend a 9×9 trapping grid, which precludes many camera trap studies. Alternatively, they recommend reducing the spacing between cameras, or increasing the number of traps to increase capture rates. An experimental evaluation of this approach (Parmenter et al. 2003) indicated poor performance when sample sizes are small, which highlights the limitation of this approach for typical camera trap studies in which trapping arrays include fewer than 80 points.

Wilson and Anderson (1985) explored the problem of correcting sampling area of trapping arrays using one-half of the mean of the maximum distances

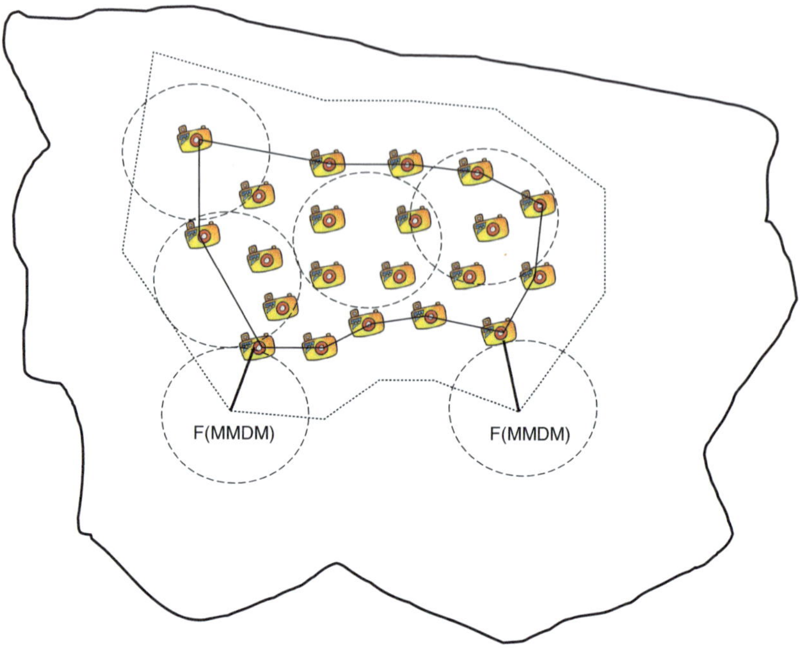

Fig. 6.4 Schematic array of camera traps and effective sampling area. A convex polygon of area A_{tp} is defined by connecting the camera trap points. Circular home ranges are calculated with radii expressed as some function $F(F=½, 1)$ of the mean maximum distance moved by individuals between camera traps for individuals captured more than once. F(MMDM) defines a length W specifying an area A_w that is added as a buffer around the convex polygon under the assumption that individuals with home ranges centered outside the convex polygon of trap points but within the buffer area fall within the sampled population. The effective sample area is then defined as $A(W)=A_{tp}+A_w$

between capture locations. This approach is known as the "mean maximum distance moved" or MMDM. The choice of one-half the MMDM is arbitrary, but follows the recommendation of Dice (1938). Here, I refer to the Wilson and Anderson method as the ½MMDM to distinguish it from the recommendations of Parmenter et al. (2003) who use the actual or full MMDM (Fig. 6.4).

Let d_i be the maximum distance moved for the ith animal in a sample of m animals caught at least twice in a study. Then:

$$\bar{d} = \left(\sum_{i=1}^{m} d_i \right) / m, \text{ and } \hat{W} = \bar{d} / 2, \tag{6.14}$$

and

$$\text{var}\left(\bar{d}\right) = \sum_{i=1}^{m} (d_i - \bar{d}^2) / (m(m-1)), \text{ and } \text{var}(\hat{W}) = \text{var}\left(\bar{d}\right) / 4. \tag{6.15}$$

$\text{Var}(A(\hat{W}))$ is then calculated using the delta method and $\text{var}(\hat{D})$ as in (6.12). This approach includes an underlying assumption that ½MMDM is greater than or equal to the radius of a circle of area equal to the average home range of the study animal so that animals whose (circular) home range centers fall within *W* are likely to be included in the sampling area.

The Wilson and Anderson (1985) modeling exercise found good support for the ½MMDM; the relative percent bias was good at low densities and still reasonable at high densities. Wilson and Anderson caution that the ½MMDM approach requires that a sufficient number of recaptures is needed for reliable estimates of MMDM. Parmenter et al. (2003) found that use of ½MMDM consistently overestimated true densities in a controlled experiment comparison of density estimation using different *W* estimation techniques with CR models, and web-based DISTANCE sampling.

Parmenter et al. (2003) examined a number of approaches to estimate density of rodents using trapping grids with CAPTURE and trapping webs with DISTANCE. They found that using the full MMDM, $\hat{W} = \bar{d}$, provided the most accurate grid-based density estimates for the area calculations considered. They found that the full MMDM density estimators tended to underestimate at low densities and overestimate at high densities, but that the bias declined as the cumulative capture probabilities increased. One encouraging result for camera trap studies is that the most accurate estimates were of low density populations with large MMDM values, a situation typical of many species targeted by camera traps.

Which form of MMDM to choose depends on whether a radius or a diameter provides a better estimate of movements within a home range. The use of either form of MMDM requires additional considerations about movement patterns, effect of social interaction on movements, home range size and shape, the degree that home ranges overlap the grid and constraints on the upper limit of "apparent" distance moved due to trap spacing (Parmenter et al. 2003).

Factors influencing movement patterns are likely to vary by study and species, and this makes generalizations of the "best" method to estimate sample area difficult. Many confounding factors can be controlled by careful design (i.e., restricting analysis to certain age classes, planning seasonal studies when dispersal is likely to be low). Perhaps the biggest constraint on the use of either MMDM approach, however, is that camera trap studies yield relatively small numbers of inter trap movement distances. When inter-trap distances are used, MMDM usually increases as the number of recaptures increases (Stickel 1954). Small sample sizes are expected to yield a constrained set of inter-trap distances, tend to underestimate movement distances and, on average, result in underestimates of full MMDM, ½MMDM and "true" *A*(*W*) for a trapping grid. For example, in a study of Sumatran tigers, O'Brien et al. (2003) found only four recaptures and based their *A*(*W*) estimate on the maximum inter-trap distance because the MMDM was unrealistically small for tigers. While most published camera trap abundance studies have 10 or more capture occasions, few have more than 30 estimates of maximum distances. Jett and Nichols (1987), recognizing that maximum distances tend to increase with the number of recaptures, suggested a method for determining the maximum movement for animals observed a large number of times. Let $\bar{d}_i$ = the MMDM for animals caught exactly i times, and d^*=the expected MMDM for animals that move an indefinitely large number of times. The equation;

$$E\left(\bar{d}_i\right)=[1-\mathrm{e}^{-(i-1)b}]d^*, \tag{6.16}$$

where b is a model parameter, may be solved by nonlinear least squares, and d^* used as an estimate of MMDM. Again, this method requires an adequate number of recaptures. Williams et al. (2002) recommend a minimum of 10 trapping occasions when using MMDM methods, or the use of the expected maximum movement (d^*) of animals rather than the average.

Ideally, the best way to estimate W is to use information on actual movement patterns collected during the capture period. Karanth (1995) was the first to combine radio telemetry information on home ranges with camera trap abundance estimates to determine density. He used the estimated non-overlapping home ranges of five resident, female tigers that he detected in camera traps as the effective size of his sampling area. Later estimates of sample area for this study, based on ½ MMDM, were only 55% as large as the home range estimate. Soisalo and Cavalcanti (2006) used radio telemetry in conjunction with a camera trap study of jaguars to compare different methods of deriving the effective area and density estimates. They considered ½MMDM, full MMDM, a MMDM calculated from telemetry movement data and ½ diameter of the average home range. They found that the ½MMDM estimate of W from camera traps was much smaller than ½ diameter of the home range based on telemetry. As a result, the density estimates based on ½ MMDM from camera traps were much higher than the telemetry-based estimates (Fig. 6.5). The full MMDM density estimates were similar, though slightly lower, than the telemetry based estimates. Interestingly, the full MMDM in both years was greater than the telemetry-based MMDM, probably due to the small sample size of the telemetry data set (n=6, compared to 25 and 31 for the full MMDM estimates). Recent work on a known leopard population in South Africa, however, supports the use of a ½ MMDM based on camera traps for estimating effective sampling area using camera traps (Balme et al. 2009). These contradictory results highlight the problems of density estimation when the sampling area must be estimated.

Recognizing the problems that arise from edge effects and the *ad hoc* estimation of W from trapping data, Efford (2004) proposed a combination of Monte Carlo simulation and inverse prediction methods (Pledger and Efford 1998) to estimate jointly population density and two parameters of individual detection probability directly from CR data. The model assumes stationary home ranges that are of equal sizes with centers Poisson-distributed with density D. The probability that an animal is detected in a trap at distance r from its home range center is described by a 2-parameter spatial detection function $g(r)$, usually a half-normal distribution with parameters g_0 when r=0 and σ, a scale measure of home range size. D, g_0 and σ define the detection process. Monte Carlo simulation is used to match the parameters to the statistics calculated from the CR study ($\hat{N}$, $\hat{p}$, $\bar{d}$). An inverted linear model is then used

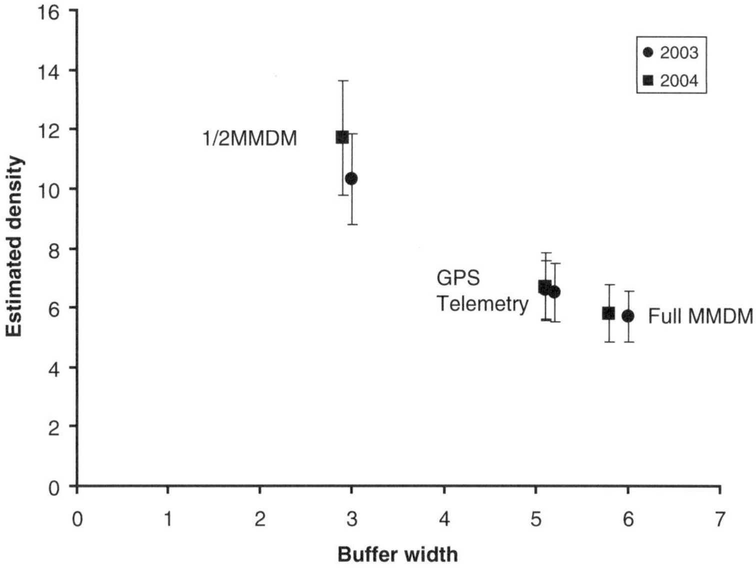

Fig. 6.5 Estimates of jaguar density from Soisalo and Cavalcanti (2006) using the ½MMDM, full MMDM based on camera trap data and GPS telemetry data. Telemetry estimates include an actual MMDM estimate for collared animals and an estimate based on the radius of the home range. Estimates based on ½MMDM are significantly larger that estimates based on telemetry and full MMDM

to predict the values of the parameters (D, g_0, σ) from the field data. In addition to the normal closed population assumptions, DENSITY-based estimation assumes that animals occupy stable home ranges and that capture does not affect the probability of recapture. The method is robust to the choice of an abundance estimator (including the number of individuals caught [M_{t+1}]; Efford et al. 2004, 2005), choice of trapping configuration and number of traps, and produces unbiased estimates when assumptions are met. Relative precision was density dependent (Efford 2004) between 0.5 and 5.0 ha^{-1} but showed only minor variation at higher densities. Relative precision also was dependent on the number of recaptures (Efford et al. 2004). A software program, DENSITY (Efford et al. 2004) implements the method.

Density estimation using DENSITY has been tested with birds (Efford et al. 2004), brushtail possums *Trichosurus vulpecula* (Efford et al. 2005), and performed well, producing unbiased and precise estimates compared to population estimates. A test of DENSITY methods to estimate density of ship rats *Rattus rattus* (Wilson et al. 2007) produced estimates that were less precise than desired for the preferred model, but this was attributed to sampling design rather than inherent problems with the estimation methods. Borchers and Efford (2008) have developed likelihood-based model fitting and model selection procedures and Royle et al. (2009) have applied a Bayesian extension of the methods.

6.4 Relative Abundance Indices

Perhaps one of the most contentious issues among wildlife researchers is the extent to which we can rely on indices of abundance for monitoring populations and making comparisons over space and time (Nichols and Pollock 1983; Conroy 1996; Gibbs 2000; Anderson 2001, 2003; McKelvey and Pearson 2001; Pollock et al 2002; Engeman 2003; Conn et al. 2004). An index can be thought of as the metric of last resort. They should only be used when it is not possible to conduct a proper population estimation study or a study that includes estimation of the detection probability. This issue is especially acute for sampling cryptic or rare species, typical target species in camera trap surveys (Carbone et al. 2001, 2002; Jennelle et al. 2002). An index can be any count of animals or sign that is expected to vary directly with population size (Caughley 1977). Indices typically are used when (1) the target species is difficult to capture or observe, (2) it is logistically difficult or too expensive to implement surveys capable of determining detection probability, or (3) there is historical precedent for index surveys. Index surveys are almost always easier and less expensive to implement than surveys based on formal abundance estimation, but yield weaker inferences compared to surveys based on CR or line transect or any other sampling approach that incorporates detection probability. Because indices rely on (often untested) assumptions about the relationship to abundance, their use should be based on a careful consideration of trade-offs between strength of inference, feasibility of other methods to estimate abundance, and cost. Indices based on camera trap data include the counts of individually recognized animals in photographs (C or M_{t+1}), and catch per unit effort indices (CPUE indices), usually some variation of the number of camera trapped animals/100 trapdays.

A fundamental problem with use of indices is the recognition that counts are underestimates of population size when $p<1$ and that p is not likely to be uniform across populations being compared. If we assume that detectability, represented here by β, is constant between two time periods, then $\beta_l=\beta_2=\beta$. Population counts C_1 and C_2, should represent patterns in population abundance irrespective of possible bias associated with the counts, and we can also assume that changes in counts reflect changes in population size because the bias stays constant and can be discounted. For the estimate C_2-C_1:

$$\mathrm{E}(C_2 - C_1) = N_2\beta_2 - N_1\beta_1 = (N_2 - N_1)\beta, \tag{6.17}$$

when $\beta_l=\beta_2=\beta$. Changes in the count reflect changes in the population and the bias attributable to imperfect detection does not affect the interpretation of patterns in population size (Williams et al. 2002). When $\beta_1\neq\beta_2$, changes in the count statistic may arise from population change or from changes in detection, and it is impossible to know which process is occurring without additional information on detectability. In general, use of an index may be justified so long as the emphasis is on relative abundance rather than changes in absolute abundance and assumptions about the relationships between relative and absolute abundance are met.

A more common relative abundance index treats differences in relative abundance between time 1 and time 2 as a proportional abundance (Skalski and Robson 1992; Williams et al. 2002). Following Williams et al. (2002), let $T = C_2/C_1$ and:

$$\mathrm{E}(T) = \mathrm{E}(C_2 / C_1) \approx \mathrm{E}(C_2) / \mathrm{E}(C_1).$$

Because C_1 and C_2 are related to N_1 and N_2 through β_1 and β_2, we can express E(*T*) as

$$\mathrm{E}(T) \approx \beta_2 N_2 / \beta_1 N_1. \tag{6.18}$$

When $\beta_1 = \beta_2 = \beta$, C_2/C_1 is an unbiased estimator of N_2/N_1. Furthermore, when β is constant between sample occasions, ratios of count statistics have smaller sampling error than ratios of abundance estimates because the abundance estimate incorporates sampling error in the count statistic plus variation due to estimating β. Skalski and Robson (1992. p. 64) provide a test of equal detection probability for surveys in which individuals are distinctive. MacKenzie and Kendall (2002) provide an equivalence testing approach and a model-averaging approach to the same problem. Note that testing for similar detection probabilities requires collection of data for estimation of, and inference about, detection probabilities.

The use of nonproportional indices *requires* the demonstration of a functional relationship between the index and abundance over the desired range of inference, a calibration of the relationship, and an evaluation of the precision of the relationship. The basic assumption is that the index, denoted by *I*, has a monotonic relationship to *N*, i.e. $\mathrm{E}(I) = \beta_0 + \beta_1 N$, to ensure that inferences based on the index relate to changes in *N*. Usually, β_0 is assumed to equal 0 and the relationship becomes the familiar $\mathrm{E}(I) = \beta N$. Most ecologists also assume the index and *N* share a positive relationship that is constant across habitats and time. This often is an untested and incorrect assumption (Conroy 1996; Link and Sauer 1998; Gibbs 2000). An index may have a lower limit of *N* below which the index is insensitive to changes in *N*. In terms of a positive linear relationship, the index may have a negative *y*-intercept, $\beta_0 < 0$, crossing the *x*-axis at $(N^*, 0)$, with $N^* > 0$. This relationship is particularly common in surveys of rare species. Similarly, an index may become insensitive to changes in population size when *N* is large. This is manifested as an asymptotic curve in which the relationship between the index and *N* approaches 0 as *N* increases. An index that is essentially flat for large values of *N* may still be useful for monitoring a population at low and intermediate levels of abundance, whereas an index that is insensitive when *N* is small is unlikely to be of much use. If the slope of an index-abundance relationship is shallow, the index will not be good predictor because large changes in *N* may be accompanied by small changes in the index that may be masked by sampling error. Finally, if detection varies by habitat or time, then a single value of the index may represent multiple population sizes N_i that depend on habitat or time *i*.

Camera trap surveys offer some promise for standardization, at least for indices that track changes in the population over time within a survey site. Proper sample design and trap placement can reduce large-scale habitat effects that might influence

p while ensuring that the objectives of the survey are met by the sampling effort. Since the detection of animals is automated, observer error is limited to variation in placement of cameras, rather than the actual observation of individuals. Typically standardization in camera trap surveys involves scaling the index to a standard unit of effort, usually 100 camera trap days and conducting surveys in the same season. Skalski and Robson (1992) argue that attempts to make inferences about populations using an index require careful standardization in any experimental design to minimize variation in p. It is unlikely however, that even standardization would suffice to allow comparisons of camera trap indices across sites or across regions.

In the situation where detection probability is thought to vary due to uncontrollable effects, the use of covariates that relate to sources of variation in p may help. Choice of covariates, however, is important. Covariates that influence both detection probability and abundance, however, cannot be used as a covariate for an index since it will be impossible to disentangle the effect of the covariate on N and p. Habitat type is an example of a covariate that is likely to affect both p and N. Environmental variation, such as rainfall, or temperature are likely to affect p and not N. Typically, analysis of covariates is used to deal with variation in the index due to suspected, covariate-induced differences in p. Unfortunately, it is nearly impossible to foresee and control for the many possible factors that affect detection. Residual variation in p due to unmodeled variation adds noise to the index values that is simply relegated to sampling error under the untestable assumption that this error is small, unimportant, and unrelated to the dimensions (i.e., space, time) of comparison.

An index should be validated through periodic calibration with independently derived estimates of abundance or through complementary surveys. The use of a double sampling design can help calibrate an index. O'Brien et al. (2003) used a combination of line transect and CR estimates to test the relationship between a relative abundance index based on independent photographic events per 100 trap nights and found a reasonably strong relationship ($r^2=0.79$) for several species. An analysis using line transect and point count estimates of three ungulate species preyed upon by Asiatic cheetah (*Acinonyx jubatus venaticus*) to calibrate photographic indices of prey abundance yielded similar results (T. O'Brien, unpublished data). However, as Williams et al. (2002) note, most indices are neither calibrated nor validated. Borrowing a calibrated index from another site or another species and assuming the same relationship exists is not a reliable form of calibration or validation and is not recommended. Local conditions can affect p and/or N, making the demonstrated relationship at site A unreliable or meaningless at site B unless the effects on p and N can be disentangled.

When analyzing time trends at a site, use of relative abundance indices may be more easily justified. Gibbs (2000) points out that trend analysis is more concerned with the signal in the index data and that should be apparent despite noise in the data, although there is an assumption of a common distribution for p so $E(p)$ is constant. Basically, variation in detection probability of a species is likely to be lower over time within a study area compared to variation between study areas. Karanth and Nichols (2002) agree that use of relative abundance to detect trends is "safer" than using an index to infer change over space. As shown above (Skalski and

Robson 1992), a proportional difference in abundance, N_2/N_1, may be approximated by C_2/C_1 when $\beta_1=\beta_2$. If $\beta_1\neq\beta_2$, the uncertainty in the count ratio is minimized only to the degree that the differences among β_i are minimized. Researchers need to be on guard, however, because human-related covariates may still cause changes in p over time causing inferential problems for detecting time trends using indices. Much of work on use of abundance indices to monitor changes in populations over time has been developed to minimize the effect of variation in p and address the shortcomings of count data (Link and Sauer 1997, 1998, 2002). Because of the weak inference associated with abundance indices, any use of indices should be accompanied by a discussion of the relationship of the index used to the population size, and the limitations of the method.

Recently, binomial mixture models have been proposed for estimating abundance and detection probability from count data (Dodd and Dorazio 2004; Royle 2004; Kéry et al. 2005). These models do not necessarily rely on individual identification for the estimation of detectability, making them potentially useful for species lacking distinguishing characteristics and appearing in camera traps. The main requirement is that count surveys are replicated at a number of sampling locations over a period for which populations are closed. If we treat a camera trap survey as a sampling protocol deployed at a large number of very small, spatially distinct sampling units, then the photographic record of a single camera may be treated as a point count survey of local abundance replicated over d days. The size of the population at each camera trap point is expected to be small and often 0. The observed counts for a single day at a trap also are likely to be small (1 or 2 individuals), and probably include many 0's. Abundance at the different trap sites will vary spatially (N_i) and repeated counts at site i may be treated as binomial random variables with parameters N_i (local abundance) and p_i (detection probability). Under these circumstances, estimates of N_i may be poor, but the estimate of total or average point abundance, $\hat{N}$, at the trap sites may be good (Royle and Nichols 2003). The average point abundance can then be used as a detectability-corrected abundance estimate for purposes of tracking changes over time and space. I include this approach in the relative abundance section because the average point abundance is expected to change as N changes; therefore changes in ratios of average point abundances $\hat{N}_2/\hat{N}_1$ should be an unbiased estimate of changes in N over time and space.

One obstacle to implementing this approach is the potential for double counting the same individual during a sample or at multiple sites. Careful scrutiny of consecutive photographs of the same species, or multiple photographs of a single species during a single sampling period, can usually discern differences in individuals, even if the individuals cannot be identified reliably from day to day, reducing the chance of double counts. This method will work best when applied to species that live singly or in small groups. It is also appropriate for territorial species when cameras are spaced such that coverage is approximately one camera per territory. For example, in five replicate camera trap surveys from Indonesia (T. O'Brien, unpubl. data), four of the most commonly observed species in 24-h samples were usually photographed only one time per day and as an individual (range 58–89% of samples). For red muntjac (*Muntiacus muntjac*), 89% of non-zero samples were of a single animal. For Great Argus Pheasants (*Argusianus argus*), 80.4% of samples were of individuals and two-third of the samples with > 1 individual were photographs of two to three individuals in a single frame. Even group-living wild pigs (*Sus scofa*) and pig-tailed macaques (*Macaca nemistrina*), were often discernable to individual in consecutive photographs; 68% and 58% of samples, respectively, were classified as a single animal per sample.

Williams et al. (2002), Pollock et al. (2002) and others stress their discomfort with the use of indices to monitor populations because of the formidable challenge of meeting model assumptions and, often, the inability to test these assumptions. Link and Sauer (1998) defend the use of indices but caution that any analysis that relies on indices should recognize that counts are not necessarily good surrogates for population size. The vast majority of pictures/data collected during a camera trap survey are species whose members are difficult to reliably identify as individuals, requiring innovative analytical techniques and interpretations. While the concerns regarding the use of indices are real, the resolution of problems related to unidentifiable individuals should be a major area of inquiry in camera trap studies. Finding ways to improve the reliability of abundance indices and extend the use of count statistics is required if the use of camera trap studies in conservation and ecology is to move beyond the limited and rather special case of abundance estimation for species with recognizable individuals.

Properly designed studies of abundance and density using camera traps will always pose logistical challenges to investigators, but the payoff can be great considering the insights gained into populations of rare and endangered species (e.g., Karanth et al. 2004, 2006). Where possible, camera trap CR studies should be replicated over time to take advantage of open model estimation of demographic parameters. For species that are not individually recognizable, it may be worthwhile to capture and tag a portion of the population and use resighting of marked animals in camera traps along with counts of unmarked animals to provide the data to estimate detection probabilities. The use of point abundance estimators has potential for territorial or solitary species as well as those that live in small groups.

Finally, the development of unbiased estimators of density is probably the most important issue facing practitioners of camera trapping for abundance estimation. Density estimates make the comparison of abundances in space and time possible. Abundance estimates based on CR methods are confounded with variation in spatial and non-spatial components of detection and with the configuration of the camera trap layout (Efford et al. 2004). Abundance estimates without an associated area may be as difficult to interpret as count-based indices (Anderson 2003). The lack of consensus on the best practice for density estimation and the potential for misinterpreting the results of an abundance analysis suggest that researchers exercise caution in reporting and comparing abundance and density estimates. Although the methods of Efford (2004) offer a promising answer to the problem of density estimation, there is still need for investigation of the effects of transients in the population, heterogeneous home range size, minimum data requirements and extension to open population models. Clearly, advances in abundance and density estimation are creating new questions and new opportunities for application in research and monitoring.

Acknowledgements I thank K.U. Karanth, M.F. Kinnaird, J.D. Nichols, and A.F. O'Connell for many helpful comments that greatly improved the content and structure of the manuscript. I also thank The Wildlife Conservation Society and J. Ginsberg, E. McBean, A. Rabinowitz and J. Robinson for support during the preparation of this manuscript and, more importantly, for support of efforts to expand the application of camera trap methods in conservation biology.

References

Anderson, D. R. 2001. The need to get the basics right in wildlife field studies. Wildlife Society Bulletin 29:1294–1297

Anderson, D. R. 2003. Response to Engeman: index values rarely constitute reliable information. Wildlife Society Bulletin 31:288–291

Bailey, L. L., T. R. Simons, and K. H. Pollock. 2004a. Estimating detection probability parameters fro plethodon salamanders using the robust capture-recapture design. Journal of Wildlife Management 68:1–13

Bailey, L. L., T. R. Simons, and K. H. Pollock. 2004b. Spatial and temporal variation in detection probability of plethodon salamanders using the robust capture-recapture design. Journal of Wildlife Management 68:14–24

Balme, G. A., L. T. B. Hunter, and R. Slowtow. 2009. Evaluating methods for counting cryptic carnivores. Journal of Wildlife Management 73:433–441

Blair, W. F. 1940. Home ranges and populations of the meadow vole in southern Michigan. Journal of Wildlife Management 4:149–161

Borchers, D. L. and M. G. Efford. 2008. Spatially explicit maximum likelihood methods for capture–recapture studies. Biometrics 64(2):377–385

Brownie, C., J. R. Hines, and J. D. Nichols. 1986. Constant-parameter capture–recapture models. Biometrics 42:561–574

Buckland, S. T., K. P. Burnham, and N. H. Augustin. 1997. Model selection: an integral part of inference. Biometrics 53:603–618

Burnham, K. P. and D. R. Anderson. 2002. Model selection and multimodal inference: a practical information-theoretic approach, Second edition. Springer, New York

Burnham, K. P., D. R. Anderson, G. C. White, C. Brownie, and K. H. Pollock. 1987. Design and analysis methods for fish survival experiments based on release-recapture. American Fishery Society Monograph 5:1–437

Burt, W. H. 1943. Territoriality and home range concepts as applied to mammals. Journal of Mammalogy 24:346–352

Carbone, C., S. Christie, K. Conforti, T. Coulson, N. Franklin, J. R. Ginsberg, M. Griffiths, J. Holden, K. Kawanishi, M. Kinnaird, R. Laidlaw, A. Lynam, D. W. MacDonald, D. Martyr, C. McDougal, L. Nath, T. O'Brien, J. Seidensticker, D. Smith, M. Sunquist, R. Tilson, and W. N. Wan Shahruddin. 2001. The use of photographic rates to estimate densities of tigers and other cryptic mammals. Animal Conservation 4:75–79

Carbone, C., S. Christie, K. Conforti, T. Coulson, N. Franklin, J. R. Ginsberg, M. Griffiths, J. Holden, K. Kawanishi, M. Kinnaird, R. Laidlaw, A. Lynam, D. W. MacDonald, D. Martyr, C. McDougal, L. Nath, T. O'Brien, J. Seidensticker, D. Smith, M. Sunquist, R. Tilson, and W. N. Wan Shahruddin. 2002. The use of photographic rates to estimate densities of cryptic mammals: response to Jennelle et al. Animal Conservation 5:121–123

Caughley, G. 1977. Analysis of vertebrate populations. Wiley, New York

Chao, A. and R. M. Huggins. 2005a. Classical closed-populations capture–recapture models. Pages 22–35 *in* S. C. Amstrup, T. L. McDonald, and B. F. J. Manly, editors. Handbook of capture–recapture analysis. Princeton University Press, Princeton, NJ

Chao, A. and R. M. Huggins. 2005b. Modern closed-populations capture-recapture models. Pages 58–87 *in* S. C. Amstrup, T. L. McDonald, and B. F. J. Manly, editors. Handbook of capture–recapture analysis. Princeton University Press, Princeton, NJ

Cochran, W. G. 1997. Sampling techniques. Wiley, New York
Conn, P. B., L. L. Bailey, and J. R. Sauer. 2004. Indexes as surrogates to abundance for low-abundance species. Pages 59–74 *in* W. L. Thompson, editor. Sampling rare or elusive species. Island Press, Washington, DC
Conroy, M. J. 1996. Abundance indices. Pages 179–192 *in* D. E. Wilson, F. R. Cole, J. D. Nichols, R. Rudran, and M. S. Foster, editors. Measuring and monitoring biological diversity: standard methods for mammals. Smithsonian Institution Press, Washington, DC
Cooch, E. and G. White, editors. 2006. Program MARK: a gentle introduction. Fifth edition. Cornell University Press, Ithaca, NY, http://www.phidot.org/software/mark/docs/book/
Crosbie, S. F. and B. F. J. Manly. 1985. Parsimonious modeling of capture–mark–recapture studies. Biometrics 41:385–398
Darroch, J. N. 1959. The multiple-recapture census: II. Estimation when there is no immigration or death. Biometrika 46:343–359
Dice, L. R. 1938. Some census methods for mammals. Journal of Wildlife Management 2:119–130
Dodd, K. C. and R. M. Dorazio. 2004. Using counts to simultaneously estimate abundance and detection probabilities in a salamander community. Herpetologica 60:468–478
Efford, M. 2004. Density estimation in live-trapping studies. Oikos 106:598–610
Efford, M. G., D. K. Dawson, and C. S. Robbins. 2004. DENSITY: software for analyzing capture-recapture data from passive detector arrays. Animal Biodiversity and Conservation 27:217–228
Efford, M. G., B. Warburton, M. C. Coleman, and R. J. Barker. 2005. A field test of two methods for density estimation. Wildlife Society Bulletin 33:731–738
Efron, B. and R. Tibshirani. 1986. Bootstrap methods for standard errors, confidence intervals, and other measures of statistical accuracy. Statistical Science 1:54–77
Engeman, R. M. 2003. More on the need to the basics right: population indices. Wildlife Society Bulletin 31:286–287
Gibbs, J. P. 2000. Monitoring populations. Pages 213–247 *in* L. Boitani and T. Fuller, editors. Research techniques in animal ecology: controversies and consequences. Columbia University Press, New York
Hayne, D. W. 1949. Calculations of home range size. Journal of Mammalogy 30:1–18
Holdenried, R. 1940. A population study of the long-eared chipmunk (*Eutamias quadrimaculatus*) in the central Sierra Nevada. Journal of Mammalogy 21:405–411
Jennelle, C. S., M. C. Runge, and D. I. Mackenzie. 2002. The use of photographic rates to estimate densities of tigers and other cryptic mammals: a comment on misleading conclusions. Animal Conservation 5:119–120
Jett, D. and J. D. Nichols. 1987. A field comparison of nested grid and trapping web density estimators. Journal of Mammalogy 68:888–892
Jolly, G. M. 1965. Explicit estimates from capture-recapture data with both death and immigration – stochastic model. Biometrika 52:225–247
Karanth, K. U. 1995. Estimating tiger *Panthera tigris* populations from camera trap data using capture-recapture models. Biological Conservation 71:333–338
Karanth, K. U. and J. D. Nichols. 2000. Ecological status and conservation of tigers in India. Final technical report to the Division of International Conservation, US Fish and Wildlife Service and Wildlife Conservation Society, Washington, DC
Karanth, K. U. and J. D. Nichols, editors. 2002. Monitoring tigers and their prey: a manual for researchers, managers, and conservationists in tropical Asia. Centre for Wildlife Studies, Bangalore, India, 193 pp
Karanth, K. U., J. D. Nichols, N. S. Kumar, W. A. Link, and J. E. Hines. 2004. Tigers and their prey: predicting carnivore densities from prey abundance. Proceedings National Academy of Sciences USA 101:4854–4858
Karanth, K. U., J. D. Nichols, N. S. Kumar, and J. E. Hines. 2006. Assessing tiger population dynamics using photographic capture–recapture sampling. Ecology 87:2925–2937
Kendall, W. L. 1999. Robustness of closed capture-recapture methods to violations of the closure assumption. Ecology 80:2517–2525

Kendall, W. L. and J. D. Nichols. 1995. On the use of secondary capture–recapture samples to estimate temporary emigration and breeding proportions. Journal of Applied Statistics 22:751–762

Kendall, W. L., K. H. Pollock, and C. Brownie. 1995. A likelihood-based approach to capture-recapture estimation of demographic parameters under the robust design. Biometrics 51: 293–308

Kendall, W. L., J. D. Nichols, and J. E. Hines. 1997. Estimating temporary emigration using capture–recapture data with Pollock's robust design. Ecology 78:563–578

Kéry, M., J. A. Royle, and H. Schmid. 2005. Modeling avian abundance from replicated counts using binomial mixture models. Ecological Applications 15:1450–1461

LaPlace, M. 1786. Sur les naissances, les marriages et les mortes. Histoire de l'Acad´emie Royale des Sciences, Ann´ee 1783:693–702

Lincoln, F. C. 1930. Calculating waterfowl abundance on the basis of banding returns. U.S. Dept. Agric. Circ. No. 118:1–4

Link, W. A. and J. R. Sauer. 1997. Estimation of population trajectories from count data. Biometrics 53:63–72

Link, W. A. and J. R. Sauer. 1998. Estimation of population change from count data: application to the North American Breeding Bird Survey. Ecological Applications 8:258–268

Link, W. A. and J. R. Sauer. 2002. A hierarchical analysis of population change with application to Cerulean warblers. Ecology 83:2832–2840

MacKenzie, D. I. and W. L. Kendall. 2002. How should detection probability be incorporated into estimates of relative abundance? Ecology 83:2387–2393

MacKenzie, D. I., J. D. Nichols, J. A. Royle, K. H. Pollock, L. L. Bailey, and J. E. Hines. 2006. Occupancy estimation and modeling: inferring patterns and dynamics of species occurrence. Academic, New York

McKelvey, K. S. and D. E. Pearson. 2001. Population estimation with sparse data: the role of estimators versus indices revisited. Canadian Journal of Zoology 79:1754–1765

Mohr, C. O. 1947. Table of equivalent populations of North American small mammals. American Midland Naturalist 37:223–249

Nichols, J. D. and K. H. Pollock. 1983. Estimation methodology in contemporary small mammal capture–recapture studies. Journal of Mammalogy 4:253–264

Norris, J. L. and K. H. Pollock, 1996. Nonparametric MLE under two closed-capture models with heterogeneity. Biometrics 59:639–649

O'Brien, T. G., M. F. Kinnaird, and H. T. Wibisono. 2003. Crouching tigers, hidden prey: Sumatran tiger and prey populations in a tropical forest landscape. Animal Conservation 6:131–139

Otis, D. L., K. P Burnham, G. C.White, and D. R. Anderson. 1978. Statistical inference from capture data on closed animal populations. Wildlife Monographs 62:1–135

Parmenter, R. R., T. L. Yates, D. R. Anderson, K. P. Burnham, J. L. Dunnum, A. B. Franklin, M. T. Friggins, B. C. Lubow, M. Miller, G. S. Olson, C. A. Parmenter, J. Pollard, E. Rexstad, T. M. Shenk, T. R. Stanley, and G. C. White. 2003. Small-mammal density estimation: a field comparison of grid-based vs. web-based density estimators. Ecological Monographs 73:1–26

Peterson, C. G. J. 1896. The yearly immigration of young plaice into the Limfjord from the German Sea. Danish Biological Station Report 6:1–48

Pledger, S. 2000. Unified maximum likelihood estimates for closed capture–recapture populations using mixtures. Biometrics 56:434–442

Pledger, S. and M. Efford. 1998. Correction of bias due to heterogeneous capture probability in capture–recapture studies of open populations. Biometrics 54:888–898

Pollock, K. H. 1982. A capture–recapture design robust to unequality of capture. Journal of Wildlife Management 46:757–760

Pollock, K. H., J. E. Hines, and J. D. Nichols. 1985. Goodness-of-fit tests for open capture-recapture models. Biometrics 41:399–410

Pollock, K. H., J. D. Nichols, C. Brownie, and J. E. Hines. 1990. Statistical inference for capture-recapture experiments. Wildlife Monograph 107:1–97

Pollock, K. H., J. D. Nichols, T. R. Simons, G. L. Farnsworth, L. L. Bailey, and J. R. Sauer. 2002. Large scale wildlife monitoring studies: statistical methods for design and analysis. Environmetrics 13:105–119

Pradel, R. 1996. Utilization of capture–mark–recapture for the study of recruitment and population growth rate. Biometrics 52:703–709

Royle, J. A. 2004. N-mixture models for estimating population size from spatially replicated counts. Biometrics 60:108–115

Royle, J. A. and J. D. Nichols, 2003. Estimating abundance from repeated presence absence data or point counts. Ecology 84:777–790

Royle J. A., J. D. Nichols, K. U. Karanth, and A. M. Gopalaswamy. 2009. A hierarchical model for estimating densities in camera-trap studies. Journal of Applied Ecology 46:118–127

Schwarz, C. J. and A. N. Arnason. 1996. A general methodology for the analysis of capture–recapture experiments in open populations. Biometrics 52:860–873

Schwarz, C. J. and W. T. Stobo. 1997. Estimating temporary migration using the robust design. Biometrics 53:178–194

Seber, G. A. F. 1965. A note on the multiple-recapture census. Biometrika 52:249–259

Seber, G. A. F. 1982. The estimation of animal abundance and related parameters. Second edition. Macmillan, New York

Silver, S. C., L. E. T. Ostro, L. K. Marsh, L. Maffei, A. J. Noss, M. J. Kelly, R. B. Wallace, H. Gómez, and G. Ayala. 2004. The use of camera traps for estimating jaguar *Panthera onca* abundance and densityusing capture/recapture analysis. Oryx 38:148–154

Skalski, J. R. and D. S. Robson. 1992. Techniques for wildlife investigations. Academic, San Diego

Smith, M. H., R. Blessing, J. G. Chelton, J. B. Gentry, F. B. Golley, and J. T. McGinnis. 1971. Determining density for small mammal populations using a grid and assessment lines. Acta Theriologica 16:105–125

Soisalo, M. K. and S. M. C. Cavalcanti. 2006. Estimating the density of a jaguar population in the Brazilian Pantanal using camera-traps and capture–recapture sampling in combination with GPS radio-telemetry. Biological Conservation 129:487–496

Stanley, T. R. and K. P. Burnham. 1998. Information-theoretic model selection and model averaging for closed-population capture–recapture studies. Biometrical Journal 40:475–494

Stickel, L. F. 1954. A comparison of certain methods of measuring ranges of small mammals. Journal of Mammalogy 35:1–15

Thompson, S. K. 1992. Sampling. Wiley, New York

Wegge, P., C. P. Pokheral, and S. R. Jnawali. 2004. Effects of trapping effort and trap shyness on estimates of tiger abundance from camera trap studies. Animal Conservation 7:251–256

White, G. C. 2005. Correcting wildlife counts using detection probabilities. Wildlife Research 32:211–216

White, G. C. and K. P. Burnham. 1999. Program MARK: survival rate estimation from both live and dead encounters. Bird Study 46(suppl.):S120–S139

Williams, B. K., J. D. Nichols, and M. J. Conroy. 2002. Analysis and management of animal populations. Academic, San Diego

Wilson, K. R. and D. R. Anderson. 1985. Evaluation of two density estimators of small mammal population size. Journal of Mammalogy 66:13–21

Wilson, D. J., M. G. Efford, S. J. Brown, J. F. Williamson, and G. J. McElrea. 2007. Estimating density of ship rats in New Zealand forests by capture–mark–recapture trapping. New Zealand Journal of Ecology 31:47–59

Yoccoz, N. G., J. D. Nichols, and T. Boulinier. 2001. Monitoring of biological diversity in space and time. Trends in Evolution and Ecology 16:446–453

Chapter 7
Estimating Tiger Abundance from Camera Trap Data: Field Surveys and Analytical Issues

K. Ullas Karanth, James D. Nichols, and N. Samba Kumar

7.1 Introduction

7.1.1 Camera Trap Studies of Tigers: Natural History and Science

Automated photography of tigers *Panthera tigris* for purely illustrative purposes was pioneered by British forester Fred Champion (1927, 1933) in India in the early part of the Twentieth Century. However, it was McDougal (1977) in Nepal who first used camera traps, equipped with single-lens reflex cameras activated by pressure pads, to identify individual tigers and study their social and predatory behaviors. These attempts involved a small number of expensive, cumbersome camera traps, and were not, in any formal sense, directed at "sampling" tiger populations.

Karanth (1995) first employed camera traps as a population sampling tool, using tiger photos to generate capture histories that were then used to estimate population size (abundance) in a closed model capture–recapture (CR) framework (Otis et al. 1978; White et al. 1982). Although this post-hoc analysis partially shoe-horned data into a CR framework, it did lead to identification of key issues related to trap-spacing, population closure, model selection and density estimation. These issues were addressed in subsequent refinements introduced by Karanth and Nichols (1998), and were elaborated in a technical manual (Karanth and Nichols 2002). Thereafter, several camera trap studies have tried to estimate tiger abundance using the Karanth–Nichols approach (Karanth et al. 2004a,b, 2006; Kawanishi and Sunquist 2004; Simcharoen et al. 2007) or variations on it (O'Brien et al. 2003; Wegge et al. 2004; Johnson et al. 2006). The material in this chapter will be based on the approach advocated by Karanth and Nichols (1998, 2002). This approach is sound and can be implemented with existing software at this time.

K.U. Karanth (✉) and N.S. Kumar
Wildlife Conservation Society – India Program, Centre for Wildlife Studies,
26-2, Aga Abbas Ali Road (Apt: 403), Bangalore, Karnataka 560042, India
e-mail: ukaranth@gmail.com

J.D. Nichols
U.S. Geological Survey, Patuxent Wildlife Research Center, Laurel, MD 20708, USA

A.F. O'Connell et al. (eds.), *Camera Traps in Animal Ecology: Methods and Analyses*,
DOI 10.1007/978-4-431-99495-4_7, © Springer 2011

However, we note that a new approach to density estimation using spatially explicit CR models (Efford 2004; Borchers and Efford 2008; Royle and Young 2008; Royle et al. 2009 a, b) holds a great deal of promise and may become the method of choice in many situations.

Overall, these studies highlight the potential for monitoring populations of this highly endangered big cat using camera traps. However, to generate reliable estimates of tiger abundance and density, camera trap studies must overcome several daunting challenges posed by the ecology of this scarce, elusive animal. In this chapter, we cover key issues including equipment needs, resource requirements, field survey protocols and analytical methods, and illustrate them with examples from tiger surveys. We also try to evaluate common problems associated with the rigorous conduct of camera trap studies of tigers. We have not tried to explain here the general concepts underlying animal abundance estimation (Seber 1982; Thompson et al. 1998; Williams et al. 2002). We refer the reader to the extensive literature on CR sampling for closed populations (Otis et al. 1978; White et al. 1982; Chao and Huggins 2005a, b). Thompson (2004) provides a good overview of statistical issues in sampling populations of rare or elusive animals, wherein application of camera traps to this task is covered by Karanth et al. (2004c).

7.1.2 Tiger Ecology in Relation to Abundance Estimation Issues

Despite the tiger's formerly vast range across Asia shrinking by 93% during historical times, the cat is still distributed widely but patchily across an extensive area of 1.1 million km^2 (Sanderson et al. 2006). However, tigers occur at relatively low densities of 1–20 individuals per 100 km^2, compared to, say large ungulates that attain densities of 100–10,000 individuals in an equivalent area (Karanth et al. 2004a). Furthermore, tigers are secretive, nocturnal animals that possess extraordinarily keen senses that they effectively employ to avoid encounters with humans. Tigers can move over long distances (5–25 km) overnight, but also stay localized around prey carcasses for 3–10 days at a time (Sunquist 1981; Smith 1993; Karanth and Sunquist 2000). Home ranges of individual tigers vary widely in size (15–1,500 km^2), depending on local ecology and social status of the individual. Tigers are solitary animals, except for mother-cub associations that may last for 18–24 months and mating pairs that associate for 3–7 days.

Because of these ecological factors, tiger population size cannot be estimated using methods based on visual detection and counting, such as distance sampling (Buckland et al. 2001). Invasive techniques that involve physical capture and radio-telemetry, which are useful for tiger behavioral studies (Sunquist 1981; Smith 1993), pose impossible logistical challenges when the goal is to adequately sample populations to estimate numbers.

Because of such problems rooted in tiger ecology, non-invasive "camera trapping" has emerged as an attractive alternative methodology for assessing tiger numbers. In this context, the fact that tigers are "naturally marked" confers a major advantage for investigators interested in abundance estimation using CR methods (Nichols 1992). Moreover, none of the alternatives proposed can boast of comparable theoretical development, analytical rigor, empirical testing, and software development (Williams et al. 2002; Amstrup et al. 2005). Furthermore, except for careful advance planning, there are virtually no extra costs or difficulties associated with rigorous photographic CR sampling when compared to other ad hoc approaches (Karanth and Nichols 2002). However, we do emphasize that challenges posed by tiger biology that defeat alternative counting methods (see Karanth et al. 2003) also pose problems for photographic CR surveys. Therefore, planning of surveys keeping these problems in view is the key to successful camera trapping.

7.2 Equipment and Field Practices

7.2.1 Camera Traps and Related Equipment

Camera traps are now manufactured by a number of commercial firms and can even be assembled at home. However, because deployment and maintenance of camera traps at a scale required by tiger surveys is a difficult and expensive exercise, we strongly advise against use of unreliable home-made equipment. Among various types of commercially available camera trap units (see Swann et al., Chap. 3), choice in a particular context is governed by several factors.

As we shall see later, deployment of dozens or even hundreds of camera traps is necessary for generating sufficient capture data for reliable abundance estimation. Further we recommend that every trap contains *two cameras* to photograph both flanks of the tiger (Karanth et al. 2002). Abundance estimation is still possible with single-flank photographs, but the additional uncertainty associated with individual identification typically combines with small sample size to yield very imprecise estimates.

Because typical tiger surveys are funded only to the extent of a few thousand dollars, investigators must choose cheaper units over expensive ones with more frills. This is an overarching constraint under which the following additional factors come into play.

Active infrared camera trapping units (Swann et al., Chap. 3) typically have a shorter recycling time (~5 s) between pictures, thus enabling photo-capture of a second or even a third tiger when mother and her pre-dispersal offspring, siblings, or mating pairs are involved. Although active units take longer to set up, they are easier to target and frame the tigers clearly, thus facilitating easier individual identification. Active units perform more reliably than passive ones, particularly in extreme cold weather or if the temperature difference between the tiger's body and surrounding environment is small.

However, in wet or rainy weather passive infrared units perform more reliably, which is a major advantage for surveying tigers in tropical regions. Passive units are less often triggered by non-target objects. Moreover, passive traps tend to be substantially cheaper compared to active ones, conferring another major advantage. Although some types of active units can trip more than one camera, saving on cost of a tripping unit, the overall relative cost advantage of passive infrared units over active ones is still substantial.

Camera traps are sometimes damaged by inclement weather or animals such as elephants, and, rarely by bears or tigers. More often, traps are stolen or vandalized by humans. Providing a rugged protective unit (such as the one described in Karanth and Nichols 2002, p. 184–186) is one option. However, strong protective shells that can thwart elephants or determined thieves tend to be heavy and cumbersome. An alternative option is to hide or mask the camera traps, which becomes particularly difficult with flashlight photography. Careful choice of trap sites increases tiger capture probabilities (Plate 1), and the use of a metal shell can reduce theft or animal damage.

However, it is our general view that most presently available commercial camera traps are of inadequate quality and/or too expensive to deploy in sufficient numbers necessary to sample tiger populations at the right spatial scales. Tigers Forever, a project sponsored by Wildlife Conservation Society and Panthera Foundation is currently working on developing an entirely new camera trap for large mammal surveys that works on a cellphone platform, which aims to overcome these constraints (Ed Yarmchuk and Alan Rabinowitz, Panthera Foundation, pers. commun).

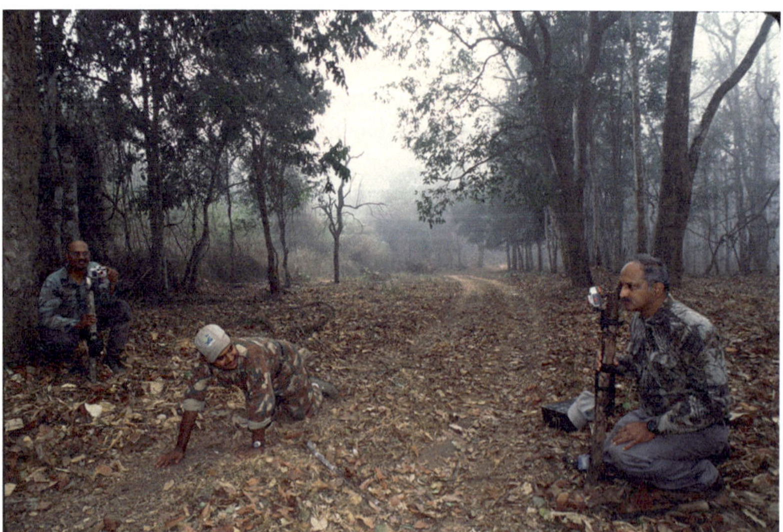

Plate 1 Researchers setting up a camera trap on a frequently used tiger travel route in Nagarahole, India. Photo credit: Eleanor Briggs

7.2.2 Choice of Trap Sites

Because tigers live at relatively low densities even in the best of habitats, the probabilities of their encountering camera traps are very low. Therefore, the most important survey design goal should be to maximize probabilities of photo-capturing tigers by increasing the per sample detection probability "*p*" (Otis et al. 1978; Williams et al. 2002; Nichols and Karanth 2002). An important secondary design goal is to expose the different individuals in the surveyed area to similar capture probabilities and to minimize variation among individuals in the probability of capture. Increased capture probabilities lead to captures of more individuals in the sampled population and increase recapture rates of previously caught tigers. Therefore, camera traps must be set on trails and paths that are most likely to be frequented by tigers. Attempts to place traps systematically by using a grid-like pattern typical of rodent trapping or to randomize selection of trap locations (see Rowcliff et al. 2008) are likely to detract from selection of the best trap sites. Such practices may, in fact, drastically reduce tiger capture probabilities. We note, however, that if the goal of the survey is not solely to maximize captures of tigers, alternative ways of choosing trap locations may be justified (see Kéry, Chap. 12; O'Brien et al., Chap. 13). It must be noted that tiger cubs <1 year age are usually camera trap-shy with extremely low capture rates, rendering it almost impossible to use photographic CR analyses to estimate their abundance. For example, in a 10-year data set of 366 tiger photo-captures from Nagarahole, India (Karanth et al. 2006), only two photo-captures were of cubs less than 1 year in age.

Typically, tigers move along well-used travel routes (Smith et al. 1989), which can be identified by skilled trackers based on tiger signs such as tracks, scent marks and scat-deposits. Optimal camera trap sites are found on such routes, particularly where trails of ungulate prey species converge. Although we have not used baits or lures in our tiger studies, these can potentially increase photo-capture probabilities in studies of big cats and should be used in such cases.

The survey area should be thoroughly reconnoitered beforehand using the best available local knowledge, maps and survey tools. This reconnaissance should have the *specific goal* of identifying approximately double the number of camera trap sites than are likely to be used later (as explained in the next section). These extra trap sites provide flexibility to locate traps optimally in the final survey design, without lowering capture probabilities or compromising on trap-spacing needs. Investigators should seek the assistance of local naturalists for such reconnaissance surveys: after all, the goal of "maximizing tiger captures in traps" is the same one pursued by illegal hunters who are often the most knowledgeable with respect to tiger natural history locally.

7.2.3 Accurately Recording Data

Regardless of the type of camera trap equipment used, trouble-shooting – fixing problems with cameras, film, tripping devices, batteries, cables, etc. – has to be done as often as possible, preferably once a day if the survey logistics permit. We recommend developing a clear check-list and data form for maintaining trouble-shooting records to ensure cost-effectiveness and data integrity.

Each tiger photo-capture should have the following ancillary data for abundance estimation: the location of trap-site (for computing distances moved, buffer area, and for use in spatially explicit modeling); the date, and if possible the time, of each capture event, for assigning it to the correct sampling occasion. Therefore, it is efficient to have pre-designed data forms (Karanth and Nichols 2002, p. 183). We cannot emphasize strongly enough the absolute need to uniquely number each photo-frame (by marking each film canister with a unique indelible identification number *before* loading it into the camera). Similarly the location and date of each photo must be recorded accurately. If errors creep in while recording these details, the capture-histories generated, as well as the resulting analyses, will be flawed.

Tigers are not difficult animals to reliably identify from good photographs based on stripe patterns on their flanks, limbs, face and even the tail (Plate 2). Poor quality photos might lead to uncertain individual identifications, which cannot be used

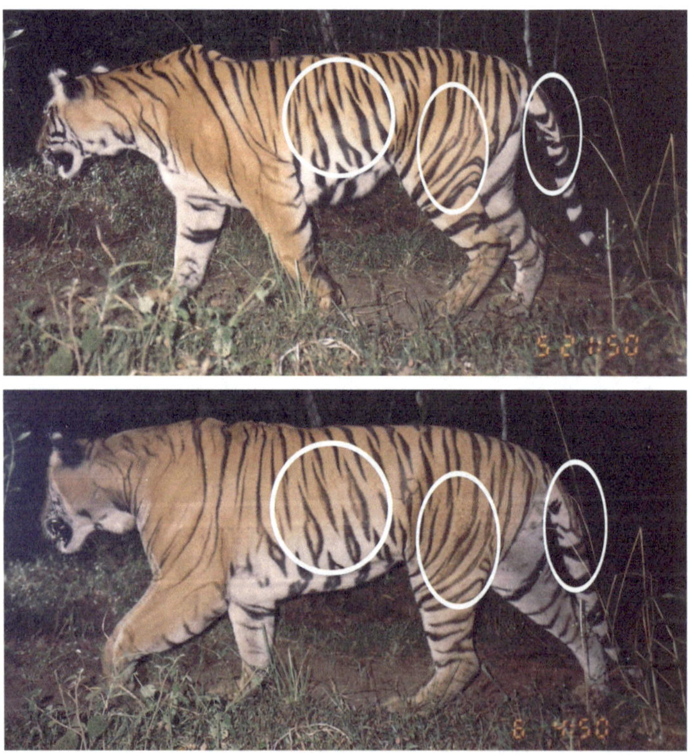

Plate 2 Camera trap photos of flanks enable unambiguous individual tiger identifications from stripe patterns as well as additional information on the sex and broad age-category for some of the captured tigers. The top picture shows an adult tigress uniquely identified as NHT-115 and the bottom picture shows the male tiger NHT-004, both photographed on successive nights at the same trap site in Nagarahole, India during January 1995. Photo credit: Ullas Karanth, Wildlife Conservation Society

Table 7.1 Capture histories of individual tigers photographed in Panna Tiger Reserve, Central India, on 15 sampling occasions during February–April 2002

Individual identification number	1	2	3	4	5	6	7	8	9	10	11	12	13	14	15
PAT-101	0	0	0	0	1	0	0	0	0	0	0	0	0	0	0
PAT-102	0	0	1	1	1	0	0	0	0	0	0	0	0	0	0
PAT-103	0	0	0	0	0	1	0	0	0	0	0	0	0	0	0
PAT-104	0	1	0	0	0	0	0	0	0	0	0	0	0	0	0
PAT-105	0	0	0	1	0	0	0	0	0	0	0	1	1	0	1
PAT-106	0	0	0	0	0	1	0	0	0	0	0	0	0	0	0
PAT-107	0	0	0	0	0	0	0	0	0	0	0	0	1	0	0
PAT-108	1	0	0	0	0	0	0	0	0	0	0	0	0	0	0
PAT-109	0	1	0	1	0	0	0	0	0	0	0	0	0	0	0
PAT-110	0	0	0	0	0	0	0	0	0	1	0	0	0	0	0
PAT-111	0	0	0	0	0	0	0	0	0	0	0	0	1	0	0

efficiently in CR analyses. In longer term studies, number of individuals captured increases over the years, and, newly caught animals must be compared to all previously caught ones. An automated pattern-matching freeware computer program specifically designed for identifying tigers, developed by Hiby et al. (2009; ExtractCompare v 1.8, http://www.conservationresearch.co.uk/tigers/tigers1.htm) can now facilitate rapid ranking and short-listing of most likely matches from a database of tiger photos. This automated process permits investigators to concentrate on the final visual identifications, thus greatly reducing effort.

The photographic capture data analysis requires "capture histories" of individual tigers in the "x matrix" format (Table 7.1). Capture histories are simply strings of 1's and 0's indicating capture and no capture, respectively, at each specific sampling period (see O'Brien, Chap. 6). For example, a capture history of 01001 indicates that the animal was captured only on sampling periods 2 and 5 of a 5-occasion study. Such histories are constructed after a careful comparison of photographs to identify individuals from stripe patterns on both flanks of the animals. A single pair of photographs of both flanks of a tiger (obtained simultaneously) is needed to "link" the left and right profiles and identify the animal permanently. Thereafter, the animal can be identified from the photograph of any one flank. However, this in effect means that every camera trap should have two cameras. There is no avoiding this investment.

7.3 Survey Design Considerations

7.3.1 Season, Survey Duration and Population Closure

The survey duration is the total number of days taken to sweep *the entire area of interest* with cameras, two or more times (sampling periods). Abundance estimation assuming a closed population (Otis et al. 1978; White et al. 1982; Williams et al. 2002) dictates that the duration of such a multi-sample CR survey should be as "short" as possible in relation to likely turnover of the targeted animal population as a result of births, deaths, immigration and emigration. Tiger populations have high mortality, recruitment and turnover rates (Karanth et al. 2006), but assuming closure over periods of 30–60 days may be justifiable. However, an even shorter duration is always preferable! Keeping cameras going for months is likely to violate the closure assumption, and requires different types of analyses specific to demographically open populations (Pollock et al. 1990; Kendall et al. 1995; Karanth et al. 2006, see Karanth et al., Chap. 9; O'Brien, Chap. 6).

The choice of the survey season is dictated by several factors: weather conditions necessary to effectively run the camera traps, accessibility of the entire area to survey personnel, incidence of human activities that may help or hamper the work, the feasibility of deploying traps to maximize tiger capture probabilities, and administrative factors such as availability of personnel, permits and equipment.

7.3.2 Spacing and Placement of Traps

Closed model CR analyses are based on the "ball-and-urn" conceptual model (White et al. 1982, p. 4). Their major underlying statistical assumption is that samples consisting of a few tigers are being repeatedly drawn from a single population consisting of all tigers. Therefore, each tiger must have "some" probability of being photo-captured (although not all animals may in fact be caught during the actual survey). In other words, the entire sampled area should be covered by camera traps, without leaving any large "holes" in which a tiger could spend the entire survey period without any probability of encountering a camera trap (Fig. 7.1).

The maximum spacing between two traps necessary to avoid a "hole", and ensure that every individual tiger is potentially exposed to trapping, depends on the expected minimum home range size of tigers. Typically, breeding tigresses holding territories have the smallest home ranges (Sunquist 1981; Smith 1993). The size of these female ranges will vary from 10 to 500 km^2 depending primarily on prey abundance. Correspondingly, the maximum spacing between traps can be set at about 2–10 km depending on expected female range size specific to that habitat. Another rule of thumb is to set at least two traps per female home range, although more traps are always better than fewer. By using female range size to apply this minimum rule, it is automatically ensured that other social classes of tigers such as post-dispersal offspring and adult males, whose home ranges may be 3–10 times larger (Smith 1993), get a reasonable number of traps set in their ranges.

Investigators face a dilemma with regard to trap spacing: given a fixed number of traps, increased trap density (traps being placed more closely) is likely to increase CR rates for tigers exposed to traps, thus increasing capture probabilities; on the other hand, reduced trap density (traps placed further apart) can potentially sample a larger number of animals, thus catching a larger sample of individual

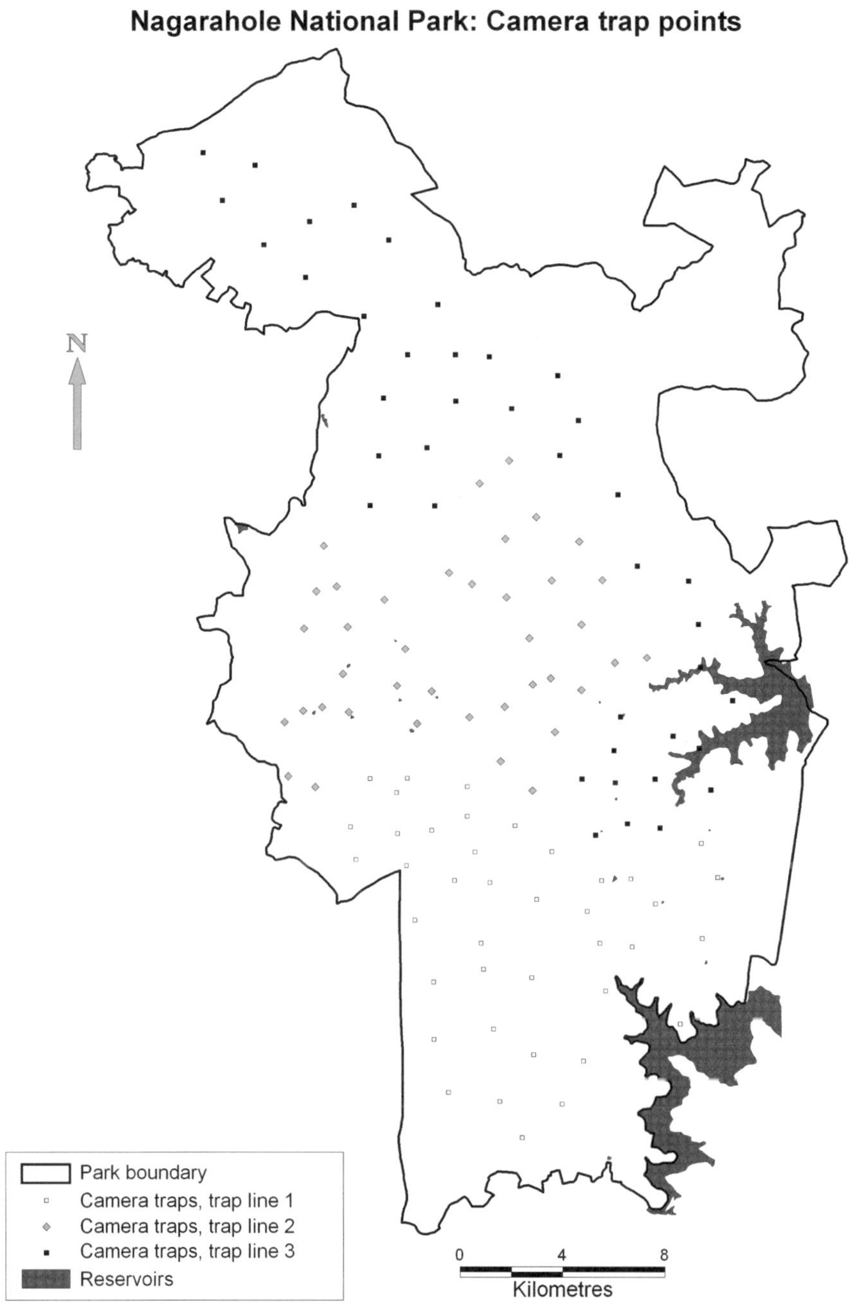

Fig. 7.1 Camera trap locations used for survey of the tiger population at Nagarahole, India

Table 7.2 The relationship between trap spacing, capture probability (p), number of tigers caught (M_{t+1}) and estimated population size from a data set in Nagarahole, India, 2004. Number of camera traps: 40; Number of sampling occasions: 10; Model: M_h. Captures from trap sites were successively eliminated to create survey designs with different trap spacing

Trap spacing (km)	Area of camera trap polygon (km^2)	Estimated capture probability per occasion $\hat{p}$	Number of tigers captured M_{t+1}	Estimated population size $\hat{N}$
~1.5	98.7	0.0840	16	25
~2.5	249.6	0.0383	16	47
~3.5	373.3	0.0255	23	94

tigers (M_{t+1}) and increasing this component of sample size (Otis et al. 1978; White et al. 1982). Table 7.2 shows this relationship between trap-spacing, capture probability (p), number of tigers caught (M_{t+1}) and estimated population size from a data set from Nagarahole, India.

Tiger densities are primarily dependent on densities of ungulate prey (Karanth et al. 2004a). In Table 7.3 we show prey density estimates, as well as predicted and estimated tiger densities at 11 sites in India. Potential number of tigers in an area to be sampled can be predicted from prey density data because there is approximately one tiger for 500 ungulate prey animals.

If there are no prior data on tiger movement and home range use in the area, our practical recommendation is to potentially expose at least 10–30 individual tigers to trapping, ensuring there are no "holes" in the sampled area. Thereafter, if more traps become available, trap densities can be increased to improve capture rates.

7.3.3 Adequate Coverage of the Sampled Area

After decisions are made on the survey season, duration, trap-spacing, number of trap locations, and geographic area to be covered, a second set of decisions has to be made based on the total number of camera traps available and logistical constraints.

For illustration, we can assume that 100 trap locations have to be sampled in a 30-day closed survey period. If the investigator has 100 functioning traps and sufficient logistical capacity to run all 100 traps simultaneously, a maximum of 30 sampling periods (these are also called "samples" or "sampling occasions") will be available to construct tiger "capture histories."

The resulting closed model CR analyses can have a minimum of 2 sampling periods to a maximum of 30 sampling periods. In the first case, all tiger capture data from the two 15-day periods (days 1–15, 16–30) are treated as the two samples, whereas in the second case all captures during each day are treated as one sample (total 30 samples). With reasonable numbers of tiger captures, generally, the ability to fit more complex models to the data increases if there are more sampling periods. On the other hand, decisions about sample periods should not

Table 7.3 Prey density estimates, predicted tiger densities, camera trapping effort, number of tigers photo-captured, estimated capture probabilities, estimated tiger abundance with standard error, estimated area sampled and estimated tiger densities at 11 sites in India (based on Karanth et al. 2004a)

Location	Ungulate prey densities (per km^2)	Predicted tiger densities (per 100 km^2)	Camera trap effort (number of trap-days)	Number of tigers captured M_{t+1}	Capture probability per sample $\hat{p}$	Estimated abundance $\hat{N}(SE[\hat{N}])$	Estimated area sampled $\hat{A}(\hat{w})(km^2)$	Estimated density $\hat{D}(SE[\hat{D}])$ (per 100 km^2)
Melghat	5.3	1.04	896	15	0.058	24 (6.09)	360	6.67 (1.85)
Tadoba	13.1	2.61	706	10	0.174	12 (1.97)	367	3.27 (0.59)
Pench-MR	16.2	3.24	715	14	0.108	20 (4.41)	274	7.29 (2.54)
Bhadra	16.8	3.36	587	7	0.220	9 (1.93)	263	3.42 (0.84)
Panna	30.9	6.18	914	11	0.039	29 (9.65)	418	6.94 (3.23)
Bandipur	35.2	7.04	946	16	0.055	34 (9.90)	284	11.97 (3.71)
Nagarahole	56.1	11.22	938	25	0.120	29 (3.77)	243	11.92 (1.71)
Kanha	57.3	11.46	803	26	0.180	33 (4.69)	282	11.70 (1.93)
Kaziranga	58.1	11.62	544	22	0.190	28 (4.51)	167	16.76 (2.96)
Ranthambore	60.6	12.12	840	16	0.115	28 (7.29)	244	11.46 (4.20)
Pench-MP	63.8	12.76	788	5	0.220	6 (1.41)	122	4.94 (1.37)

produce many sample periods with zero captures, as this can create numerical problems with estimation software and simply makes the modeling of capture probability more difficult. Therefore, in the above case, it may be preferable to construct capture histories with 10 samples of 3 days each or perhaps 6 samples of 5 days each.

Usually, investigators will not have enough camera traps to cover all the identified locations simultaneously. In such a case, trap locations are spatially segregated into logistically convenient "blocks" or "trap lines" (see Fig. 7.1) and sampled successively. An additional logistical constraint is that camera traps cannot be moved easily and set up every day. In the above case, if we had only 25 camera traps, the 100 locations would be segregated into four trap lines. Cameras in each line would be trapped continuously for 7 days and then moved to the next line, to complete the survey in 28 days. This approach would result in a capture history matrix with seven sampling periods, each containing tiger photo-captures from one day selected from each one of the four trap lines (Nichols and Karanth 2002, design 4, p. 134). For example, the first day of trapping in each line could be combined to form sample period 1, the second day from each line combined to form sample period 2, etc.

The key point is that tiger capture data going into each sampling period should be generated from across the entire study area, to ensure that each individual tiger in the population has some probability of being exposed to traps during each period. If the approach described above involving trap lines and movement of camera traps is employed during the survey, abundance estimation models that incorporate influence of time (the sampling occasion) on capture probabilities typically should not be used in the analyses. However, we believe that in short duration surveys, such time-related influence on capture probability may not be important for tigers. In this context, we note that the spatially explicit CR model recently developed by Royle et al. (2009a, b), does not require sampling periods to be structured in this manner and can effectively deal even with the presence of potential "holes" in the sampled area at any sampling period.

The area over which camera traps are deployed typically attains the shape of an irregular polygon of some sort (Fig. 7.1). In practice, the shape of this "trap polygon" may be dictated by a variety of influences, such as: local tiger distribution and potential densities; geographical and logistical factors that affect trap deployment; and social factors such as legal access and need to protect traps against theft or vandalism. Although the computation of the sampled area is covered later in this chapter, the following points should be considered when selecting the trap locations.

The more a trapped area resembles a complete circle, the better it is from an analytical point of view. Shapes that have high periphery-to-area ratios (such as "doughnuts", narrow strips, radiating projections) are to be avoided if possible. Furthermore, if the tiger habitat in the sampled area has a "hard-edge" that borders non-tiger habitat, instead of locating traps right on the edge, placing traps inside will avoid coverage of areas where tigers do not occur, and thus free up camera traps.

Tigers are exposed to traps as they move along their travel routes looking for prey, water, shade, or social contact with other tigers. They do not choose travel routes through a "random process" that might be best sampled by a regular grid of intersecting points at which to set traps. Therefore, constraining the choice of potential camera trap sites indirectly by using some sort of a pre-planned "grid" to guide trap setting will almost certainly reduce tiger capture probabilities. The best possible set of trap locations for photographing tigers is unlikely to be selected through such self imposed constraints on the investigator's part. Therefore, investigators should avoid using grid-like patterns or randomly selected GPS coordinates as a basis for setting camera traps for tigers. The primary reason for use of such systematic or random trap locations is to provide similar capture probabilities for all animals in the sampled area. However, if animal movement is restricted to a small subset of all possible areas (i.e., along trails and roads), then restriction of traps to this subset can accomplish this objective. Given the numerous ecological and logistical challenges investigators already face in getting sufficient photo-captures of this elusive species, selecting trap locations "randomly" is a luxury they can ill-afford.

7.4 Data Analysis: Issues and Examples

7.4.1 The Approach to Analysis of Tiger Photo-Capture Data

It is our view that given major analytical advantages of using "marked animals" for studying animal population dynamics (Nichols 1992; Williams et al. 2002), tiger ecologists should not forgo these advantages by using approaches based on simple trapping rates to make inferences about abundance (e.g., Carbone et al. 2001). For estimating tiger abundance from short-duration surveys, we recommend use of sound closed CR models which are being continually developed, evaluated and refined (for details, see reviews by Otis et al. 1978; Williams et al. 2002; Chao and Huggins 2005a, b; Royle et al. 2009a, b).

Relevant software, such as the reliable old war-horse, program CAPTURE (White et al. 1982; Rexstad and Burnham 1991), as well as the more recent and versatile program MARK (White and Burnham 1999), is available at no cost off the internet (http://www.mbr-pwrc.usgs.gov/software.html; http://www.warnercnr.colostate.edu/~gwhite/mark/mark.htm). Tiger biologists using these modern approaches are also supported by helpful list servers (http://www.phidot.org/forum) and detailed manuals (Karanth and Nichols 2002).

The analysis of tiger photo-capture data for estimating abundance and density using the above software involves the following steps:

(1) Testing the assumption that the tiger population was closed, using the observed tiger capture frequencies, (2) Selecting the CR model most likely to have generated the observed tiger capture history data, by using the discriminant function model selection statistics and comparing the relative fit and number of parameters of various plausible models, (3) Estimating capture probabilities per sampling

period (p) and tiger abundance (N) using the most appropriate model for the data on hand, and (4) Estimating the area effectively sampled by camera traps and computing the tiger population densities based on abundance estimates.

7.4.2 Testing for Population Closure

The capture history data are first tested to check if the assumption of population closure during the "short" survey is justified. In program CAPTURE the null hypothesis of a closed population (and heterogeneity in capture probabilities) is statistically tested using an approach based on the observed times between first and last captures for all individuals captured twice or more (Otis et al. 1978; Williams et al. 2002). This test is sensitive to certain time-related patterns in capture probabilities (e.g., low probabilities at the beginning or end of a study), behavioral responses to capture, and temporary emigration. Program MARK (White and Burnham 1999) can be used to test the null hypothesis of complete closure (and time-specific variation in capture probabilities) against the alternative hypothesis of a completely open population, with both mortality and recruitment (Stanley and Burnham 1999a). This test is somewhat sensitive to behavioral responses and individual heterogeneity, but not to temporal variation in capture probabilities. If the closure assumption cannot be rejected, analyses can proceed. However, if the tiger population appears to be open, investigators may have to apply suitable open model analyses (Pollock et al. 1990; Williams et al. 2002; Karanth et al., Chap. 9) such as the Jolly-Seber model (Jolly 1965; Seber 1965) and more recent parameterizations of this model (Pradel 1996; Schwarz and Arnason 1996). However, closed models are generally more useful for abundance estimation from short duration surveys.

7.4.3 Model Selection and Estimation of Tiger Abundance

The next step in the analysis requires comparisons among plausible CR models that attempt to approximate the processes that generated the observed capture histories. In program CAPTURE these comparisons are performed using a series of between-model and goodness-of-fit hypothesis tests, and the resulting statistics are used to compute an overall score based on a discriminant function developed using simulated data (Otis et al. 1978; White et al. 1982). In program MARK, likelihood-based model selection criteria such as Akaike's Information Criterion (AIC; Burnham and Anderson 2002) may be used to compare models. Different closed CR models consider the potential effects on capture probabilities of such potential sources of variation as behavioral response of tigers to camera trapping (e.g., trap-avoidance), time-specific variation (e.g., weekly weather changes), and heterogeneity among individual tigers (e.g., caused by factors such as territorial status of the animal or trap access). More complex models that incorporate the combined effects of these factors on capture probabilities are also available.

Typically the following models (Otis et al. 1978; White et al. 1982; Williams et al. 2002; Chao and Huggins 2005a, b) are considered in the analyses of tiger capture data:

- M_o – Capture probability is the same for all tigers and is not influenced by behavioral response, time or individual heterogeneity.
- M_h – Capture probabilities are heterogeneous among individual tigers, but are not affected by trap response or time.
- M_b – Capture probabilities differ between previously caught and uncaught tigers due to trap-response behavior, but are not influenced by heterogeneity or time.
- M_t – Capture probability is the same for all individual tigers, but varies from one sampling period to another due to time-specific factors.

The model selection process also considers more complex models such as M_{bh}, M_{th}, M_{tb} and M_{tbh} that incorporate effects of heterogeneity, trap-response, and time, in various combinations.

Generally, because tigers are territorial animals and the number of traps in each animal's range is likely to vary, models that incorporate either heterogeneity alone (M_h) or in combination with trap-response (M_{bh}) are appropriate for analyses of tiger capture data. In short duration surveys, time (sampling period) is unlikely to influence capture probabilities. For estimation of capture probabilities and abundance based on models M_h or M_{bh}, several specific estimators have been implemented in programs CAPTURE and MARK. These are reviewed in detail by several authors (Otis et al. 1978; Williams et al. 2002; Chao and Huggins 2005a, b). While tigers may in some cases develop a trap-shy response, as pointed out by Wegge et al. (2004), we note that this problem can be effectively handled by using behavioral response models (e.g., model M_{bh}), provided sufficient capture data are available.

In cases where there is a substantial amount of uncertainty about appropriate models, it may be useful to compute model-averaged estimates (e.g., Buckland et al. 1997; Stanley and Burnham 1998, 1999b). Specifically, a weighted mean abundance estimate is computed, with weights based on AIC (Burnham and Anderson 2002) or whatever other model selection statistic is computed (Stanley and Burnham 1998; 1999b). The variance of this weighted abundance estimate includes two components: the conditional variance of each abundance estimate based on its model and the squared deviations of the model-specific abundance estimates from the weighted mean (Buckland et al. 1997; Stanley and Burnham 1998, 1999b). The latter variance component reflects model uncertainty. Model averaging was found to perform well for closed population CR models in simulation studies by Stanley and Burnham (1998, 1999b).

The analyses using programs CAPTURE or MARK yield estimates of capture probability per sample occasion, *p*, and population size, *N* (abundance). Table 7.3 shows tiger abundance estimates from camera trap studies in 11 reserves across India (Karanth et al. 2004a).

7.4.4 Estimating the Sampled Area Size and Tiger Densities

While abundance estimates are useful for tracking changes in a tiger population at a given site over years, estimates of tiger density (usually expressed as number of tigers per 100 km^2) permit spatial comparisons across different sites, habitat types and larger regions. However, to derive tiger densities it is necessary to estimate area sampled by the camera trap survey. In our view, this problem remains somewhat of a weak link in analyses of CR data of big cats. Indeed, the issue of geographic closure extends beyond large cats and can cause problems with CR estimation in general (e.g., White et al. 1982).

For estimating the area actually sampled by camera traps, Nichols and Karanth (2002) proposed a refinement of a technique originally suggested by Dice (1938) and found to perform well in simulation studies by Wilson and Anderson (1985). This approach uses the mean of the maximum distance moved between captures for each individual tiger (MMDM) as an estimate of the average home range diameter, and attaches a buffer strip width of half this home range diameter around the polygon formed by outermost camera traps to estimate the area sampled (described in detail by Karanth and Nichols 1998; Karanth et al. 2004b). Application of the Wilson and Anderson (1985) method in field studies has generated tiger density estimates that appear to be very reasonable when compared to densities predicted from prey abundance (Table 7.3).

Nichols and Karanth (2002) proposed that instead of the delta approximation method (e.g., Seber 1982) used for estimation of variance of the buffer distance, which assumes a trap polygon roughly circular in shape (Karanth and Nichols 1998), a GIS-based bootstrap simulation using the actual camera-trapped area may perform better. However, this refinement has not been tried out in any field studies.

Soisalo and Cavalcanti (2006) studied radio-collared jaguars in Brazil and suggested that the Wilson and Anderson (1985) approach may underestimate jaguar home range diameter and, thereby overestimate densities. In their study, use of the full trap-based MMDM (rather than half of this distance) provided a better approximation to trap-based home range radius. It is not possible to judge the generality of this result. If MMDM is based on small numbers of recaptures (this is frequently true with tigers), then it may be wise to implement an estimation approach that accounts for the fact that greater distances tend to be associated with larger numbers of recaptures. For example, Jett and Nichols (1987) proposed use of an exponential model in which MMDM was modeled as an increasing function of number of captures and then estimated model parameters using data on MMDM for animals recaptured once, twice, etc. This approach permits estimation of the asymptotic MMDM that would be expected if there was a large number of recaptures of every animal.

We like the idea of incorporating radio telemetry data into analyses directed at density estimation (e.g., Nichols and Karanth 2002). Unfortunately, in most cases, radio-telemetry data on tigers (particularly on the same group of animals that are camera-trapped) are not available. Furthermore, if radio-telemetry data are indeed available at a site where tigers are being camera trapped, we believe that careful thought should be devoted to the manner in which they are used in the density estimation process. For example, Powell et al. (2000) did not focus on density estimation but developed a joint likelihood for telemetry and CR data to investigate temporary emigration. A similar approach based on the superpopulation concept (Kendall et al. 1997) could be developed for combining camera trap and telemetry data. Under the new spatial CR approach of Royle et al. (2009a, b), the telemetry data could be used to inform the spatial parameter relating an animal's capture probability at a specific trap to distance between the home range center and the trap. If studies employing radio telemetry and camera traps become more common, we believe there is much interesting work to be done investigating the best ways to combine these data sources.

7.5 Camera Trapping Tigers: Some General Comments

Having examined various constraints faced by investigators, the complexity of the answer to the common question "how many camera traps should be used in a study of tigers?" can be appreciated. The short answer is: "as many as you can afford." However, all too often investigators photo-trapping tigers for abundance estimation (as opposed to just getting tiger pictures) deploy too few cameras, but with far too many expectations about the data they can generate. Unfortunately, closed model estimates of tiger abundance generated from less than half a dozen tiger capture histories may not be very reliable. While one may be forced to begin the pilot work with a small number of cameras, ultimately, for reliable abundance estimation, there is no getting around the fact that a substantial number of camera traps will be required. This is simply a function of the relatively small numbers of study animals, the extensive spatial scale, and the short study duration that investigators studying tigers are forced to deal with.

In spite of the above problems, investigators using camera traps sometimes try to answer fundamental questions related to CR methodology, simply by chopping up their small capture data sets into even smaller and smaller sub-samples. We believe that many basic statistical and modeling issues of CR methodology are better addressed through appropriate analytical studies (e.g., Seber and Whale 1970; Carothers 1973; Williams et al. 2002, p. 293–295) or simulations (Otis et al. 1978; Menkens and Anderson 1988; Lee and Chao 1994; Rosenberg et al. 1995) or even field studies in special situations where population size is known (e.g., Greenwood et al. 1985; Manning et al. 1995). Some of these studies are published in specialized statistical journals (e.g., *Biometrics*) that investigators studying tigers sometimes tend to overlook.

While the easy availability of flexible software such as CAPTURE and MARK has been helpful, sometimes the output from these programs is not carefully studied, understood and reported by investigators. For the reader to understand the reliability and value of tiger abundance estimates, components of the analyses such as closure test results, the model selection process, model comparison results, and the relevant parameter estimates under more likely models should be reported clearly.

Sometimes, investigators studying tigers appear to become disheartened by the large variances associated with the abundance and density estimates generated by CR analyses. In response, they either abandon standard estimation protocols or make unjustifiable modifications to these. We point out that large variances of CR estimates are consequences of uncertainties resulting from small sample sizes of tigers caught and low levels of effort. They also arise partly because of a scientific philosophy that tries to explicitly model relevant parameters such as capture probability, instead of simply pretending that these don't exist. "Seat-of-the-pants" approaches may provide investigators an escape from the complexities of modeling tiger ecology, but they do not yield the kinds of reliable inferences that should emerge from studies that require nontrivial investments of research or conservation funds. Tiger studies conducted across extensive spatial (Karanth et al., 2004a) and temporal (Karanth et al. 2006) scales demonstrate that these uncertainties around estimates of tiger abundance are best dealt with by increases in the quality and quantity of the photographic capture data, and by overall integration of such data using improved modeling and estimation techniques now available in the general realm of marked animal studies.

We submit that camera trap studies of tigers will be more useful for science and conservation when they employ well-developed statistical methodologies, and when investigators make efforts to ensure that assumptions of these methods are satisfied in their field studies. Given the acute scarcity of demographic data on tigers and the many existing questions about tiger population responses to management and conservation efforts, we see great potential for investigations using camera traps. Because of the substantial effort and resources required to conduct a good camera trap study of tigers, we recommend that future studies be closely focused on either informing conservation efforts or addressing scientific questions (see recommendations in Nichols et al., Chap. 4). Such integration of estimation and monitoring into larger programs of science or conservation is especially important when estimation is such an expensive endeavor.

References

Amstrup, S. C., T. L. McDonald, and B. F. J. Manly. 2005. Handbook of capture–recapture analysis. Princeton University Press, Princeton, NJ

Borchers, D. L. and M. G. Efford. 2008. Spatially explicit maximum likelihood methods for capture–recapture studies. Biometrics 64:377–385

Buckland, S. T., K. P. Burnham, and N. H. Augustin. 1997. Model selection: an integral part of inference. Biometrics 53:603–618

Buckland, S. T., D. R. Anderson, K. P. Burnham, J. L. Laake, D. L. Borchers, and L. Thomas. 2001. Introduction to distance sampling. Oxford University Press, Oxford

Burnham, K. P. and D. R. Anderson. 2002. Model selection and multi-model inference: a practical information-theoretic approach. Springer, New York

Carbone, C., S. Christie, T. Coulson, N. Franklin, J. Ginsberg, M. Griffiths, J. Holden, K. Kawanishi, M. Kinnaird, R. Laidlaw, A. Lynam, D. W. Macdonald, D. Martyr, C. McDougal, L. Nath, T. Obrien, J. Seidensticker, D. Smith, M. Sunquist, R. Tilson, and W. N. W. Shahruddin. 2001. The use of photographic rates to estimate densities of tigers and other cryptic mammals. Animal Conservation 4:75–79

Carothers, A. D. 1973. The effects of unequal catchability on Jolly–Seber estimates. Biometrics 29:79–100

Champion, F. W. 1927. With a camera in tiger-land. Chatto & Windus, London

Champion, F. W. 1933. The jungle in sunlight and shadow. Chatto & Windus, London

Chao, A. and R. M. Huggins. 2005a. Classical closed population models. Pages 22–36 *in* S. Amstrup, T. McDonald, and B. Manly, editors. The handbook of capture–recapture analysis. Princeton University Press, Princeton, NJ

Chao, A. and R. M. Huggins. 2005b. Modern closed population models. Pages 58–86 *in* S. Amstrup, T. McDonald, and B. Manly, editors. The handbook of capture–recapture analysis. Princeton University Press, Princeton, NJ

Dice, L. R. 1938. Some census methods for mammals. Journal of Wildlife Management 2:119–130

Efford, M. 2004. Density estimation in live-trapping studies. Oikos 106:598–610

Greenwood, R. J., A. B. Sargeant, and D. H. Johnson. 1985. Evaluation of mark-recapture for estimating striped skunk abundance. Journal of Wildlife Management 49:332–340

Hiby, L., P. Lovell, N. Patil, N. S. Kumar, A. M. Gopalaswamy, and K. U. Karanth. 2009. A tiger cannot change its stripes: using a three dimensional model to match images of living tigers and tiger skins. Biology Letters 5:383–386

Jett, D. A. and J. D. Nichols. 1987. A field comparison of nested grid and trapping web density estimators. Journal of Mammalogy 68:888–892

Johnson, A., C. Vonkhameng, M. Hedemark, and T. Saithgodham. 2006. Effect of human-carnivore conflict on tiger (*Panthera tigris*) and prey populations in Lao PDR. Animal Conservation 9:421–430

Jolly, G. M. 1965. Explicit estimates from capture–recapture data with both death and immigration – stochastic model. Biometrika 52:225–247

Karanth, K. U. 1995. Estimating tiger *Panthera tigris* populations from camera-trap data using capture–recapture models. Biological Conservation 71:333–338

Karanth, K. U. and J. D. Nichols. 1998. Estimation of tiger densities in India using photographic captures and recaptures. Ecology 79:2852–2862

Karanth, K. U. and J. D. Nichols, editors. 2002. Monitoring tigers and their prey: a manual for researchers, managers and conservationists in tropical Asia. Centre for Wildlife Studies, Bangalore, India

Karanth, K. U. and M. E. Sunquist. 2000. Behavioral correlates of predation by tiger, leopard and dhole in Nagarahole, India. Journal of Zoology (London) 250:255–265

Karanth, K. U., N. S. Kumar, and J. D. Nichols. 2002. Field surveys: estimating absolute densities of tigers using capture–recapture sampling. Pages 139–152 *in* K. U. Karanth and J. D. Nichols, editors. Monitoring tigers and their prey: a manual for researchers, managers and conservationists in tropical Asia. Centre for Wildlife Studies, Bangalore, India

Karanth, K. U., J. D. Nichols, J. Seidensticker, E. Dinerstein, J. L. D. Smith, C. McDougal, A. J. T. Johnsingh, R. S. Chundawat, and V. Thapar. 2003. Science deficiency in conservation practice: the monitoring of tiger populations in India. Animal Conservation 6:141–146

Karanth, K. U., J. D. Nichols, N. S. Kumar, W. A. Link, and J. E. Hines. 2004a. Tigers and their prey: predicting carnivore densities from prey abundance. Proceedings of the National Academy of Sciences USA 101:4854–4858

Karanth, K. U., R. S. Chundawat, J. D. Nichols, and N. S. Kumar. 2004b. Estimation of tiger densities in the tropical dry forests of Panna, Central India, using photographic capture–recapture sampling. Animal Conservation 7:285–290

Karanth, K. U., J. D. Nichols, and N. S. Kumar. 2004c. Photographic sampling of elusive mammals in tropical forests. Pages 229–247 *in* W. L. Thompson, editor. Sampling rare or elusive species: concepts, designs and techniques for estimating population parameters. Island Press, Washington, DC

Karanth, K. U., J. D. Nichols, N. S. Kumar, and J. E. Hines. 2006. Assessing tiger population dynamics using photographic capture–recapture sampling. Ecology 87:2925–2937

Kawanishi, K. and M. E. Sunquist. 2004. Conservation status of tigers in a primary rainforest of Peninsular Malaysia. Biological Conservation 120:329–344

Kendall, W. L., K. H. Pollock, and C. Brownie. 1995. A likelihood-based approach to capture–recapture estimation of demographic parameters under the robust design. Biometrics 51:293–308

Kendall, W. L., J. D. Nichols, and J. E. Hines. 1997. Estimating temporary emigration and breeding proportions using capture–recapture data with Pollock's robust design. Ecology 78:563–578

Lee, S. M. and A. Chao. 1994. Estimating population size via sample coverage for closed capture–recapture models. Biometrics 50:88–97

Manning, T., W. D. Edge, and J. O. Wolff. 1995. Evaluating population-size estimators: an empirical approach. Journal of Mammalogy 76:1149–1158

McDougal, C. 1977. The face of the tiger. Rivington, London

Menkens, G. E., Jr., and S. H. Anderson. 1988. Estimation of small-mammal population size. Ecology 69:1952–1959

Nichols, J. D. 1992. Capture–recapture models: using marked animals to study population dynamics. BioScience 42:94–102

Nichols, J. D. and K. U. Karanth. 2002. Statistical concepts: estimating absolute densities of tigers using capture–recapture sampling. Pages 121–138 *in* K. U. Karanth and J. D. Nichols, editors. Monitoring tigers and their prey: a manual for researchers, managers and conservationists in tropical Asia. Centre for Wildlife Studies, Bangalore, India

O'Brien, T. G., M. F. Kinnaird, and H. T. Wibisono. 2003. Crouching tigers, hidden prey: Sumatran tiger and prey populations in a tropical forest landscape. Animal Conservation 6:131–139

Otis, D. L., K. P. Burnham, G. C. White, and D. R. Anderson. 1978. Statistical inference from capture data on closed animal populations. Wildlife Monographs 62:1–135

Pollock, K. H., J. D. Nichols, C. Brownie, and J. E. Hines. 1990. Statistical inference for capture–recapture experiments. Wildlife Monographs 107:1–97

Powell, L. A., M. J. Conroy, J. E. Hines, J. D. Nichols, and D. G. Krementz. 2000. Simultaneous use of mark-recapture and radiotelemetry to estimate survival, movement, and capture rates. Journal of Wildlife Management 64:302–313

Pradel, R. 1996. Utilization of capture–mark–recapture for the study of recruitment and population growth rate. Biometrics 52:703–709

Rexstad, E. and K. P. Burnham. 1991. Users' guide for interactive program CAPTURE. Abundance estimation of closed animal populations. Colorado State University, Fort Collins, CO

Rosenberg, D. K., W. S. Overton, and R. G. Anthony. 1995. Estimation of animal abundance when capture probabilities are low and heterogeneous. Journal of Wildlife Management 59: 252–261

Rowcliff, J. M., J. Field, S. T. Turvey, and C. Carbone. 2008. Estimating animal density using camera traps without the need for individual recognition. Journal of Applied Ecology 45: 1228–1236

Royle, J. A. and K. V. Young. 2008. A hierarchical model for spatial capture–recapture data. Ecology 80:2281–2289

Royle, J. A., J. D. Nichols, K. U. Karanth, and A. M. Gopalaswamy. 2009a. Hierarchical models and inference for spatial capture–recapture data: estimating tiger density from camera trap arrays. Journal of Applied Ecology 46:118–127

Royle, J. A., K. U. Karanth, A. M. Gopalaswamy, and N. S. Kumar. 2009b. Bayesian inference in camera-trapping studies for a class of capture-recapture models. Ecology 90:3233–3244

Sanderson, E., J. Forrest, C. Loucks, J. Ginsberg, E. Dinerstein, J. Seidensticker, P. Leimgruber, M. Songer, A. Heydlauff, T. O'Brien, G. Bryja, S. Klenzendorf, and E. Wikramanayake. 2006.

Setting priorities for the conservation and recovery of wild tigers: 2005–2015. The technical assessment. WCS, WWF, Smithsonian, and NFWF-STF, New York, Washington, DC

Schwarz, C. J. and A. N. Arnason. 1996. A general methodology for the analysis of capture–recapture experiments in open populations. Biometrics 52:860–873

Seber, G. A. F. 1965. A note on the multiple-recapture census. Biometrika 52:249–259

Seber, G. A. F. 1982. The estimation of animal abundance and related parameters. MacMillan, NewYork

Seber, G. A. F. and J. F. Whale. 1970. The removal method for two and three samples. Biometrics 26:393–400

Simcharoen, S., A. Pattanavibool, K. U. Karanth, J. D. Nichols, and N. S. Kumar. 2007. How many tigers *Panthera tigris* are there in Huai Kha Khaeng wildlife sanctuary, Thailand? An estimate using photographic capture–recapture sampling. Oryx 41:447–453

Smith J. L. D. 1993. The role of dispersal in structuring the Chitwan tiger population. Behavior 124:165–195

Smith J. L. D., C. McDougal, and D. Miquelle. 1989. Scent marking in free ranging tigers, *Panthera tigris*. Animal Behavior 37:1–10

Soisalo, M. K. and S. M. C. Cavalcanti. 2006. Estimating the density of a jaguar population in the Brazilian Pantanal using camera-traps and capture–recapture sampling in combination with GPS radio-telemetry. Biological Conservation 129:487–496

Stanley, T. R. and K. P Burnham. 1998. Information-theoretic model selection and model-averaging for closed-population capture–recapture studies. Biometrical Journal 40:475–494

Stanley, T. R. and K. P Burnham. 1999a. A closure test for time-specific capture–recapture data. Environmental and Ecological Statistics 6:197–209

Stanley, T. R. and K. P Burnham. 1999b. Estimator selection for closed-population capture–recapture. Journal of Agricultural, Biological, and Environmental Statistics 3:131–150

Sunquist M. E. 1981. Social organization of tigers (*Panthera tigris*) in Royal Chitwan National Park, Nepal. Smithsonian Contributions to Zoology 336:1–98

Thompson, W. L., editor. 2004. Sampling rare or elusive species: concepts, designs and techniques for estimating population parameters. Island Press, Washington, DC

Thompson, W. L., G. C. White, and C. Gowan. 1998. Monitoring vertebrate populations. Academic, New York

Wegge, P., C. Pokheral, and S. R. Jnawali. 2004. Effects of trapping effort and trap shyness on estimates of tiger abundance from camera trap studies. Animal Conservation 7:251–256

White, G. C. and K. P. Burnham. 1999. Program MARK: survival rate estimation from both live and dead encounters. Bird Study 46:120–139

White, G. C., D. R. Anderson, K. P. Burnham, and D. L. Otis. 1982. Capture–recapture and removal methods for sampling closed populations. Los Alamos National Laboratory Publication LA-8787-NERP. Los Alamos, NM

Williams, B. K., J. D. Nichols, and M. J. Conroy. 2002. Analysis and management of animal populations. Academic, San Diego, CA

Wilson, K. R. and D. R. Anderson. 1985. Evaluation of two density estimators of small mammal population size. Journal of Mammalogy 66:13–21

Chapter 8 Abundance/Density Case Study: Jaguars in the Americas

Leonardo Maffei, Andrew J. Noss, Scott C. Silver, and Marcella J. Kelly

8.1 Introduction

Since camera traps were first used to estimate the density of tiger *Panthera tigris* populations in India (Karanth 1995; see also Karanth et al. this volume), this methodology has been widely used to study a variety of species: leopards *Panthera pardus* (Henschel and Ray 2003; Karanth et al. this volume; Kostyria et al. 2003), snow leopards *Panthera uncia* (Jackson et al. 2006), pumas *Puma concolor* (Kelly et al. 2008), ocelots *Leopardus pardalis* (Di Bitetti et al. 2006, 2008; Dillon and Kelly 2007, 2008; Maffei et al. 2005; Trolle and Kéry 2003, 2005), and Geoffroy's cats *Oncifelis geoffroyi* (Cuéllar et al. 2006; Pereira et al. 2006). However, jaguars *Panthera onca* have probably received the most attention with respect to using camera traps to estimate the abundance and density of populations that cover the species' entire Neotropical range (Cullen et al. 2005; Kelly 2003; Maffei et al. 2004b; Miller and Miller 2005; Silver et al. 2004; Soisalo and Cavalcanti 2006). To date, at least 83 different camera trapping efforts have been carried out to survey jaguars, from southern Arizona in the north to northern Argentina in the south. In this chapter, we describe the details of this methodology – summarizing information on survey design and methodologies, results, data manipulation and analyses – and discuss how future surveys can be refined to allow for more robust inferences.

L. Maffei (✉)
Jaguar Conservation Program, Wildlife Conservation Society, New York, USA
e-mail: lmaffei@wcs.org

A.J. Noss
Latin America and Caribbean Program, Wildlife Conservation Society, New York, USA

S.C. Silver
Queens Zoo and Jaguar Conservation Program, Wildlife Conservation Society, New York, USA

M.J. Kelly
Virginia Polytechnic Institute and State University, Blacksburg, VA, USA

A.F. O'Connell et al. (eds.), *Camera Traps in Animal Ecology: Methods and Analyses*,
DOI 10.1007/978-4-431-99495-4_8, © Springer 2011

8.2 Study Sites

The studies have been carried out in 14 countries and 12 major habitat types that range from dry and moist forests to grasslands (Fig. 8.1 and Table 8.1). Most of them were conducted inside designated Jaguar Conservation Units (Sanderson et al. 2002; Zeller 2007). The surveys have covered portions of at least 19 national parks or other protected areas, one Biosphere Reserve, three state or provincial parks, six private reserves, three wildlife sanctuaries or management areas, four indigenous territories, 15 cattle ranches, 11 forestry reserves or concessions, and one private conservation concession (Table 8.1). Additional surveys are underway or planned (for example, by V. Quiroga in the Argentine Chaco, by WCS-Ecuador in Yasuní National Park), the most ambitious of which is Mexico's national jaguar census (CENJAGUAR) to be completed during 2008–2009, with the participation of 18 researchers, and the support of more than 10 institutions led by the Universidad Nacional Autónoma de México (Chávez et al. 2006).

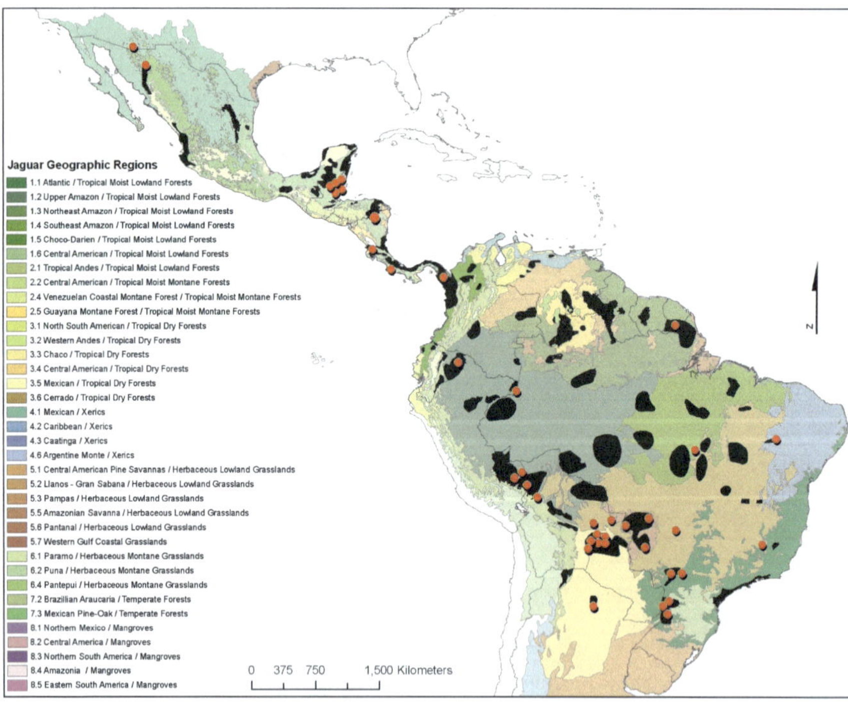

Fig. 8.1 Jaguar Conservation Units and points where systematic jaguar camera trapping surveys have been carried out (map adapted from Zeller 2007). Not all sites are represented at this scale – some single points represent more than one site in Argentina, Belize, Bolivia, Colombia, Costa Rica, and Peru

Table 8.1 Camera trap surveys for jaguars: number of surveys per site, ecoregion and land use

Country	Study site	Surveys	Type of forest-ecoregion	Land use
Argentina	Copo	1	Chaco/tropical dry forests	National Park
Argentina	Impenetrable ChacoAboriginal Reserve	1	Chaco/tropical dry forests	Indigenous Territory
Argentina	Iguazú	2	Atlantic/tropical moist lowland forests	National Park and Forestry Reserve
Argentina	Urugua-í	1	Atlantic/tropical moist lowland forests	Provincial Park and Private Reserve
Argentina	Yabotí	1	Atlantic/tropical moist lowland forests	Forestry Reserve
Belize	Chiquibul	5	Central America/tropical moist lowland and submontane forests	Forest Reserve and National Park
Belize	Cockscomb Basin	6	Central America/tropical moist lowland forests	Wildlife Sanctuary
Belize	Fireburn	1	Central America/tropical moist lowland forests	Private Reserve, Forest Corridor, Mesoamerican Biological Corridor
Belize	Gallon Jug Estate	2	Central America/tropical moist lowland forests	Private protected area
Belize	Rio Bravo	1	Central America/tropical moist lowland forests	Conservation and Management Area
Belize	Mountain Pine Ridge	6	Central American/tropical pine forests	Forest Reserve
Bolivia	Alto Madidi	2	Tropical Andes/tropical moist lowland forests	National Park
Bolivia	Cerro Cortado, Kaa-Iya	2	Chaco/tropical dry forests	National Park and Indigenous communal lands
Bolivia	El Encanto CIMAL	2	Cerrado/tropical dry forests (Chiquitano dry forest)	Certified forestry concession
Bolivia	Estación Isoso, Kaa-Iya	2	Chaco/tropical dry forests (transitional Chaco-Amazon)	National Park
Bolivia	Guanacos, Kaa-Iya	2	Chaco/tropical dry forests (grasslands)	National Park and cattle ranches
Bolivia/ Paraguay	Palmar, Kaa-Iya	2	Chaco/tropical dry forests (transitional Chaco-Chiquitano)	National Park, private reserve, and cattle ranch
Bolivia	Puestos Ganaderos	1	Chaco/tropical dry forests (transitional Chaco-Chiquitano)	Cattle ranches

(continued)

Table 8.1 (continued)

Country	Study site	Surveys	Type of forest-ecoregion	Land use
Bolivia	Ravelo, Kaa-Iya	2	Chaco/tropical dry forests (transitional Chaco-Chiquitano)	National Park
Bolivia	Rio Heath, Madidi	2	Tropical Andes/tropical moist lowland forests, tropical grasslands	National Park
Bolivia	Rios Tuichi and Hondo, Madidi	3	Tropical Andes/tropical moist lowland forest	National Park
Bolivia	San Matias	1	Pantanal/herbaceous lowland grasslands	Cattle ranch and National Integrated Management area
Bolivia	San Miguelito	2	Cerrado/tropical dry forests (Chiquitano dry forest)	Private reserve and cattle ranch
Bolivia	Tucavaca, Kaa-Iya	3	Chaco/tropical dry forests (transitional Chaco-Chiquitano)	National Park
Brazil	Emas National Park, Goiás	1	Cerrado/tropical dry forests	National Park
Brazil	Fazenda Cauaia	1	Cerrado/tropical dry forests	Cattle ranch
Brazil	Fazenda Santa Fé and Cantão State Park, Tocantins	1	Amazon/tropical moist forests – Cerrado/tropical dry forests ecotone	Cattle ranch, State Park
Brazil	Fazenda Sete	2	Pantanal/herbaceous lowland grasslands	Cattle ranch
Brazil	Moro do Diablo	1	Atlantic/tropical moist lowland forest	National Park
Brazil	Serra da Capivara	1	Caatinga/xerics	National Park
Brazil	SESC Pantanal	1	Pantanal/herbaceous lowland grasslands	Private reserve
Brazil	Varzeas do Rio Ivinhema	1	Atlantic/tropical moist lowland forest/varzea	State Park
Colombia	Amacayacu National Park and Ticoya Indigenous Territory	1	Amazon/tropical moist lowland forest	National Park and indigenous territory
Colombia	Calderón river valley	1	Amazon/tropical moist lowland forest	National Forestry Reserve (unprotected) and indigenous territory
Costa Rica	Corcovado	1	Central American/tropical moist lowland forest	National Park

(continued)

Table 8.1 (continued)

Country	Study site	Surveys	Type of forest-ecoregion	Land use
Costa Rica	Golfo Dulce, Golfito	1	Central American/ tropical moist lowland forest	Private ranches, Forest Reserve, Wildlife Reserve
Costa Rica	Golfo Dulce	1	Central American/ tropical moist lowland forest	Forest reserve
Costa Rica	Santa Rosa, Guanacaste, San Cristobal	3	Central American/ tropical dry forest	National Parks and biological corridor
Ecuador	Yasuní and Waorani Ethnic Reserve	2	Amazon/tropical moist lowland forest	National Park and indigenous territory
French Guiana	Counami forest	1	Amazon/tropical moist lowland forest	Unprotected
Guatemala	Carmelita-AFISAP	1	Central America/tropical moist lowland forest	Forestry concessions
Guatemala	La Gloria-Lechugal	1	Central America/tropical moist lowland forest	Forestry concession, multiple use zone
Guatemala	Rio Azul	1	Central America/tropical moist lowland forest	National Park
Guatemala	Tikal	1	Central America/tropical moist lowland forest	National Park
Mexico	Sonora	1	Mexican xerics/tropical thorn scrub	Private Reserve and cattle ranches
Nicaragua	Bosawas	1	Central America/tropical moist lowland forest	Biosphere Reserve
Panama	Darien	2	Central America/tropical moist lowland forest	National Park
Peru	Los Amigos	2	Tropical Andes/tropical moist lowland forest	Conservation concession
Peru	Bahuaja Sonene, Tambopata	1	Tropical Andes/tropical moist lowland forest	National Parks
United States	Southern Arizona	1[a]	Mexican xerics/tropical thorn scrub	National Forest, National Wildlife Refuge, private ranches

[a]McCain and Childs (2008) established a grid system of camera traps to monitor the southern Arizona borderlands continuously from 2001 through 2007.

8.3 Survey Design and Data Analysis

Two approaches have been used to set camera traps for jaguar surveys: (1) placing traps in a single grid for the entire sample period, or (2) shifting traps to a different area within the study period for a length of time equal to the initial sample. The second approach is used when the number of cameras available cannot cover the entire study area in a single sample period. In the second case, the sample period is considered the length of time the camera traps are operable in a single location.

For either of the sampling approaches described, jaguar surveys have followed a systematic survey design that typically follows some defined travel route (Silver et al. 2004; http://savingwildplaces.com/media/file/SilverJaguarCamera-TrappingProtocol.pdf) to accommodate low jaguar densities and capture probabilities high enough to run capture–recapture (CR) models (but see Discussion). In some cases, researchers clear trails specifically for the survey in order to reach inaccessible areas and distribute the camera traps throughout the study area, as well as providing a feature to attract jaguars. Once trails are established, they are routinely cleared to maintain travel routes for jaguars. Figure 8.2 shows a selection of camera trapping grids that take advantage of available roads, trails, and rivers at various survey sites.

Cameras usually are set 30–40 cm above the ground to accommodate the height of the target species. Jaguars can be active day or night, and therefore camera traps are programmed to take pictures 24 h per day. The time delay for activation is usually between 30 s and 5 min, but in places with high traffic of non-target species/objects (e.g., roads with trucks, trails or salt licks with people or wildlife), a longer time delay can be used. We also note that on several occasions different male jaguars have been photographed within 2 min of each other at one camera station (M. Kelly, Virginia Tech University, Blacksburg, VA, unpublished data). In locations with low traffic, camera traps are typically checked only once every 10–14 days, whereas they are checked every two or three days at sites with high traffic, in order to avoid running out of film. Pilot surveys are useful in determining the frequency with which cameras and film need to be checked and replaced, as well as in evaluating the optimum sites for photographing jaguars (Rosas-Rosas 2006). Scents or attractants are not known to be necessary or even effective for increasing capture probabilities, but jaguars are known to occasionally investigate scents. This can result in multiple photographs from different angles, facilitating identification in some cases. In areas with abundant livestock, researchers have protected cameras with fencing that permits wildlife and especially jaguar movement, but keeps livestock away from the cameras (Rosas-Rosas 2006).

Surveys are based upon the standard procedures used in CR sampling of closed populations (see Karanth and Nichols 1998; 2002) using cameras in place of live traps, and using the natural markings of the jaguar to recognize individuals and "recaptures" in photographs. The objective of our CR (in this case, photograph/rephotograph) surveys was to estimate the number of individuals within a sample area. In general terms, this estimate is obtained by first estimating capture probability based on the capture histories of individuals that are caught at least once. The number of animals in the sampled area is then estimated by dividing the total number of animals caught by the estimated average probability of catching an animal at least once. The technique does not have to be based on a random sampling of the area, but rather, cameras are set up systematically in a pattern designed to maximize capture probability for all animals in the sampled area (Silver 2004). The method estimates the efficiency of the survey to photograph all the individuals in the survey area. The more jaguars that are photographed, and subsequently the more often they can be rephotographed, the more robust the abundance estimate

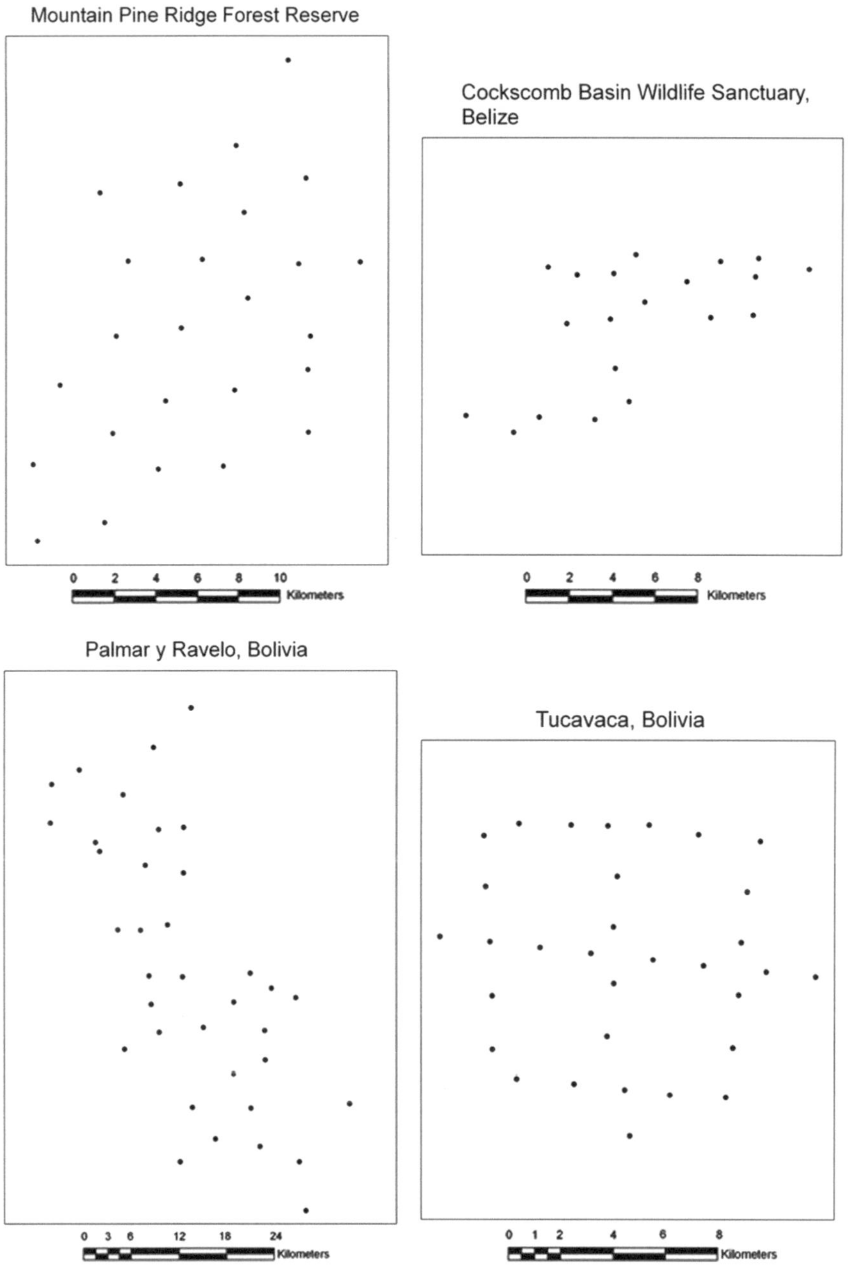

Fig. 8.2 Camera trap placement patterns for jaguar surveys (*dots* camera trap positions)

will be for the study period. With the date and time stamped on the photographs, researchers can measure days or blocks of days as discrete sampling events.

Single CR surveys assume a closed population (i.e., no births, deaths, immigration or emigration of individuals) within the study area during the survey. In reality, few animal populations are actually closed, so in practice researchers try to meet this assumption by limiting the duration of the survey. A short survey length, relative to the lifespan of the animal, decreases the likelihood of violating this assumption. Since jaguars, like tigers, are long-lived, most jaguar surveys follow the convention established by Karanth and Nichols (1998) in using no longer than a 4-month time period to gain photographs to conduct CR and yet still satisfy the assumption of a closed population. Similarly, surveys on African leopards have typically used two to three months (Henschel and Ray 2003). Although there are few life history data available for jaguars, it is reasonable to assume the same duration is satisfactory. Most jaguar surveys have used three months or less as a data collection period. The most commonly used software for estimating jaguar abundance through camera photographs is the program CAPTURE (Otis et al. 1978; White et al. 1982), available online from the Patuxent Wildlife Research Center website (http://www.mbr-pwrc.usgs.gov/software/capture.html). This program uses different models to generate abundance estimates based on the number of individuals captured and the proportion of recaptures. The models differ in their assumed sources of variation in capture probability, including variation among individuals (e.g., sex, age, ranging patterns, dominance, activity), variation over time, behavioral responses to having been captured, and various combinations of these factors. Specifically, the model M(o) indicates that the probability of capture is the same for every animal at every occasion; M(h) incorporates heterogeneity, a unique capture probability for each individual; M(t) is characterized by differences in capture due to time; and M(b) applies where animals have different reaction to the camera traps such as being trap-happy or trap-shy. A series of models also combines the aforementioned factors. The majority of jaguar surveys have used M(h) as the best fitting model based on our knowledge of individual animal behavior and ecology, individuals – especially territorial carnivores – that have different capture probabilities (Karanth and Nichols 1998). Occasionally, however, M(o) may be the model that CAPTURE recommends; but we recommend caution when this is the case. The M(h) model uses the jackknife estimator, which is much more robust than the maximum likelihood estimator that the other models use.

Collapsing data from a long survey into fewer trapping occasions (e.g., a 70-day survey into ten 7-day trapping occasions), increases the capture probability per trapping occasion, and may ameliorate violations of closure. If sampling generates multiple recaptures of multiple individuals, collapsing the number of trapping occasions does not generally affect the abundance estimate and may reduce the standard error of the estimate. CAPTURE uses a discriminant analysis function in its model selection procedure to determine which model best fits the available data. It should be noted that CAPTURE is also a built-in feature of the program MARK (http://welcome.warnercnr.colostate.edu/~gwhite/mark/mark.htm).

The second important assumption is that every jaguar inhabiting the survey area has at least some probability of being photographed (i.e., one camera trap within each animal's home range for the duration of the survey). This assumption dictates distance between camera traps and determines the maximum size of an area to be sampled by at least one camera trap. Thus, the estimated minimum home range of a jaguar in the study area ultimately determines the local minimum camera trap density. Ideally, there should be no gaps between camera trap stations large enough to encompass a jaguar's home range. A conservative approach to satisfy this assumption is to adopt the smallest home range estimate documented locally for jaguars. In practice, most jaguar surveys have spaced cameras 2–3 km apart using the smallest home range of 10 km^2 for a female jaguar in Belize (Rabinowitz and Nottingham 1986). This spacing may not be applicable for other areas where jaguars have larger home ranges.

Once we have the abundance estimate, the next step is to calculate the area surveyed. This has been one of the most problematic issues for estimating jaguar population density based on camera trap surveys. The classical way to estimate the sampling area is to calculate the mean maximum distance moved (MMDM) as a proxy for home range diameter (Wilson and Anderson 1984), sum the maximum distances moved by every individual captured in at least two different locations (but see Dillon and Kelly 2007 regarding animals repeatedly captured at one location), calculate the average, diameter, divide by two (radius), and apply this as a buffer around the camera traps. In the scientific literature, the buffer has been applied two ways: as a strip around the polygon formed by connecting the camera trap locations (polygon buffer), or as a circular buffer surrounding each camera trap location (point buffer). The first method is more subjective because different researchers (and software programs) create different polygons depending on the way they connect the camera locations. The second method is not subject to an interpretation of polygon-drawing because it generates the same area surveyed each time and is the one commonly used in jaguar surveys. However, some argue that buffering each camera location individually does not conform to the idea of a single jaguar population being sampled under the "ball-and-urn concept" (White et al. 1982), where individual jaguars represent the "balls" within a single population or "urn." All areas determined by creating circular buffers and dissolving those buffers have resulted in a continuous sampling area. It is important to note, however, that this may not always be the case; for example, when using camera trap data from a jaguar study to estimate buffers for animals with a smaller home range such as the ocelot. The estimation of the buffer, which in turn determines the area effectively sampled, is the weakest link in density estimation. The MMDM can vary widely even between surveys (in the same location); thus, when data are available from multiple surveys in the same location, we can opt to use one half of a cumulative MMDM. This cumulative MMDM averages the maximum distances moved by all individuals across multiple surveys in different years. This increases the sample size and reduces the variance associated with the MMDM, and gives a more precise estimate of the effective sample area (Dillon and Kelly 2007).

However, even this does not improve the estimate of MMDM if the overall sample area is too small relative to the ranging patterns of the individuals.

New approaches are being developed to address the deficiencies in density estimation procedures (Borchers and Efford 2008; Efford et al. 2004; Royle et al. 2009).

8.4 Results

Both the camera trap polygons and the effective survey area (including a ½ MMDM buffer around the camera traps) vary considerably across surveys, from 24–555 km^2 to 54–938 km^2 respectively (Table 8.2). In the case of private reserves, cattle ranches, and relatively small reserves, the cameras can be distributed across 30–100% of the land use unit: Moro do Diablo National Park and Fazenda Sete ranch in Brazil, San Miguelito Private Reserve in Bolivia, Gallon Jug Estate in Belize. At the opposite extreme is the Kaa-Iya del Gran Chaco National Park, where surveys at six different sites add up to barely 1% of the park's land area. Considering the area effectively surveyed, most surveys cover at least 35% of the land area, again with the exception of the Bolivian parks such as Kaa-Iya where the all surveys total only 4% of the area. Other surveys fall between the two extremes. For example, the largest camera polygons achieved in any study, 550 km^2 at Iguazú and Yabotí, represented 21% of the protected areas in each case. Including the buffer, the effective survey areas in these two studies covered 35% of the protected areas. Density estimations also varied considerably across study areas (Table 8.2): from 0 to >11 individuals per 100 km^2. Some of the highest density estimates were reported from private properties: a cattle ranch in the Brazilian Pantanal (Soisalo and Cavalcanti 2006) and a private reserve in Belize (Miller and Miller 2005). Another unexpectedly high density estimate comes from forestry concessions that are under heavy pressure from non-timber forest product harvesters and hunters (McNab et al. 2008).

Several surveys using camera traps in specific areas have not photographed jaguars despite documentation of individuals by other means. We can attribute these results to a number of issues: (1) camera failure, (2) low jaguar densities, (3) camera trapping period was not long enough to photograph an individual, and (4) lack of local knowledge about routes jaguars travel combined with a failure to place camera traps in such areas. Problems with density estimation also arise when too few individuals are photographed, without recaptures or with very few recaptures. Nevertheless, these data do comprise a minimum confirmed population based on the number of individuals positively identified. Camera trapping data has been used to calculate a "capture frequency" based on the number of photographs recorded per 100 or 1,000 trap-nights. Overall capture frequency has been found to correlate with abundance of target animals (see O'Brien, Chap. 6), but population density estimates based on individual identification and CR analysis provide the only reliable comparisons across studies and species when measuring abundance

Table 8.2 Capture frequency, population density estimates (average when multiple surveys were conducted at the same location), and survey area

Country	Study site	Reference	Captures per 100 trap nights	Density ± SE (ind/100 km^2)	Effective survey area using ½ MMDM buffer (km^2)	Camera polygon (km^2)	Protected area size (km^2)	% surveyed[c]
Argentina	Copo	1	0	0	–	388	1,140	34
Argentina	Impenetrable Chaco Aboriginal Reserve	39	0	0	–	449	1,500	30
Argentina	Iguazú	2	0.5–1.5	1.12 ± 0.30	576–958	209–555	670–2,594[b]	**31, 21** / 86, 37
Argentina	Urugua-í	2	0.1	0.3[a]	368	81	872	**11** / 42
Argentina	Yabotí	2	0.2	0.2[a]	1,001	549	2,600	**21** / 39
Belize	Cockscomb basin	3,	3.1–8	3.5–11	196–322	80	400	**20** / 40–80
Belize	Chiquibul	3, 6	3.5	7.48 ± 2.74[d]	107–405	89–146	1,670	**5–9** / 6–24
Belize	Fireburn	27	1.2	5.3 ± 1.76	132	55	8 priv res	**100+** / 100+
Belize	Gallon Jug Estate	4	3.3–4.7	10.05 ± 2.47	170–195	130–165	520 ranch	**32** / 38
Belize	Mountain Pine Ridge	35	3.3–7.1	2.32–5.35	302–345	105–140	430	**24–33** / 74–80
Bolivia	Cerro Cortado, Kaa-Iya	7	1.0–2.0	5.24 ± 2.46	137–149	49–52	34,400	**<1** / <1
Bolivia	El Encanto	8	0.4	5.66 ± 2.33	106	36	876 concession	**4** / 12
Bolivia	Estación Isoso, Kaa-Iya	9	2.2–3.2	2.91 ± 0.33	153–158	48–51	34,400	**<1** / <1
Bolivia	Guanacos, Kaa-Iya	10	1.1–2.9	2.28 ± 0.66	191–243	49–62	34,400	**<1** / <1
Bolivia/Paraguay	Palmar, Kaa-Iya	9, 25	2.4–2.9	1.13 ± 0.13	230–1,068	71–434	34,400	**1** / 3
Bolivia	Puestos Ganaderos	32	0	0	–	217	270	**80**
Bolivia	Ravelo, Kaa-Iya	12	1.2–1.5	1.92 ± 0.76	309–319	100	34,400	**<1** / <1
Bolivia	Rios Tuichi and Hondo, Madidi	3	0.9	2.84 ± 1.78	458	200	15,000 lowlands	**1** / 3
Bolivia	San Matias	13	0	0	–	125	29,200	**<1**
Bolivia	San Miguelito	14	1.2–3.2	4.23 ± 1.43	54–142	24–53	24 priv res, 600 ranch	**100** / 100priv res, **9** / 24 ranch

(continued)

Table 8.2 (continued)

Country	Study site	Reference	Captures per 100 trap nights	Density ± SE (ind/100 km^2)	Effective survey area using ½ MMDM buffer (km^2)	Camera polygon (km^2)	Protected area size (km^2)	% surveyed[c]
Bolivia	Tucavaca, Kaa-Iya	15	0.8–1.3	3.41 ± 1.21	125–272	49–130	34,400	**<1** / <1
Brazil	Emas National Park	28	4.56	2.00	500[f]	–[e]	1,320	**38**
Brazil	Fazenda Cauaia	26	0	0	–	16	17 ranch	**94**
Brazil	Fazenda Santa Fé	29	4.02	2.59 ± 1.03	425	80	570[g]	**14** / 75
Brazil	Fazenda Sete	16	13.6–16.4	11.0 ± 1.73	274–360	110–165	460 ranch	**36** / 78
Brazil	Moro do Diablo	17	3.0	2.22 ± 1.33	300	330	370	**89** / 81
Brazil	Serra da Capivara	37	6.5	2.67 ± 1.06	524	157	1,291	**41** / 12
Brazil	SESC Pantanal	18	0	0	–	54	1,063 priv res	**5**
Colombia	Amacayacu	22	0.56	4.2	120	32	2,930 park, 1,406 indig territory	**1** / 4 park,**2** / 9 indig territory
Colombia	Calderón river valley	34	0.62	2.5	242	70	–	–
Costa Rica	Corcovado	19	1.9	6.98 ± 2.36	86	29	425	**7** / 20
Costa Rica	Golfo Dulce / Golfito	38	0.5	2 ± 1.49	218	102	630	**16** / 35
Costa Rica	Golfo Dulce	30	0	0	–	24	617	**4**
Costa Rica	San Cristobal	20	1.1	6.7	60	134	40 biol corridor	**100** / 100 biol corridor
Ecuador	Yasuní-Waorani	36	0.3	1.38 ± 0.60	218	94	9,820	**<1** / 2
Guatemala	Carmelita-AFISAP	33	3.1	11.28 ± 3.51	115	51	1,056	**5** / 11
Guatemala	La Gloria-Lechugal	34	1.5	1.54 ± 0.85	390	128	911	**14** / 43
Guatemala	Rio Azul	5	2.9	10.5	95	50	1,169	**4** / 8
Guatemala	Tikal	21	5.9	6.63 ± 2.46	121	39	575	**7** / 21
Mexico	Sonora	23	0.9	1.0 ± 1.30	140	100	400	**25** / 35
Nicaragua	Bosawas	24	0.3	3.7	127	52	19,928	**<1** / <1

Panama	Darien	11	0.8	1.8–4.4	213–274	67–110	5,790	**1–2** / 4–6
Peru	Los Amigos	31	1.0–1.6	9.6 ± 2.35	130–141	56	1,460	4 / 10
Peru	Bahuaja-Sonene, Tambopata	31	0.5	11.4 ± 19.8	105	52	13,830	**<1** / <1

[a]Density estimates from observed individuals only
[b]Combined area of National Parks in Brazil and Argentina and San Jorge Forest Reserve
[c]Figure in bold is camera polygon as proportion of protected area, other figure is effective survey area as proportion of protected area
[d]Densities declining since the high of this first survey
[e]Does not apply, as cameras were placed along a trap line
[f]Only 30% of the total area was considered suitable jaguar habitat
[g]Total ranch area, including 200 km² permanent pasture, and over 50% deciduous forest

References: 1, Denapole (2007); 2, Paviolo et al. (2008); 3, Silver et al. (2004); 4, Miller (2005); 5, Miller and Miller (2005); 6, Kelly (2003); 7, Maffei et al. (2003); 8, Arispe et al. (2007); 9, Romero-Muñoz (2008); 10, Cuéllar et al. (2004); 11, Moreno (2006); 12, Cuéllar et al. (2003); 13, Maffei (2005); 14, Arispe et al. (2005); 15, Maffei et al. (2004a); 16, Soisalo and Cavalcanti (2006); 17, Cullen et al. (2005); 18, Trolle and Kéry (2005); 19, Salom-Pérez et al. (2007); 20, Amit (2007); 21, García et al. (2006); 22, Payan (2008); 23, Rosas-Rosas (2006); 24, Polisar (2006); 25, Montaño et al. (2007); 26, Trolle et al. (2007); 27, Miller (2006); 28, Silveira (2004); 29, L. Silveira and N.M. Negrões (Jaguar Conservation Fund / Instituto Onça-Pintada, Mineiros, Brazil) (unpublished data); 30, Carrillo et al. (2007); 31, S. Carrillo-Percastegui, M. Tobler and G. Powell (unpublished data); 32, Arispe et al. (2006); 33, McNab et al. (2008); 34, Moreira et al. (2007); 35, M. Kelly, Virginia Tech University, Blacksburg, VA (unpublished data); 36, S. Espinosa, University of Florida, Gainesville, FL (unpublished data); 37, Silveira et al. (2009); 38, Bustamante (2008); 39, V. Quiroga (CONICET, Mendoza, Argentina) (unpublished data).

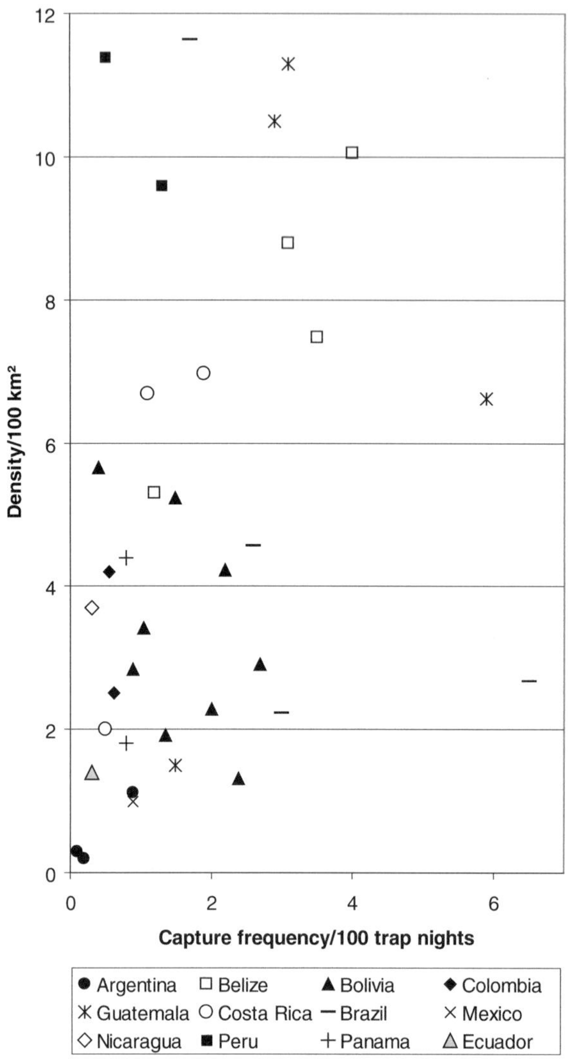

Fig. 8.3 Jaguar population density estimates relative to capture frequency (excluding the Fazende Sete site, Brazil, with a capture frequency of 13–16 photographs per 100 trap nights and a population density of 11 individuals per 100 km^2)

or density. Using capture frequency as an index of abundance, therefore, remains controversial (Carbone et al. 2001, 2002; Jennelle et al. 2002). Figure 8.3 confirms that population density and capture frequency do not correlate across sites: the sites with the highest capture frequencies in Belize and Bolivia are not the same as the sites with the highest population densities. The Atlantic forest site, Moro do Diablo, in Brazil has a capture frequency as high as the Belize/Guatemala moist forest sites, but a considerably lower population density in the range of the dry forest sites in Bolivia. Furthermore, excluded from Fig. 8.3 is the case of the Fazenda Sete in the Brazilian Pantanal. This area had this area recorded an average capture frequency of 15 photographs per 100 trap nights across two surveys, a figure >2.5 times the next highest capture frequency recorded anywhere. However, the population density was similar to the highest Belize estimate (Table 8.2).

Sex ratios also vary across camera trap surveys (Table 8.3), but most surveys have recorded more males than females: from 3:2 (Maffei et al. 2004a, Soisalo and

Table 8.3 Adult sex ratios by jaguar survey site (cumulative where multiple surveys conducted), and locations where cubs/juveniles were photographed

Study (reference)	Males	Females	Unsexed	Cubs/ juveniles
Argentina Iguazú (Paviolo et al. 2008)	4	6	0	Yes
Argentina Urugua-í (Paviolo et al. 2008)	1	0	0	
Argentina Yabotí (Paviolo et al. 2008)	1	0	0	
Belize Chiquibul (M. Kelly [Virginia Tech University, Blacksburg, VA] unpublished data)	15	6	0	Yes
Belize Cockscomb (Silver et al. 2004)	9	0	2	
Belize Fireburn (Miller 2006)	3	0	2	
Belize Gallon Jug (Miller and Miller 2005)	9	7	4	
Belize Mountain Pine Ridge (M. Kelly [Virginia Tech University, Blacksburg, VA] unpublished data)	14	7	0	Yes
Bolivia Cerro Cortado (Maffei et al. 2003)	6	2	1	Yes
Bolivia CIMAL (Arispe and Venegas [WCS/Fundación para la Conservacion del Bosque Chiquitano, Santa Cruz, Bolivia], unpublished data)	2	4	0	Yes
Bolivia El Encanto (Arispe et al. 2007)	4	0	0	
Bolivia Estación Isoso (Romero-Muñoz 2008)	4	1	0	Yes
Bolivia Guanacos (Cuéllar et al. 2004)	2	2	2	Yes
Bolivia Palmar (Romero-Muñoz 2008; Montaño et al. 2007)	7	2	0	
Bolivia Ravelo (Cuéllar et al. 2003)	5	2	0	Yes
Bolivia Río Tuichi/Río Hondo (Silver et al. 2004)	5	3	1	Yes
Bolivia San Miguelito (Arispe et al. 2005; Rumiz et al. 2003)	5	5	1	Yes
Bolivia Tucavaca (Maffei et al. 2004a)	5	3	1	Yes
Brazil ENP (Silveira 2004)	2	1	5	
Brazil Fazenda Santa Fé and Cantão State Park (L. Silveira and N.M. Negrões [Jaguar Conservation Fund/Instituto Onça-Pintada, Mineiros, Brazil], unpublished data)	6	0	2	
Brazil Fazenda Sete (Soisalo and Cavalcanti 2006)	15	10	6	Yes
Brazil Moro do Diablo (Cullen et al. 2005)	2	3	1	Yes
Brazil Serra da Capivara (Astete 2008)	6	4	3	Yes
Colombia Amacayacu (Payan 2008)	3	1	0	

(continued)

Table 8.3 (continued)

Study (reference)	Males	Females	Unsexed	Cubs/ juveniles
Colombia Calderón river valley (Payan 2008)	2	1	1	
Costa Rica Corcovado (Salom-Pérez et al. 2007)	3	1	0	
Costa Rica Corcovada buffer zone (Bustamante 2008)	4	0	0	
Costa Rica San Cristobal (Amit 2007)	0	3	1	
Ecuador Yasuní-Waorani (S. Espinosa [University of Florida, Gainesville, FL], unpublished data)	3	0	0	
Guatemala Carmelita-AFISAP (McNab et al. 2008)	7	3	0	
Guatemala La Gloria-Lechugal (Moreira et al. 2007)	4	2	0	
Guatemala Río Azul (Miller and Miller 2005)	6	0	1	
Guatemala Tikal (García et al. 2006)	3	1	3	
Mexico Sonora (Rosas-Rosas 2006)	4	1	0	Yes
Nicaragua Bosawas (Polisar 2006)	3	0	1	
Panama Darien (Moreno 2006)	1	3	0	
Peru Los Amigos (S. Carrillo-Percastegui, M. Tobler and G. Powell [Arizona State University, Tucson, AZ], unpublished data)	6	3	1	
Peru Bahuaja-Sonene, Tambopata (S. Carrillo-Percastegui, M. Tobler and G. Powell [Arizona State University, Tucson, AZ], unpublished data)	5	1	1	
United States (McCain and Childs 2008)	4	0	0	

Cavalcanti 2006) to 4:1 (Kelly 2003, Wallace et al. 2003) and up to 9:0 (no animals positively identified as females, Silver et al. 2004). One exception is in the Darien, and two others in Atlantic forest: Iguazú and Moro do Diablo National Parks. In the latter two cases, the protected areas are islands of forest surrounded by heavily transformed landscapes and may provide breeding refuges for jaguars. Most radio telemetry studies report that males have larger home ranges than females (Crawshaw 1995; Cullen et al. 2005; Rabinowitz and Nottingham 1986; Scognamillo et al. 2002, 2003; Soisalo and Cavalcanti 2006), so we would assume that more females than males are present in any given area where there is a resident breeding population. However, males may have a higher capture probability because of larger home ranges that are presumably include relatively more cameras. In addition, males tend to walk more than females (Rabinowitz and Nottingham 1986) and use human trails/roads (where camera traps are almost always set) more than females (Salom-Pérez et al. 2007). Both radio telemetry and camera trapping studies suggest that multiple males and females overlap in their ranging patterns. Sites where females and cubs are present clearly represent conservation priorities. On the other hand, the failure to photograph females does not mean that they are absent from an area, but only that such areas should be evaluated more carefully to determine whether they function principally as corridors or dispersal areas, and whether they potentially represent population sinks.

Finally, the camera trap methodology can provide considerable information about jaguars besides density estimation (see other chapters of this volume), including activity patterns, reproduction data (number of cubs, seasonality) and information on prey

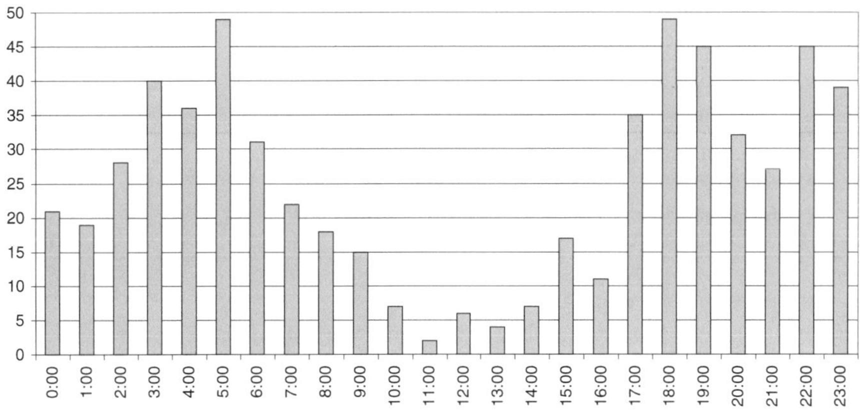

Fig. 8.4 Sample activity patterns for jaguars based on camera trapping records from Bolivia dry forest sites ($N=605$ records)

(species present, relative abundance from capture frequency, activity patterns). Figure 8.4 presents jaguar activity budgets derived from camera trapping records at a sample of survey sites. Camera trap photos suggest that jaguars can be active at any time of day, but are principally crepuscular-nocturnal in their habits. Cubs are occasionally photographed (single cubs only in the Chiquibul in Belize; and Cerro Cortado, Guanacos, Tucavaca in Bolivia), and juvenile animals occur in the company of their mothers more frequently (a pair of juveniles together in the Chiquibul in Belize and Cerro Cortado in Bolivia; single juveniles only in Estación Isoso, Ravelo, San Miguelito, Tucavaca in Bolivia). This type of information provides preliminary information on reproductive patterns: one to two cubs born during the rainy season, December–May in Chaco dry forests, and maternal care until the juveniles approach adult size.

Conducting multiple surveys at the same site can validate density estimates, ranging patterns, and document rough turn-over rates of individuals within specific populations. For example, Table 8.4 suggests which individuals may be resident (females T2 and T4, males T5 and T6) vs. transient (possibly males T1, T3, T10, and T9 [unknown sex]). This information must be viewed in context, however, because alternatively, the latter group could have been photographed at the edge of their ranges, thereby incorrectly categorizing as transients resident individuals whose ranges overlap minimally with the camera layout. For example, the cub of female T2 and the juvenile offspring of female T7 were not subsequently photographed, suggesting that they dispersed outside the survey area if they survived. Karanth et al. (2006) go much further to estimate rate of change, survival, recruitment, temporary emigration, etc., based on 12 years of data. Though jaguar researchers have not estimated these rates to date, the longest running surveys are currently six years and these estimates should be possible in the near future. Camera trap surveys have also documented transboundary movements of jaguars between the United States and Mexico (McCain and Childs 2008), between Argentina and Brazil (Paviolo et al. 2006), and between Bolivia and

Table 8.4 Turn-over of individual jaguars according to multiple camera trap surveys at Tucavaca, Kaa-Iya del Gran Chaco National Park (Maffei et al. 2004a)

	T1	T2	T2	T3	T4	T5	T6	T7	T8	T9	T10	
	M	F	Cub	M	F	M	M	F	J	?	M	Total
Preliminary May–Dec, 2001		14	2	1				3		2		20
Survey I Jan–Mar, 2002	11	5		1	2	3	1	1				24
Survey II Apr–Jun, 2003		3					4	3	2			12
Survey III Mar–May, 2004						8	2	1			3	14

Paraguay (Romero-Muñoz et al. 2007). Such information is invaluable for promoting international conservation efforts.

8.5 Discussion

Another approach when there are too few detections to calculate abundance or when grid trapping is not possible is to use detection-non detection data at each camera site to model detection probabilities and the proportion of area occupied (MacKenzie and Kendall 2002; MacKenzie et al. 2003, MacKenzie et al. 2006, see O'Connell and Bailey, Chap. 11). In this way, detection-nondetection data (often referred to as presence-absence) data can be used as a surrogate for abundance for cryptic or low density species. The underlying logic is that changes in the proportion of occupied sites will be correlated with changes in the population size, provided sites are defined at an appropriate spatial scale (MacKenzie 2005; MacKenzie et al. 2006). So far, this approach has not yet been applied to jaguars, but holds promise for future studies.

Although new camera trapping techniques are developing that use random camera placement, combined with information on species' day range, to address spatial variability (Rowcliffe et al. 2008), random placement is unrealistic for most jaguar field studies because capture probabilities would be impossibly low. Given that capture probability is already low even in studies that target jagaurs (~2 per 100 trap nights), the increased effort required to obtain captures using random placement is probably not realistic. The study approach for jaguars – systematic, regularly-spaced, traps set to target jaguars (i.e., on roads, trails, games trails, riverbeds, etc.) – violates the random placement of traps which has proven to be necessary to generate unbiased estimates as in the gas model approach of Rowcliffe et al. (2008). However, increasing the capture probability is also necessary to obtain enough recaptures to conduct CR surveys. Perhaps a compromise approach of random placement with directed sampling will be fruitful. Alternatively, the approach of Borchers and Efford (2008) used capture locations to estimate animal locations and spatially referenced capture probabilities. With this technique, density is evaluated in a maximum likelihood framework, based on spatial and temporal co-variates. This approach has not yet been applied to jaguars.

An underlying problem for all jaguar camera trap surveys is that we do not actually know the true densities of the target population and therefore cannot judge whether we are underestimating or overestimating true densities. Calibrating the camera trapping technique would require conducting a camera survey in an area with known densities. This may be possible for other animals such as lions *Panthera leo* where all study animals in an area are known (C. Packer [University of Minnesota] pers. comm.), but it is unlikely to be the case for any area in the jaguar's range.

The systematic camera trapping methodology was originally developed for tigers in India, where many protected areas are relatively small islands and where surveys can cover large proportions or even all of the area, and where the target species may have difficulty moving outside the protected area. Similar conditions may exist for jaguars in parts of their range, for example in much of Central America and in Atlantic forest patches in Brazil. However, in many other landscapes and particularly in South America, we are often surveying only tiny portions of vast protected areas or potential habitat, exceeding 10,000 km^2, through which jaguars can move freely beyond the boundaries of a 100–500 km^2 camera trap survey. The density estimate is then crucial because it provides information on the status of the species within this wider landscape. However, it should only be used tentatively and cautiously to extrapolate and estimate total populations (Maffei et al. 2004b) for wider protected areas or regions. Carnivore densities may vary significantly even under natural conditions with no or minimal human interventions (Karanth et al. 2004; Sunquist et al. 1999).

Density estimates are extremely sensitive to the calculation of the effective survey area, which depends on the size of the buffer surrounding traps. Camera trap spacing, total survey area, and degree of concordance between home range radius and ½ MMDM from cameras have arisen as three important factors impacting density estimation (Dillon and Kelly 2007; 2008). Increased camera spacing can lead to decreases in density estimates because MMDM increases (Dillon and Kelly 2007). Maffei and Noss (2008) suggest that MMDM may not be an appropriate proxy for home range diameter when camera survey areas are small compared to home range areas of the target species because the small area leads to an underestimate of maximum distance moved. While the use of ½ MMDM as a proxy for home range radius has a long history in the literature (Dice 1938) and has performed well in simulation studies (Wilson and Anderson 1985), its use has recently been called into question. Parmenter et al. (2003) found that small numbers of capture locations produce severe underestimates of home range size and movement distances. Most jaguar studies use 30 or fewer camera stations, undoubtedly a small number of capture locations. And while Parmenter et al. (2003) found that using the full rather than the ½ MMDM performed very well empirically in their small mammal studies, they caution against using MMDM at all due to the large number of underlying assumptions about animal movement. They instead suggest substituting known movement distances derived from radio telemetry.

A few studies have done this. Soisalo and Cavalcanti (2006), who followed jaguars with radio collars simultaneously with camera trapping efforts in the Pantanal, found that distances moved with radio collars were as much as twice the distance estimated with camera traps. Based upon comparisons between the ranging behavior of the collared jaguar and their MMDM, they recommended using the full MMDM to buffer camera locations rather than ½ MMDM (following Parmenter et al. 2003). Recent research on ocelots with simultaneous camera trapping and radio telemetry has proven equivocal with one study finding similar results to Soisalo and Cavalcanti (2006) (Dillon and Kelly 2008) and the other finding ½ MMDM a good proxy for home range radius (Maffei and Noss 2008). Habitat types were different in the two ocelot studies pointing to flexibility in wild cat movements patterns from one subpopulation to another.

In order for the MMDM to be an accurate characterization of ranging patterns in surveyed jaguar populations (and therefore an accurate tool in estimating the effective sample area), the camera trapping grid must be large enough to account for the long distances the jaguars are likely to travel during the survey. Obviously, having camera trap arrays with cameras spread only 15 km apart will not allow an accurate ranging characterization of animals that travel > 15 km. Thus, investigators designing camera trap surveys need to make some a priori assumptions about the minimum dimensions of a camera trap grids.

In Central America, radio telemetry studies have reported the following home range sizes for jaguars: 10–40 km^2 in the tropical moist lowland forests of Belize (Rabinowitz and Nottingham 1986), 32–59 km^2 in tropical moist lowland forests of Mexico (Ceballos et al. 2002), and 25–65 km^2 in Mexican dry forests (Núñez et al. 2002). Applying the recommendation that camera trap surveys encompass at least four average home ranges of the target species (Maffei and Noss 2008), jaguar surveys in Central America should, at a minimum, cover areas in the range of 100–180 km^2. Several of the Belize surveys have met this requirement, as well as the San Cristobal survey in Costa Rica and the second Darien survey in Panama (Table 8.2). The low population densities and wide ranging patterns of jaguars in the Mexico–USA border region require that even extensive areas be surveyed (McCain and Childs 2008).

In South America, average home ranges are considerably larger than in Central America: 52–176 km^2 in Pantanal grasslands (Crawshaw and Quigley 1991; Soisalo and Cavalcanti 2006), 43–177 km^2 in Atlantic tropical moist lowland forest (Crawshaw 1995; Cullen et al. 2005), 48–130 km^2 in Venezuelan Llanos grasslands (Scognamillo et al. 2002, 2003), and 69–1,200 km^2 in the Chaco (McBride et al. 2004, 2005; Romero-Muñoz et al. 2007). Again, applying the tentative rule suggested by Maffei and Noss (2008), jaguar surveys in South America should ensure that cameras cover a minimum of 500–600 km^2. The Yabotí and second Iguazú surveys (Argentina – Paviolo et al. 2008) do so, each covering around 550 km^2, which is equivalent to 21% of the protected area in each case. The Moro do Diablo study (Brazil) comes close to doing so, coincidentally also covering 90% of the island protected area, and with telemetry information to confirm the camera trap density estimation (Cullen et al. 2005). The second Palmar survey in Bolivia also comes close to doing so, but covers barely 3% of the immense Kaa-Iya National Park (Montaño et al. 2007).

We recommend that density estimates from camera trapping surveys, particularly when they cover only small portions of vast protected areas or potential jaguar habitat, be treated only as preliminary until the methodology can be tested further by conducting camera trap surveys with cameras spread ≥ 500 km^2. If it is logistically impossible to insure that the area covered by the camera traps include at least four average home range areas, we suggest that density estimates for jaguar populations be interpreted with great care. In addition, radio telemetry studies are needed to determine daily home ranges across similar habitats and regions that can be used as a substitute for ½ MMDM to estimate the effective area sampled by camera traps. We also recommend the development of a more theoretically sound approach, based on modeling, to estimate effective survey area.

Finally, compared to tiger surveys published in the literature, jaguar surveys have generated relatively small samples sizes (Table 8.3). Given the generally low population densities of jaguars across their range, future research should emphasize larger survey areas to confirm whether density estimates are consistent with larger sample sizes.

Acknowledgments We especially thank Alan Rabinowitz for his inspiration, leadership, and vision in consolidating range-wide efforts to study and conserve jaguars. Through the decade-long Jaguar Conservation Program of the Wildlife Conservation Society and most recently as the originator of the Jaguar Corridor with Panthera, he introduced us to the methodology, and provided technical and financial support to many of the surveys reported here. We are enormously grateful to all the researchers who generously provided information from their own work and sites: A. Paviolo, M. di Bitetti, C.D. de Angelo, Y.E. di Blanco, L. Denapole, L. Ostro, C. Miller, B. Miller, R. Arispe, E. Cuéllar, C. Venegas, D. Rumiz, R. Peña, T. Dosapey, R. Montaño, A. Romero-Muñoz, J. Barrientos. E. Ity, J. Ity, F. Mendoza, J. Segundo, G. Segundo, R. Wallace, G. Ayala, M. Trolle, T.G. de Oliveira, L. Silveira, N.M. Negroes, A.T. de Almeida Jácomo, E. Payan, E. Carrillo, R. Amit, S. Espinosa, R. McNab, R. García, H. Portillo, O. Rosas-Rosas, J. Polisar, R. Moreno, S. Astete, V. Quiroga, M. Tobler and S. Carrillo-Percastegui. Kathy Zeller graciously produced the map in Fig. 8.1. We thank all the supporters and partners (conservation organizations, local and national governments, the private sector, individuals), too numerous to name one by one, who made research possible across so many sites in so many countries. The on-going collaboration reflected here is essential for efforts on behalf of the jaguar to succeed.

References

Amit, R. R. 2007. Densidad de jaguares (*Panthera onca*), en el Sector San Cristóbal del Área de Conservación Guanacaste, Costa Rica. Undergraduate thesis. Universidad Nacional, Heredia, Costa Rica

Arispe, R., D. I. Rumiz, and C. Venegas. 2005. Segundo censo de jaguares (*Panthera onca*) y otros mamíferos con trampas cámara en la Estancia San Miguelito. Technical Report #144, Wildlife Conservation Society, Santa Cruz, Bolivia

Arispe, R., C. Venegas, D. I. Rumiz, and A. J. Noss. 2006. Estudio de mamíferos con trampas cámara en estancias ganaderas al sur de Roboré. Technical Report #166, Wildlife Conservation Society, Santa Cruz, Bolivia

Arispe, R., D. I. Rumiz, and C. Venegas. 2007. Censo de jaguares (*Panthera onca*) y otros mamíferos con trampas-cámara en la Concesión Forestal El Encanto (23 de septiembre–20 de noviembre). Technical Report #173, Wildlife Conservation Society, Santa Cruz, Bolivia

Astete, S. E. 2008. Ecología da onça pintada nos Parques Nacionais Serra da Capivara e Serra das Confusões, Piauí. M.S. thesis. Universidade da Brasília, Brasilia, Brazil

Borchers, D. L. and M. G. Efford. 2008. Spatially explicit maximum likelihood methods for capture–recapture studies. Biometrics 64:377–385

Bustamante, A. 2008. Densidad y uso de hábitat por los felinos en la parte sureste del área de amortiguamiento del Parque Nacional Corcovado, Península del Osa, Costa Rica. M.S. thesis, Universidad Nacional, Heredia, Costa Rica

Carbone, C., S. Christie, K. Conforti, T. Coulson, N. Franklin, J. R. Ginsberg, M. Griffiths, J. Holden, K. Kawanishi, M. Kinnaird, R. Laidlaw, A. Lynam, D. W. Macdonald, D. Martyr, C. McDougal, L. Nath, T. O'Brien, J. Seidensticker, D. J. L. Smith, M. Sunquist, R. Tilson, and W. N. Wan Shahruddin. 2001. The use of photographic rates to estimate densities of tigers and other cryptic animals. Animal Conservation 4:75–79

Carbone, C., S. Christie, K. Conforti, T. Coulson, N. Franklin, J. R. Ginsberg, M. Griffiths, J. Holden, M. Kinnaird, R. Laidlaw, A. Lynam, D. W. MacDonald, D. Martyr, C. McDougal, L. Nath, T. O'Brien, J. Seidensticker, J. L. D. Smith, R. Tilson, and W. N. Wan Shahruddin. 2002. The use of photographic rates to estimate densities of cryptic mammals: response to Jennelle et al. Animal Conservation 5(2):121–124

Carrillo, E., R. Salom-Pérez, J. Saénz, and C. Saénz. 2007. Análisis de distribución de poblaciones de mono araña, felinos grandes y sus presas en la Península de Osa como una herramienta para definir áreas prioritarias de conservación. Final Report presented to INBio. Instituto Nacional de Biodiversidad-Coalición Técnica del Corredor Osa, Critical Ecosystem Partnership Fund-Conservation International, Wildlife Conservation Society & Instituto Internacional en Conservación y Manejo de Vida Silvestre, San José, Costa Rica

Ceballos, G., C. Chávez, A. Rivera, C. Manterola, and B. Wall. 2002. Tamaño poblacional y conservación del jaguar en la Reserva de la Biosfera Calakmul, Campeche, México. Pages 403–417 *in* R. A. Medellín, C. Equihua, C. L. B. Chetkiewicz, P. G. Crawshaw, Jr., A. Rabinowitz, K. H. Redford, J. G. Robinson, E. W. Sanderson, and A. B. Taber, editors. El jaguar en el nuevo milenio. Fondo de Cultura Económica, Universidad Nacional Autónoma de México, Wildlife Conservation Society, Mexico City, Mexico

Chávez, C., G. Ceballos, R. Medellín, and H. Zarza. 2006. Primer censo nacional del jaguar. Pages 133–141 *in* Conservación y manejo del jaguar en México: estudios de caso y perspectivas. G. Ceballos, C. Chávez, R. List, and H. Zarza, editors. Comisión nacional para el conocimiento y uso de la biodiversidad, World Wildlife Fund, Telcel, Universidad Nacional Autónoma de México, Mexico City, Mexico

Crawshaw, P. G., Jr. 1995. Comparative ecology of the ocelot (*Leopardus pardalis*) and jaguar (*Panthera onca*) in a protected subtropical forest in Brazil and Argentina. Ph.D. dissertation. University of Florida, Gainesville, FL

Crawshaw, P. G., Jr. and H. B. Quigley. 1991. Jaguar spacing, activity and habitat use in a seasonally flooded environment in Brazil. Journal of Zoology, London 223:357–370

Cuéllar, E., T. Dosapei, R. Peña, and A. J. Noss. 2003. Jaguar and other mammal camera trap survey Ravelo II, Ravelo field camp (19° 17′ 44″S, 60° 37′ 10″W), Kaa-Iya del Gran Chaco National Park, 18 September–18 November 2003. Technical Report #103. Wildlife Conservation Society, Santa Cruz, Bolivia

Cuéllar, E., J. Segundo, G. Castro, and A. J. Noss. 2004. Jaguar and other mammal camera trap survey Guanaco II, Guanaco field camp (20° 03′ 03″S, 62° 26′ 04″W), Kaa-Iya del Gran Chaco National Park, 18 August–18 October 2004. Technical Report #108. Wildlife Conservation Society, Santa Cruz, Bolivia

Cuéllar, E., L. Maffei, R. Arispe, and A. J. Noss. 2006. Geoffroy's cats at the northern limit of their range: activity patterns and density estimates from camera trapping in Bolivian dry forests. Studies on Neotropical Fauna and Environment 41(3):169–178

Cullen, L. Jr., D. Sana, K. C. Abreu, and A. F. D. Nava. 2005. Jaguars as landscape detectives for the upper Paraná river corridor, Brazil. Natureza e Conservação 3(1):124–146

Denapole, L. 2007. Jaguar (*Panthera onca*), puma (*Puma concolor*) y presas en el Chaco Seco, Santiago del Estero, República Argentina. M.S. thesis. Universidad Nacional, Heredia, Costa Rica

Di Bitetti, M. S., A. Paviolo, and C. D. de Angelo. 2006. Density, habitat use and activity patterns of ocelots (*Leopardus pardalis*) in the Atlantic Forest of Misiones, Argentina. Journal of Zoology, London 270:153–163

Di Bitetti, M. S., A. Paviolo, C. D. de Angelo, and Y. E. di Blanco. 2008. Local and continental correlates of the abundance of a neotropical cat, the ocelot (*Leopardus pardalis*). Journal of Tropical Ecology 24:189–200

Dice, L. R. 1938. Some census methods for mammals. Journal of Wildlife Management 2:119–120

Dillon, A. and M. J. Kelly. 2007. Ocelot *Leopardus pardalis* in Belize: the impact of trap spacing and distance moved on density estimates. Oryx 41:469–477

Dillon, A. and M. J. Kelly. 2008. Ocelot home range, overlap, and density using camera trapping and radio telemetry. Journal of Zoology 275(4):391–398

Efford, M. G., D. K. Dawson, and C. S. Robbins. 2004. DENSITY: software for analysing capture–recapture data from passive detector arrays. Animal Biodiversity and Conservation 27:217–228

García, A. R., R. B. McNab, J. S. Shoender, J. Radachowsky, J. Moreira, C. Estrada, V. Méndez, D. Juárez, T. Dubón, M. Córdova, F. Córdova, F. Oliva, G. Tut, K. Tut, E. González, E. Muñoz, L. Morales, and L. Flores. 2006. Los jaguares del corazón del Parque Nacional Tikal, Petén, Guatemala. Asociación Balám and Wildlife Conservation Society, Flores, Petén, Guatemala

Harmsen, B. J., R. J. Foster, S. C. Silver, L. E. T. Ostro, and C. P. Doncaster. 2010. The ecology of jaguars in the Cockscomb Basin Wildlife Sanctuary, Belize. Pages 403–416 *in* D. Macdonald, editor. The biology and conservation of wild felids. Oxford University Press, Oxford

Henschel, P. and J. Ray. 2003. Leopards in African rainforests: survey and monitoring techniques. WCS Global Carnivore Program. Wildlife Conservation Society, New York

Jackson, R. M., J. D. Roe, R. Wangchuk, and D. O. Hunter. 2006. Estimating snow leopard population abundance using photography and capture–recapture techniques. Wildlife Society Bulletin 34(3):772–781

Jennelle, C. S., M. C. Runge, and D. I. MacKenzie. 2002. The use of photographic rates to estimate densities of tigers and other mammals: a comment on misleading conclusions. Animal Conservation 5(2):119–120

Karanth, K. U. 1995. Estimating tiger *Panthera tigris* populations from camera-trap data using capture–recapture models. Biological Conservation 71:333–36

Karanth, K. U. and J. D. Nichols. 1998. Estimation of tiger densities in India using photographic captures and recaptures. Ecology 79(8):2852–2862

Karanth, K. U. and J. D. Nichols, editors. 2002. Monitoring tigers and their prey: a manual for researchers, managers and conservationists in tropical Asia. Centre for Wildlife Studies, Bangalore, India

Karanth, K. U., J. D. Nichols, N. S. Kumar, W. A. Link, and J. E. Hines. 2004. Tigers and their prey: predicting carnivore densities from prey abundance. Proceedings of the National Academy of Sciences (PNAS) 101(14):4854–4858

Karanth, K. U., J. D. Nichols, N. S. Kumar, and J. E. Hines. 2006. Assessing tiger population dynamics using photographic capture-recapture sampling. Ecology 87(11):2925–2937

Kelly, M. J. 2003. Jaguar monitoring in the Chiquibul Forest, Belize. Caribbean Geography 13(1):19–32

Kelly, M. J., A. J. Noss, M. S. di Bitetti, L. Maffei, R. Arispe, A. Paviolo, C. D. de Angelo, and Y. E. di Blanco. 2008. Estimating puma densities from camera trapping across three study sites: Bolivia, Argentina, Belize. Journal of Mammalogy 89(2):408–418

Kostyria, A. V., A. S. Skorodelov, D. G. Miquelle, V. V. Aramilev, and D. McCullough. 2003. Results of camera trap survey of far eastern leopard population in southwest Primorski Krai, winter 2002–2003. Wildlife Conservation Society and Institute for Sustainable Natural Resource Use, Vladivostok, Russia

MacKenzie, D. I. 2005. What are the issues with presence-absence data for wildlife managers? Journal of Wildlife Management 69:849–860

MacKenzie, D. I. and W. L. Kendall. 2002. How should detection probability be incorporated into estimates of relative abundance? Ecology 83:2387–2393

MacKenzie, D. I., J. D. Nichols, J. E. Hines, M. G. Knutson, and A. B. Franklin. 2003. Estimating site occupancy, colonization, and local extinction when a species is detected imperfectly. Ecology 84:2200–2207

MacKenzie, D. I., J. D. Nichols, A. Royle, K. H. Pollock, L. L. Bailey, and J. E. Hines. 2006. Occupancy estimation and modeling: inferring patterns and dynamics of species occurrence. Academic, Boston, MA

Maffei, L. 2005. Estudios con trampas cámaras en el Área Protegida San Matias. Technical Report #162. Wildlife Conservation Society, Santa Cruz, Bolivia

Maffei, L. and A. J. Noss. 2008. How small is too small? Camera trap survey areas and density estimates for ocelots in the Bolivian Chaco. Biotropica 40(1):71–75

Maffei, L., J. Barrientos, F. Mendoza, E. Ity, and A. J. Noss. 2003. Jaguar and other mammal camera trap survey Cerro II, Cerro Cortado field camp (19° 31′ 36″S, 61° 18′ 36″W), Kaa-Iya del Gran Chaco National Park, 28 November 2002–28 January 2003. Technical Report #85. Kaa-Iya Project (Capitanía de Alto y Bajo Isoso, Wildlife Conservation Society), Santa Cruz, Bolivia

Maffei, L., B. Julio, R. Paredes, A. Posiño, and A. J. Noss. 2004a. Estudios con trampas-cámara en el campamento Tucavaca III (18° 30.97′ S, 60° 48.62′ W), Parque Nacional Kaa-Iya del Gran Chaco, 28 de marzo-28 de mayo de 2004. Technical Report #129. Wildlife Conservation Society, Santa Cruz, Bolivia

Maffei, L., E. Cuéllar, and A. Noss. 2004b. One thousand jaguars (*Panthera onca*) in Bolivia's Chaco? Camera trapping in the Kaa-Iya National Park. Journal of Zoology, London 262:295–304

Maffei, L., A. J. Noss, E. Cuéllar, and D. I. Rumiz. 2005. Ocelot (*Felis pardalis*) population densities, activity, and ranging behaviour in the dry forests of eastern Bolivia: data from camera trapping. Journal of Tropical Ecology 21(3):349–353

McBride, R., L. Zavala, and E. Buongermini. 2004. Proyecto Jaguareté. Informe Técnico 2002–2004. Faro Moro Eco Research – Conservation Force, Secretaría del Ambiente, Asunción, Paraguay

McBride, R., L. Zavala, and E. Buongermini. 2005. Proyecto Jaguareté. Informe Técnico 2005. Faro Moro Eco Research – Conservation Force, Secretaría del Ambiente, Asunción, Paraguay

McCain, E. B. and J. L. Childs. 2008. Evidence of resident jaguars (*Panthera onca*) in the southwestern United States and the implications for conservation. Journal of Mammalogy 89(1):1–10

McNab, R. B., J. Moreira, R. García, G. Ponce, V. Méndez, M. Córdova, F. Córdova, G. Ical, I. García, A. Vanegas, E. Zepeda, H. Tut, and R. Monzón. 2008. Densidad de jaguares en la Concesión Comunitaria de Carmelita y Asociación Forestal Integral San Andrés Petén, Reserva de la Biosfera Maya. Technical Report. Wildlife Conservation Society-Guatemala, Flores, Guatemala

Miller, C. M. 2005. Jaguar density in Gallon Jug Estate, Belize. Unpublished report. Wildlife Conservation Society, Gallon Jug, Belize

Miller, C. M. 2006. Jaguar density in Fireburn, Belize. Unpublished report. Wildlife Conservation Society, Gallon Jug, Belize

Miller, C. M. and B. Miller. 2005. Jaguar density in La Selva Maya. Unpublished report. Wildlife Conservation Society, Gallon Jug, Belize

Montaño, R. R., L. Maffei, and A. J. Noss. 2007. Segundo muestreo con trampas cámaras de jaguares y otros mamíferos en el Campamento Palmar de las Islas y Ravelo (Diciembre 2006–Marzo 2007). Technical Report #185. Wildlife Conservation Society, Santa Cruz, Bolivia

Moreira, J., R. B. McNab, D. Thornton, R. García, V. Méndez, A. Vanegas, G. Ical, E. Zepeda, R. Senturión, I. García, J. Cruz, G. Asij, G. Ponce, J. Radachowsky, and M. Córdova. 2007. The comparative abundance of jaguars in La Gloria-El Lechugal, Multiple Use Zone, Maya Biosphere Reserve, Guatemala. Wildlife Conservation Society, Flores, Guatemala

Moreno, R. 2006. Parámetros poblaciones y aspectos ecológicos de los felinos y sus presas en Cana, Parque Nacional Darien, Panamá. M.S. thesis. Universidad Nacional, Heredia, Costa Rica

Núñez, R., B. Miller, and F. Lindzey. 2002. Ecología del jaguar en la Reserva de la Biosfera Chamela-Cuixmala, Jalisco, México. Pages 107–126 *in* R. A. Medellín, C. Equihua, C. L. B. Chetkiewicz, P. G. Crawshaw, Jr., A. Rabinowitz, K. H. Redford, J. G. Robinson, E. W. Sanderson, and A. B. Taber, editors. El jaguar en el nuevo milenio. Fondo de Cultura Económica, Universidad Nacional Autónoma de México, Wildlife Conservation Society, Mexico City, Mexico

Otis, D. L., K. P. Burnham, G. C. White, and D. R. Anderson. 1978. Statistical inference from capture data on closed populations. Wildlife Monographs 62:1–135

Parmenter, R. R., T. L. Yates, D. R. Anderson, K. P. Burnham, J. L. Dunnum, A. B. Franklin, M. T. Friggens, B. C. Lubow, M. Miller, G. S. Olson, C. A. Parmenter, J. Pollard, E. Rexstad, T. M. Shenk, T. R. Stanley, and G. C. White. 2003. Small-mammal density estimation: A field comparison of grid-based vs. web-based density estimators. Ecological Monographs 73:1–26

Paviolo, A., C. de Angelo, Y. di Blanco, C. Ferrari, M. di Bitetti, C. B. Kasper, F. Mazim, J. B. G. Soares, and T. G. de Oliveira. 2006. Crossing the border: the need of transboundary efforts between Argentina, Brazil and Paraguay to preserve the southernmost jaguar (*Panthera onca*) population in the world. Cat News 45:12–14

Paviolo, A. J., C. D. de Angelo, Y. E. di Blanco, and M. S. di Bitetti. 2008. Jaguar population decline in the Upper Paraná Atlantic Forest of Argentina and Brazil. Oryx 42(4):554–561

Payan, C. E. 2008. The impact of indigenous hunting in jaguar, ocelot and prey populations inside and outside a national park in Colombian Amazonia. PhD dissertation, University College London and Insitute of Zoology, Zoological Society of London, London

Pereira, J., M. di Bitetti, N. Fracassi, A. Paviolo, and C. de Angelo. 2006. Densidad y patrón de actividad del gato montés (*Oncifelis geoffroyi*) en un arbustal semidesértico de Argentina central. I Congreso Sudamericano de Mastozoología. 5–8 October 2006. Ciudad de Gramado, Rio Grande do Sul, Brazil

Polisar, J. 2006. Jaguares, presas y gente en Territorios Indígenas Mayangna Sauni Bu, Reserva Biosfera BOSAWAS. The Nature Conservancy, United States Agency for International Development, Managua, Nicaragua

Rabinowitz, A. R. and B. G. Nottingham Jr. 1986. Ecology and behaviour of the jaguar (*Panthera onca*) in Belize, Central America. Journal of Zoology, London (A) 210:149–159

Romero-Muñoz, A. 2008. Densidad, patrones de actividad y comportamiento espacial de felinos en dos sitios del Gran Chaco con diferente presión de ganadería. Undergraduate thesis. Universidad Mayor de San Simon, Cochabamba, Bolivia

Romero-Muñoz, A., A. J. Noss, L. Maffei, and R. M. Montaño. 2007. Binational jaguar confirmed by camera-traps in the American Gran Chaco. Cat News 46:24–25

Rosas-Rosas, O. C. 2006. Ecological status and conservation of jaguars in northeastern Sonora, Mexico. Ph.D. dissertation. New Mexico State University, Las Cruces, New Mexico

Rowcliffe, J. M., J. Field, S. T. Turvey, and C. Carbone. 2008. Estimating animal density using camera traps without the need for individual recognition. Journal of Applied Ecology 45(4):1228–1236

Royle, J. A., J. D. Nichols, K. U. Karanth, and A. M. Gopalaswamy. 2009. A hierarchical model for estimating density in camera-trap studies. Journal of Applied Ecology 46(1):118–127

Rumiz, D. I., R. Arispe, A. J. Noss, and K. Rivero. 2003. Survey of jaguars (*Panthera onca*) and other mammals using camera traps in the San Miguelito Ranch, Santa Cruz, Bolivia. Technical Report #143. Wildlife Conservation Society, Museo de Historia Natural Noel Kempff Mercado, Santa Cruz, Bolivia

Salom-Pérez, R., E. Carrillo, J. C. Sáenz, and J. M. Mora. 2007. Critical condition of the jaguar *Panthera onca* in Corcovado National Park, Costa Rica. Oryx 41(1):1–7

Sanderson, E. W., K. H. Redford, C.-L. B. Chetkiewicz, R. A. Medellin, A. R. Rabinowitz, J. G. Robinson, and A. B. Taber. 2002. Planning to save a species: the jaguar as a model. Conservation Biology 16(1):58–72

Scognamillo, D. I., E. Maxit, M. Sunquist, and L. Farrell. 2002. Ecología del jaguar y el problema de la depredación de ganado en un hato de los llanos venezolanos. Pages 139–150 *in* R. A. Medellín, C. Equihua, C. L. B. Chetkiewicz, P. G. Crawshaw, Jr., A. Rabinowitz, K. H. Redford, J. G. Robinson, E. W. Sanderson, and A. B. Taber, editors. El jaguar en el nuevo milenio. Fondo de Cultura Económica, Universidad Nacional Autónoma de México, Wildlife Conservation Society, Mexico City, Mexico

Scognamillo, D. I., E. Maxit, M. Sunquist, and J. Polisar. 2003. Coexistence of jaguar (*Panthera onca*) and puma (*Puma concolor*) in a mosaic landscape in the Venezuelan llanos. Journal of Zoology, London 259:269–279

Silveira. L. 2004. Ecologia comparada e conservação da onça-pintada (*Panthera onca*) e onça-parda (*Puma concolor*) no Cerrado e Pantanal. PhD dissertation, University of Brasília, Brazil

Silveira, L., A. T. A. Jácomo, S. Astete, R. Sollmann, N. M. Tôrres, M. M. Furtado, and J. Marinho-Filho. 2009. Jaguar density in the Caatinga of northeastern Brazil. Oryx

Silver, S. 2004. Assessing jaguar abundance using remotely triggered cameras. Wildlife Conservation Society, New York. http://savingwildplaces.com/media/file/SilverJaguarCamera-TrappingProtocol.pdf

Silver, S. C., L. E. T. Ostro, L. K. Marsh, L. Maffei, A. J. Noss, and M. Kelly. 2004. The use of camera traps for estimating jaguar (*Panthera onca*) abundance and density using capture/recapture analysis. Oryx 28(2):148–154

Soisalo, M. K. and S. M. C. Cavalcanti. 2006. Estimating the density of a jaguar population in the Brazilian Pantanal using camera-traps and capture–recapture sampling in combination with GPS radio-telemetry. Biological Conservation 129:487–496

Sunquist, M., K. U. Karanth, and F. Sunquist. 1999. Ecology, behavior and resilience of the tiger and its conservation needs. Pages 5–18 *in* J. Seidensticker, S. Christie, and P. Jackson, editors. Riding the tiger: tiger conservation in human-dominated landscapes. The Zoological Society of London, Cambridge University Press, London

Trolle, M. and M. Kéry. 2003. Density estimation of ocelot (*Leopardus pardalis*) in the Brazilian Pantanal using capture-recapture analysis of camera-trapping data. Journal of Mammalogy 66:13–21

Trolle, M. and M. Kéry. 2005. Camera-trap study of ocelot and other secretive mammals in the northern Pantanal. Mammalia 69(3–4):405–412

Trolle, M., M. C. Bissaro, and H. M. Prado. 2007. Mammal survey at a ranch of the Brazilian Cerrado. Biodiversity and Conservation 16:1205–1211

Wallace, R. B., H. Gómez, G. Ayala, and F. Espinoza. 2003. Camera trapping for jaguar (*Panthera onca*) in the Tuichi valley, Bolivia. Mastozoología Neotropical 10(1):133–139

White, G. C., D. R. Anderson, K. P. Burnham, and D. L. Otis. 1982. Capture–recapture and removal methods for sampling closed populations. Los Alamos National Laboratory LA-8787-NERP, Los Alamos, New Mexico

Wilson, K. R. and D. R. Anderson. 1985. Evaluation of two density estimators of small mammal population size. Journal of Mammalogy 66:13–21

Zeller, K. 2007. Jaguars in the new millennium data base update: the state of the jaguar in 2006. Wildlife Conservation Society-Jaguar Conservation Program, New York

Chapter 9
Estimation of Demographic Parameters in a Tiger Population from Long-term Camera Trap Data

K. Ullas Karanth, James D. Nichols, N. Samba Kumar, and Devcharan Jathanna

9.1 Introduction

Chapter 7 (Karanth et al.) illustrated the use of camera trapping in combination with closed population capture–recapture (CR) models to estimate densities of tigers *Panthera tigris*. Such estimates can be very useful for investigating variation across space for a particular species (e.g., Karanth et al. 2004) or variation among species at a specific location. In addition, estimates of density continued at the same site(s) over multiple years are very useful for understanding and managing populations of large carnivores. Such multi-year studies can yield estimates of rates of change in abundance. Additionally, because the fates of marked individuals are tracked through time, biologists can delve deeper into factors driving changes in abundance such as rates of survival, recruitment and movement (Williams et al. 2002). Fortunately, modern CR approaches permit the modeling of populations that change between sampling occasions as a result of births, deaths, immigration and emigration (Pollock et al. 1990; Nichols 1992). Some of these early "open population" models focused on estimation of survival rates and, to a lesser extent, abundance, but more recent models permit estimation of recruitment and movement rates as well.

Given the importance of understanding long-term animal population dynamics, relatively few such studies of large mammals are available because of constraints on carrying out studies of large mammals at the appropriate spatial and temporal scales. For example, in the case of tigers, only a few studies, in Nepal (Sunquist 1981; Smith 1993; Kenny et al. 1995) and Russia (Kerley et al. 2003), have generated some ad hoc estimates of survival rates based on radio-telemetry of samples of 30–40 individuals. However, high costs and logistical difficulties severely limit the potential use of radio-telemetry for estimating tiger demographic parameters. Furthermore,

K.U. Karanth (✉), N.S. Kumar, and D. Jathanna
Wildlife Conservation Society – India Program, Centre for Wildlife Studies,
26-2, Aga Abbas Ali Road (Apt: 403), Bangalore, Karnataka 560042, India
e-mail: ukaranth@gmail.com

J.D. Nichols
U.S. Geological Survey, Patuxent Wildlife Research Center, Laurel, MD 20708, USA

A.F. O'Connell et al. (eds.), *Camera Traps in Animal Ecology: Methods and Analyses*,
DOI 10.1007/978-4-431-99495-4_9, © Springer 2011

most published estimates of vital rates used in studies of tiger population dynamics (e. g. Smith 1993; Kenny et al. 1995; Carroll and Miquelle 2006; Chapron et al. 2008) generally ignore the central problem of incorporating detection probabilities, thus weakening their final inferences.

In this chapter, we illustrate the use of non-invasive camera trap data in conjunction with open population CR models to estimate key demographic parameters, such as time-specific abundance, annual survival rate, and number of new recruits. The methodology is illustrated with examples from a 9-year study of a tiger population in Nagarahole, southern India carried out by Karanth et al. (2006) during 1991–2000.

9.2 Tiger Behavior and Demography in Relation to Monitoring Issues

9.2.1 Sampling Considerations

Efficiently "capturing" several individual tigers using camera traps (or other techniques) is an important issue since accuracy and precision of parameters estimated using CR models depends on sample size. As examined earlier in Chap. 7 camera trapping often involves a tradeoff between sampled area and sampling intensity. Fortunately, tigers often move on forest roads and trails to "patrol" their ranges or to locate prey (Karanth and Chundawat 2002). Biologists can set up camera traps along such travel routes to maximize capture probabilities (see Chap. 7 for details of selecting trap sites and other field survey design issues).

However, tigers in all age-sex classes are not equally likely to be photographed even when traps are set at optimal sites based on field assessments. Tiger spatial organization and land tenure systems pivot around breeding females, which usually establish territories in a part of, or adjacent to, their own natal territories. Typically these females first reproduce at 3–4 years age and hold their territories for the next 5–7 years. Ranges of adult males are much larger and overlap several female ranges, but male tenures are typically shorter at 2–4 years. At about 18–24 months, sub-adults disperse away, with males moving farther away from their natal ranges than females. These dispersers may move over dozens of kilometers, through several breeder territories, looking for vacant territories to settle in by trying to evict existing breeders of their own sex (Sunquist 1981; Smith et al. 1987; Smith 1993; Miquelle et al. 1999; Karanth and Sunquist 2000).

We emphasize that the problem of dealing with the nuisance parameter of detection probability looms large even in the case of open model studies. The area camera trapped by biologists might not necessarily cover the entire tiger habitat patch. Because of patterns of tiger social organization described above, tigers in any population may have heterogeneity in capture probabilities among individuals. Because cubs younger than a year may avoid being photographed, demographic parameters are usually estimated only for animals older than a year. Post-dispersal, transient tigers just passing through the sampled area will have low probability of being captured more than once. Some individuals may temporarily move out of the sampled area during some periods, but will be present and possibly captured in subsequent periods. This is known as temporary emigration. Permanent loss of a tiger from a population in a long-term study may occur as a result of post-natal dispersal or death. All of these processes interact to produce the capture history data from which population-dynamic inferences must be made, and, therefore must be at least considered when modeling CR data. Furthermore, even open model analysis must deal with trap response, behavior, and time-related variation in capture probabilities. As explained more fully under *Data Analysis Issues* below, it is important to realistically incorporate, to the degree possible, these aspects of tiger biology into models for open population camera trap studies, just as in models for closed population studies (see Chap. 7).

Sizes of tiger home ranges are primarily set by densities of large ungulate prey (Karanth and Stith 1999; Karanth et al. 2004). In prey-rich habitats such as the alluvial grasslands of Nepal and India, and the moist deciduous forests of peninsular India, home ranges of breeding female tigers tend to be small and tiger densities high, whereas in habitats where densities are lower, either naturally or because of human impacts, tiger home ranges could be larger and densities lower by as much as 10–20 times. However, even in the best of tiger habitats, mortality rates may be inherently high for all age classes, with cubs and post-dispersal transients having the lowest survival rates.

Proximate causes of tiger cub mortality include infanticide, starvation, floods, forest fires, and predation by other species. Post-dispersal transients are vulnerable to mortality from intra-specific aggression, starvation and human persecution (Karanth and Chundawat 2002). Local tiger populations also lose individuals through permanent emigration (dispersal) and death of evicted breeders (Smith 1993). Thus, even in a healthy tiger population, equilibrium densities (determined primarily by prey densities) are temporally dynamic, characterized by high turnover rates of individuals.

Spatial correlations of tiger and prey densities (Karanth and Nichols 1998; Karanth et al. 2004), as well as simulations of tiger population dynamics (Karanth and Stith 1999), predict that healthy tiger populations may be more susceptible to prey depletion (which decreases carrying capacities, as well as survival of cubs and adults) than to direct poaching of tigers themselves, since removal of tigers is likely to be at least partly compensated by increased survival and recruitment of survivors. The case study presented in this chapter tried to test the prediction that tiger populations in prey-rich habitats are demographically viable despite hypothesized high rates of annual loss.

9.2.2 Field Survey Issues

Most of the practical field survey considerations are covered in Chap. 7, and in greater detail in the manual edited by Karanth and Nichols (2002). This discussion is limited to aspects of photographic CR surveys of tigers and illustrated by observations from a long-term study of tiger population dynamics.

Selection of a study area for a long term photographic CR survey of tigers may be based on several factors, but the issues of sample size and logistics will necessarily constrain the choices. Capture–recapture estimation of tiger abundance based on camera traps has been found to work well in areas with densities above 2–3 tigers per 100 km^2 (Karanth and Nichols 2002). Areas with lower densities, or very small study areas, may not yield sufficient data to reliably model the photo-capture process. As noted in Chap. 7 selection of the actual camera trap locations should not at all be at "random," but should aim to maximize tiger capture probabilities. The trap spacing must ensure that there are no "holes" within the sampled area (Chap. 7). In a multi-year study, if the sampled area changes, time-specific abundance estimates cannot be used directly to derive estimates of population growth rates and recruitment, as changes in the sampled population will reflect both population dynamics and changes in area sampled. Expansion of study areas over time is relatively common, and estimation of population growth rate and recruitment can be accomplished by focusing on a portion of the total study area and/or by use of density (rather than abundance) estimates. Estimates of survival rates are less likely to be affected by increases in the sampled area.

9.3 Identification of Tigers and Assignment of Age-Sex Classes

Because estimates of both capture probabilities and survival rates using CR models (with camera trap data) are completely dependent on unambiguous individual identification, it is critical that (a) the same individual is never misidentified as two or more individuals, and (b) two or more individuals are never assigned the same identity. Broadside pictures of tigers have the added advantage of permitting categorization by sex and broad age-classes. If heterogeneity of estimated parameters is likely to be due to differences between age-sex classes, it may be useful to carry out a "stratified" or multistratum analysis (e.g., Williams et al. 2002) in which parameters are estimated separately for different age-sex classes of tigers. However, this is possible only if sufficient numbers of tigers are captured in all relevant age-sex categories. In the long-term study we report here, we were not able to obtain sufficient data for this type of analysis, and the parameter estimates are best viewed as averages for all tigers of all catchable classes. Because they were rarely captured (2 out of 366 tiger photo-captures), cubs <1 year are excluded from these analyses. Independent of a priori categorization into age-sex classes, some CR analyses can estimate the proportion of transient individuals in the population, as described below.

9.4 Data Analysis Issues

9.4.1 Model Framework

In closed population CR studies, the probability of observing a particular capture history (e.g., 101001; see Chap. 7) depends only on capture probabilities, which may or may not vary over time, among individuals (heterogeneity), or between newly captured and previously captured animals (trap response). The count (number of individuals captured during the study) is equal to the product of real abundance and overall capture probability (probability that an individual in the population is captured at least once). Therefore, the "nuisance parameter" of capture probability must first be estimated from the capture history data (Karanth and Nichols 2002; Williams et al. 2002) to estimate true abundance.

In open population CR studies, the probability of observing a particular capture history is dependent both on the capture probabilities and the probabilities of individual tigers surviving between sampling occasions and staying within the sampled area, since the population is open to additions and losses across time and space. This persistence is referred to as "apparent survival", and does not distinguish between losses due to death and permanent emigration. Models of capture history data thus include parameters for capture and survival probabilities. The method of maximum likelihood, the most commonly used parameter estimation method, simply finds those parameter values that are most likely, given the data (i.e., the values maximize the likelihood of actually getting the observed set of photo capture histories). Survival and capture probabilities may or may not vary over time, between groups (e.g., sex), or as a function of time- (e.g., weather) or individual-specific (e.g., weight) attributes.

The basic open population model, usually referred to as the Cormack–Jolly–Seber (CJS; Cormack 1964; Jolly 1965; Seber 1965) model, tracks recaptures of marked individuals and permits estimation of capture probabilities and apparent survival. This model does not require the assumption that marked (captured) and unmarked individuals have the same capture probabilities, and thus does not allow estimation of abundance. The Jolly–Seber (JS) model includes the CJS model within its structure. By assuming that individuals are randomly sampled from the population (i.e., captures of marked and unmarked individuals are equally probable), the JS model provides estimates of abundance in addition to capture probabilities and apparent survival (Pollock and Alpizar-Jara 2005). A major problem is that these abundance estimators are especially sensitive to violations of the assumption of homogenous capture probabilities. Heterogeneity in capture probabilities among individuals and trap-happiness tend to bias abundance estimates negatively, while trap-shyness results in positive bias. Estimates of apparent survival, on the other hand, are robust to heterogeneity of capture probabilities and trap response, though the latter may affect variance estimation (Pollock and Alpizar-Jara 2005).

One way of overcoming some of these problems is to use Pollock's (1982) robust design, where sampling is carried out at two temporal scales: primary sampling occasions are separated by relatively long time-periods, during which the population is open to losses and gains; and multiple secondary sampling occasions occur within each primary sampling occasion and are separated by relatively short periods, during which closure can be reasonably assumed (Williams et al. 2002; Nichols 2005). As originally envisaged, survival is estimated over the primary sampling occasions using CJS estimators, and abundance estimated within primary occasions using closed population estimators, based on capture histories observed over secondary sampling occasions. Recruitment into the population could then be estimated by combining estimates of survival and time-specific abundance. Kendall et al. (1995, 1997) later developed a likelihood-based approach to estimation under the robust design, where data from both primary and secondary sampling occasions are used to simultaneously estimate parameters in one step. This joint modeling of open and closed datasets has the advantage of permitting reduced parameter models, for example where some parameters are kept constant over time, effectively borrowing information across years and increasing the precision of estimated parameters.

The development of likelihood-based estimators that incorporate heterogeneity (Norris and Pollock 1996; Pledger 2000) now allows heterogeneity to be incorporated in closed population or robust design analyses. A robust design analysis also has the advantage of permitting estimation of temporary emigration. The proportion of new captures that are transients and population growth rates can be estimated using open population models and thus using robust design models as well. Temporary emigration, if unaccounted for, can lead to substantial biases in estimates of abundance and sometimes apparent survival. For example, if an individual had a capture history of 001001 over primary occasions, a model that does not permit temporary emigration will assume that the individual was present within the sampled area during occasions 4 and 5, but not captured (thus leading to potential underestimates of capture probability, as traditionally defined), while a model that permits temporary emigration will consider the possibilities that (1) the individual was present but not captured *or* (2) that the individual had temporarily moved out of the sampled area in either occasion 4, or 5, or both. Similarly, the proportion of individuals that are transients needs to be estimated, since these individuals have a near-zero probability of being captured in subsequent sampling occasions, leading to underestimation of survival for resident animals (Pradel et al. 1997).

We note here that the way "transience" and "temporary emigration" are defined in CR analyses may not correspond exactly to phenomena described by tiger biologists under these names. For instance, "transience" in CR analyses defines the proportion of newly caught tigers that have a near-zero probability of being recaptured again during the entire survey. However, biologists use the term "transients" to refer to tigers that are post-dispersal individuals yet to establish territories. Temporary emigration is likely to be a function of study area size: with small study areas, many individuals may be absent during some primary occasions, simply because parts of their home ranges lie outside the sampled area. The assumption of closure is still required across secondary sampling occasions under many robust design models, but it can be assessed using the tests described in Chap. 7.

A class of CR models known as multi-state models assigns individuals to classes (e.g., breeding vs. non-breeding, different geographic locations), and permits estimation of transition probabilities from one class to another (Arnason 1972; Brownie et al. 1993; Schwarz et al. 1993; Williams et al. 2002; Schwarz 2005). In the case of tigers, this approach is most useful in estimating rates of movement between populations, where the class memberships pertain to geographic location. Another approach known as reverse-time modeling (see Williams et al. 2002; Nichols 2005 for details) of CR data also allows population gains to be separated into those attributable to in situ reproduction and immigration (Nichols and Pollock 1990;

Nichols et al. 2000). All these analyses can be carried out using data from a robust design CR study (Nichols 2005). Open population models for spatially explicit data have the potential to deal with transience, temporary emigration and trap-induced heterogeneity within a single framework (see Royle and Gardner, Chap. 10).

9.4.2 Model Selection

Given a set of capture histories from a robust design study, a very large number of potential models is possible: capture probabilities can be modeled as constant or varying over time (primary and/or secondary occasions), constant or varying between new captures and recaptures, heterogeneous or homogenous across individuals. Similarly, survival probabilities can be varied or held constant over primary occasions, and between new captures and recaptures. Survival rates can also be allowed to vary in parallel (on a logit scale) between new captures and recaptures, or held constant for new captures while being allowed to vary for recaptures. The probability of being a temporary emigrant can be modeled as either different depending on whether the animal was or was not a temporary emigrant the previous period, or independent of previous emigration status. Temporary emigration can also be modeled as time-dependent or time-constant (see Karanth et al. 2006 for details). Considering the number of possible combinations of these parameters, clearly a very large number of potential models can be fit to the data. The models to be considered in any analysis reflect hypotheses about the process that generated the data. Our preference is to limit the models considered to as small a number as possible, including in the model set only those that we view as most plausible based on our a priori hypotheses and ecological knowledge about the system. We have no formal proof that this approach outperforms all others, but we do note that approaches that include all possible models should be much more likely to find models that fit any data set well just by chance alone. Our objective in model selection is to find a good approximating model for the process that generated the data rather than simply a model that fits well. This objective reflects our interest in estimating the values of parameters governing the generating process rather than in simply describing variation in the data.

Given a reasonable model set, the issue of model selection then involves two questions. The first question is whether the model set includes any models that are reasonable. This question is generally addressed by asking whether the most general (complicated) model fits the data adequately. The second question is then, given that the set includes at least one reasonable model, which model(s) should be selected as the basis for inference? Goodness-of-fit tests for robust design models have not yet been developed; however, it is possible to separately assess fit of the different model components. Fit is typically assessed separately for each of the closed model data sets (one set of secondary period data for each primary period) and the single open model data set obtained by combining secondary period data to indicate at least one capture or no capture at each primary period. Goodness-of-fit tests developed specifically for closed (Otis et al. 1978) and open (Pollock et al. 1985; Burnham et al. 1987) models can be used to assess fit of different components of the robust design data set (also see Williams et al. 2002; Nichols 2005). Lack of fit could be due to inappropriate model structure or to a lack of independence between fates (e.g., capture and survival) of individual animals. If the most general model in the model set fails to fit the data adequately, then a variance inflation factor, $\hat{c}$, can be estimated and used in model selection and variance estimation (see Burnham and Anderson 2002; Williams et al. 2002).

For model selection, because all models described here use likelihood-based estimation procedures, it is possible to apply objective model selection criteria such as Akaike's Information Criterion (AIC; Burnham and Anderson 2002) to trade off the greater bias associated with simple models against the lower precision of complex models. In practice, it is recommended to use the small sample correction (AIC_c) if sample sizes are moderate to low, and the quasi-likelihood correction (QAIC) based on $\hat{c}$ if there is evidence of lack of fit for the most general model. If two or more models have strong support, estimates can be derived using model averaging procedures (Burnham and Anderson 2002).

9.4.3 *Software Options*

Program MARK (White and Burnham 1999) is a flexible and powerful software program that offers a large suite of models for a range of CR data types. Closed captures, CJS, JS and robust design analyses can all be carried out using MARK. In addition, MARK implements different options for computing goodness-of-fit test statistics and $\hat{c}$, including various simulation-based approaches (White et al. 2001) and the goodness-of-fit test of program RELEASE (Burnham et al. 1987), which can be called up from the MARK interface. For the secondary sampling occasions, by comparing models where apparent survival is constrained to be = 1 with models where it is not constrained, program MARK also allows a test for violation of closure due to death or permanent emigration. In addition, programs CAPTURE (Otis et al. 1978; Rexstad and Burnham 1991) and ClosTest (Stanley and Burnham 1999) can be used to carry out tests of the closure assumption, as described in Chap. 7.

9.5 Population Dynamics of Tigers in Nagarahole, India

We carried out a 9-year long study of tiger population dynamics using camera traps, analyzing the data under the robust design (Karanth et al. 2006). Field survey protocols were as described in Chap. 7 and in Karanth and Nichols (2002). The study was conducted in the well-protected central part of Nagarahole reserve in the southern Indian state of Karnataka. This 643 km^2 reserve supports high densities of prey (~56 ungulates per km^2: Karanth et al. 2004) and, consequently, of tigers. We expected that this tiger population would be relatively stable despite suspected high annual losses from mortality and emigration. Given that our Nagarahole study area was embedded in a larger landscape consisting of other reserves, multiple use forests and agricultural land, we also expected to find a relatively high proportion of transient individuals trying to establish territories.

To test these hypotheses, we estimated time-specific abundance, survival, transience, recruitment and rates of population change. Camera trapping was first carried out over an area of 41.4 km^2, which was later expanded to 101.5 km^2, and finally to 231.8 km^2 (Fig. 9.1). Over the entire period, we invested a sampling effort of 5,725 trap-nights, photo-capturing a total of 74 adult tigers. Table 9.1 gives details of primary and secondary sampling periods, sampled areas, camera trapping effort, and number of tigers photo-captured (by primary sampling periods and cumulatively).

Analytic methods followed were based on those described above in *Data Analysis Issues* and *Model Selection* and are described in more detail by Karanth et al. (2006). We first carried out tests and preliminary analyses using program CAPTURE (Otis et al. 1978; Rexstad and Burnham 1991). The null hypothesis of closure within each primary sampling period was tested against the alternative hypothesis of an open population. These tests failed to reject the null hypothesis for any of the primary periods, although most of the z statistics were negative, suggesting some gains and losses.

We also examined the between-model tests, goodness-of-fit tests and the model selection scores (see Chap. 7 for details) provided in the CAPTURE output to help narrow down the list of candidate models to be used for the final analysis. Model M_0 was selected for most datasets, but since it is known that this estimator is not robust to violations of assumptions, we did not use it in subsequent analyses. There was evidence of heterogeneity and trap-response, with much less support for time-related variation in capture probability. Therefore, in our set of candidate models, we included heterogeneity and trap-response as modeled sources of variation across secondary sampling periods, as well as temporal variation in capture probabilities across primary sampling periods. Heterogeneity was incorporated using Pledger's (2000) finite mixture model, by including a mixing parameter and considering two groups of animals with different capture probabilities. Modeling of survival rates and temporary emigration are described in *Model Selection* above. The goodness-of-fit tests from program RELEASE indicated an adequate fit of the full CJS model to the data ($\chi^2_{16} = 16.11, P = 0.45$). However, examination of one of the tests carried out by RELEASE provided weak evidence of a transient response (χ_8^2 = 11.69, P = 0.17). Therefore, we included a transient parameterization in our candidate model set, which consisted of 30 competing models.

Examination of ΔAIC_C values and AIC_C weights (Burnham and Anderson 2002) indicated that the selected "best" model was substantially better than its closest competitor (AIC_c weight=0.68; AIC_c weight of nearest competitor=0.21). Therefore, we did not use model-averaged parameter estimates. The selected best model included heterogeneity and trap response in capture probabilities, with initial capture probabilities estimated at $\hat{p} = 0.40, \widehat{SE}(\hat{p}) = 0.067; \hat{p} = 0.15, \widehat{SE}(\hat{p}) = 0.020$, for the two groups (mixture model), respectively. Estimated recapture probabilities were

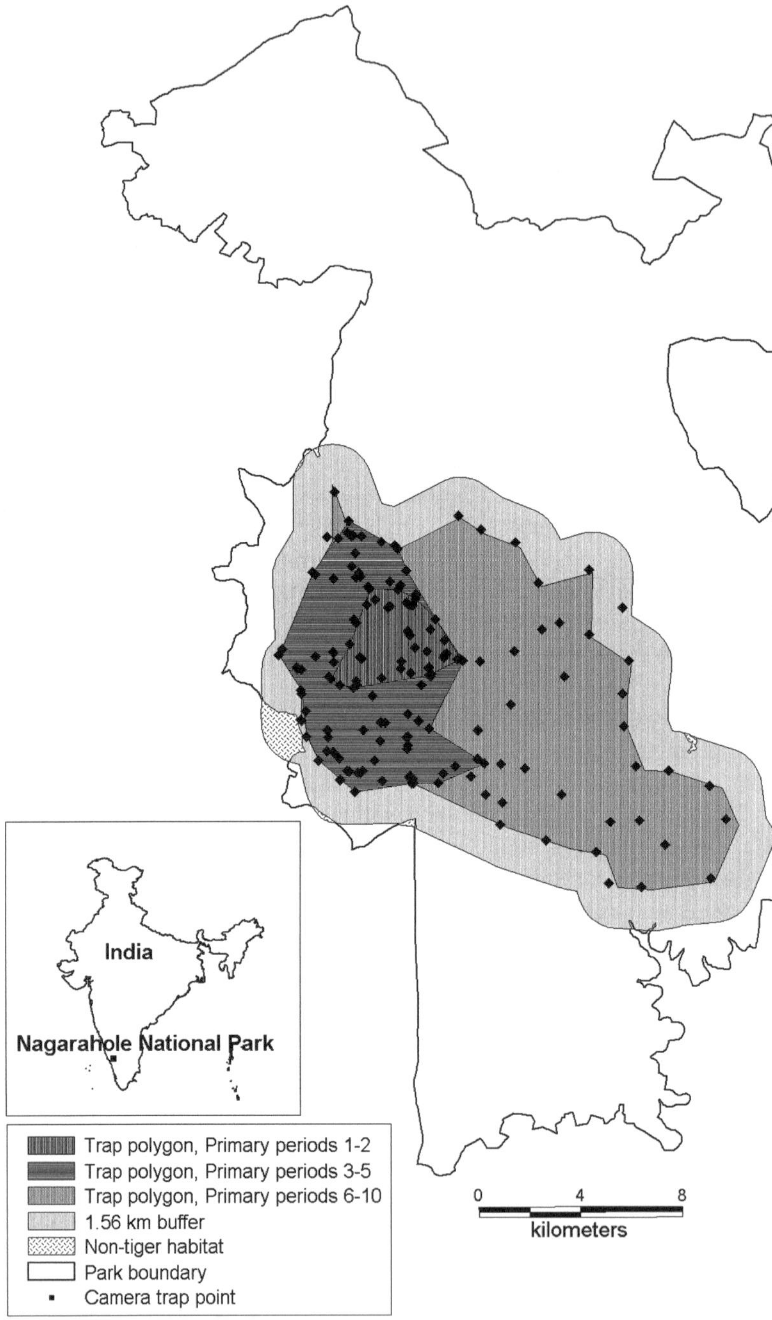

Fig. 9.1 Map of the areas sampled by camera traps in Nagarahole National Park from 1991–2000. *Inset* shows the park's location within India

Table 9.1 Details of the primary and secondary sampling periods, sampled areas, camera trapping effort and number of individual tigers photo-captured at Nagarahole, India, 1991–2000. The sampled area $\hat{A}(\hat{SE}(\hat{A}))$ has been estimated as described in Karanth and Nichols (1998)

Primary period no.	No. of second. periods	Mid point	Total days	Area $\hat{A}(\hat{SE}(\hat{A}))$ km²	Effort trap-days	No. of tigers caught	Cumulative no. of tigers caught
1	6	5/91	162	41.4 (3.3)	294	9	9
2	5	12/91	127	41.4 (3.3)	87	4	10
3	3	4/93	75	101.5 (5.2)	108	5	13
4	7	1/94	197	101.5 (5.2)	668	17	24
5	10	1/95	78	101.5 (5.2)	691	12	26
6	18	3/96	118	231.8 (7.8)	938	26	44
7	8	6/97	33	231.8 (7.8)	448	15	47
8	12	1/98	39	231.8 (7.8)	695	16	50
9	15	3/99	47	231.8 (7.8)	868	22	60
10	15	5/00	54	231.8 (7.8)	928	28	74

$\hat{c} = 0.26, \hat{SE}(\hat{c}) = 0.048; \hat{c} = 0.080, \hat{SE}(\hat{c}) = 0.010$, for the two groups, respectively, providing some evidence of trap-shyness. The proportion of individuals in each of the two groups was allowed to change over time, and there was no evidence of heterogeneous detection probabilities in some years (estimated mixing parameter approximately = 1 or 0). Temporary emigration was estimated to be constant over time, and random, rather than dependent on the previous year's status (see Kendall et al. 1997); 10% ($\hat{SE} = 0.069$) of the individuals were estimated to be temporary emigrants during any primary period.

In the selected best model, survival was modeled as differing between the first interval following detection (indicating transience) and all other intervals, but otherwise constant over time. Because the mid-points of the primary sampling periods were not separated by exactly 1 year, estimated interval survival rates were rescaled to obtain an annual rate of apparent survival of 0.77 ($\hat{SE} = 0.051$) (on average, 23% of tigers were lost annually from the population due to mortalities and permanent emigration during the study). The survival estimates for the intervals following initial detection of each animal were used to estimate the probability that a newly detected animal is a transient (0.18, $\hat{SE} - 0.11$). Abundance was modeled as varying over time (Table 9.2).

Using the parameters estimated directly by the model, we were able to compute other quantities of interest. We estimated the number of new recruits in each primary period, B_t = animals present at $t+1$ but not present in the population in any previous primary sampling period (see Karanth et al. 2006 for details). The time-specific abundance estimates also allowed us to compute time-specific finite population growth rates (Table 9.2). Growth rates and recruitment were only computed for primary periods 6–10, since the study area size increased twice between periods 1 and 5, exposing a larger number of tigers to sampling efforts and producing changes in abundance and new animals that were associated with these sample area changes as well as with population dynamics. The geometric mean of the time-specific

Table 9.2 Estimated survival, $\hat{S}_t$, recruitment, $\hat{B}_t$, abundance, $\hat{N}_t$, and rate of change in abundance, $\hat{\lambda}_t$, for primary sampling periods, for the tiger population in Nagarahole, India, 1991–2000. We report means and standard errors of various estimated parameters

Primary period (t)	Date	Time interval (Δt) between $t, t+1$	Annual survival $\hat{S}(\hat{SE}(\hat{S}))$	Interval survival $\hat{S}^{\Delta t}$ $(\hat{SE}(\hat{S}^{\Delta t}))$	Abundance $\hat{N}_t(\hat{SE}(\hat{N}_t))$	Growth rate $\hat{\lambda}_t(\hat{SE}(\hat{\lambda}_t))$	Recruitment $\hat{B}_t(\hat{SE}(\hat{B}_t))$
1	5/91	0.667	0.77 (0.051)	0.85 (0.040)	9 (0.0)	–	–
2	12/91	1.333	0.77 (0.051)	0.72 (0.061)	7 (2.6)	–	–
3	4/93	0.750	0.77 (0.051)	0.83 (0.043)	11 (5.5)	–	–
4	1/94	0.917	0.77 (0.051)	0.80 (0.048)	21 (3.2)	–	–
5	1/95	1.250	0.77 (0.051)	0.73 (0.059)	12 (0.0)	–	–
6	3/96	1.167	0.77 (0.051)	0.75 (0.056)	27 (1.4)	0.76 (0.12)	3(3.2)
7	6/97	0.583	0.77 (0.051)	0.87 (0.037)	20 (3.2)	0.86 (0.15)	0(3.0)
8	1/98	1.250	0.77 (0.051)	0.73 (0.059)	17 (1.7)	1.35 (0.15)	11(2.8)
9	3/99	1.083	0.77 (0.051)	0.77 (0.051)	23 (1.7)	1.29 (0.11)	14(2.9)
10	5/00				30 (2.1)		

Table 9.3 Estimated area sampled by camera traps, $\hat{A}$, abundance, $\hat{N}_t$, population density, $\hat{D}_t$, and rate of change in density, $\hat{\lambda}_t^D$, for primary sampling periods, for the tiger population in Nagarahole, India, 1991–2000. We report means and standard errors of various estimated parameters

Primary period (t)	Date	Time interval (Δt) between t, $t+1$	Abundance $\hat{N}_t(\hat{SE}(\hat{N}_t))$	Sampled area $\hat{A}(\hat{SE}(\hat{A}))$	Density $\hat{D}_t(\hat{SE}(\hat{D}_t)$	Density change $\hat{\lambda}_t^D(\hat{SE}(\hat{\lambda}_t^D))$
1	5/91	0.667	9 (0.0)	41.4 (3.3)	21.73 (1.7)	0.78 (0.30)
2	12/91	1.333	7 (2.6)	41.4 (3.3)	16.91 (2.6)	0.64 (0.40)
3	4/93	0.750	11 (5.5)	101.5 (5.2)	10.84 (5.4)	1.91 (1.01)
4	1/94	0.917	21 (3.2)	101.5 (5.2)	20.69 (3.3)	0.57 (0.10)
5	1/95	1.250	12 (0.0)	101.5 (5.2)	11.82 (0.6)	0.99 (0.08)
6	3/96	1.167	27 (1.4)	231.8 (7.8)	11.65 (0.7)	0.74 (0.13)
7	6/97	0.583	20 (3.2)	231.8 (7.8)	8.62 (1.4)	0.85 (0.17)
8	1/98	1.250	17 (1.7)	231.8 (7.8)	7.33 (0.8)	1.35 (0.18)
9	3/99	1.083	23 (1.7)	231.8 (7.8)	9.92 (0.8)	1.30 (0.15)
10	5/00		30 (2.1)	231.8 (7.8)	12.94 (1.0)	–

finite growth rates between 1996 and 2000 was estimated at 1.03 ($\hat{SE} = 0.020$), representing a 3% annual increase.

We also obtained primary period-specific estimates of density from the abundance estimates using the approach based on mean maximum distances moved by recaptured tigers (MMDM; see Chap. 7). These density estimates were also used to compute density-based estimates of population growth rates, which would be relatively unaffected by changes in sampled area. However, the expanded study area included new areas with lower tiger density compared to the original study site. This was reflected in the results as an apparent decline in tiger population densities over the long term, especially for the 2 years during which the area sampled was expanded (Table 9.3).

9.6 Utility of Camera Trap Data for Assessing Population Dynamics

The results of our study supported our predictions that despite substantial losses (23% loss of tigers ≥1 year age, each year), tiger populations can remain high (densities ranged between 7.3–21.7 tigers per 100 km^2) and demographically viable. This is likely a result of the high prey densities (~56 ungulates per km^2) in Nagarahole that facilitate high reproduction and recruitment rates. This result underscores the importance of controlling human hunting of prey species and not focusing solely on direct hunting of tigers. The estimated annual loss of 23% includes mortalities (human induced and natural), as well as some permanent emigration out of the area in the form of dispersing sub-adults or as evicted resident tigers (Smith 1993; Smith et al. 1999). Our prediction of finding a relatively high proportion of temporary emigrants was supported (10%), as was our expectation of high rates of transience, with the selected model estimating the transience parameter at 18%. Thus, our tiger population sampling and modeling approach using non-invasive camera traps allowed us to estimate parameters that are virtually impossible to estimate otherwise in populations of large, elusive, wide-ranging felids such as the tiger.

Our estimates of tiger abundance varied over time and are characterized by relatively high variances. This is partly a result of our analysis explicitly incorporating sources of uncertainty arising from the sampling process as well as from tiger ecology. In our opinion, tiger monitoring methods that claim to improve "precision" and "ability to detect changes" by essentially ignoring these uncertainties do not represent a valid alternative approach. We were also constrained by the limited number of camera traps we could deploy, particularly in primary periods 1–5. Increasing the numbers of camera traps as well as the sampled area will increase recapture rates and capture more individuals (See Table 9.1), improving precision of parameter estimates. In addition, sampling a greater proportion of the population of interest (e.g., the tigers of the entire Nagarahole reserve) will likely reduce the estimated proportion of temporary emigrants. These ideas are currently being tested in our ongoing field studies over a wider landscape around this site.

Some investigators have used the relatively low precision of year-specific abundance estimates to argue against the use of camera traps as a means of monitoring tiger populations. However, the population growth rate over the entire period of study was estimated relatively precisely $(\hat{\lambda} = 1.03\ [\hat{SE} = 0.02])$, at least when compared with other studies of animal populations. By using reduced parameter models that borrow information over time, we were able to estimate change in tiger population size and other parameters (e.g., annual survival) with reasonable precision, while satisfactorily dealing with various ecological and sampling-related uncertainties. Thus, in situations where a demographic monitoring program is really needed to address management or scientific questions, we believe that intermediate- to long-term camera trap studies of the type presented here can be an effective approach to the conduct of management and science on tiger populations.

Long-term studies of tigers based on radio-telemetry or visual identification of individuals (Sunquist 1981; Smith 1993; Smith et al. 1999) have traditionally classified tigers in a population as residents, pre-dispersal offspring and transients. Such post hoc categorizations are subjective, based on the detection histories themselves, and do not consider the probability of individual tigers not being detected despite their presence during some seasons. In such studies, individuals detected only once are typically categorized as transients, whereas our modeling recognizes that such individuals may also include resident tigers that die before being recaptured. Many of our models were parameterized specifically to deal with transience (see Pradel et al. 1997), and model selection results provided strong support for these models. As noted earlier, this transient parameterization does not necessarily correspond exactly to what tiger biologists (Sunquist 1981; Smith 1993; Karanth and Sunquist 2000) subjectively term as "transience," but is instead based on quantities that can be estimated from CR data.

In most large-scale ecological field studies of this nature, experimental manipulation, randomization, replication and controls are impossible to achieve. However, conservation still requires answers to questions about population responses to management and conservation actions in the face of uncertainty. Our study in Nagarahole was begun after a period of population response to effective conservation actions (Karanth et al. 1999). This history of an effective conservation program led to the prediction of a relatively large and stable tiger population. Studies that are initiated before implementation of management actions and continued afterwards can provide even stronger inferences about population responses to such actions. We believe that long-term camera trap studies using the robust design have great potential for evaluating effects of such factors as management actions and human disturbance on tiger populations.

By using reduced parameter models that borrow information over time, we were able to estimate multi-year trends reflecting changes in tiger population size and other demographic parameters with reasonable precision, while satisfactorily dealing with various ecological and sampling-related uncertainties. We believe that other camera trap studies of animal population dynamics would benefit from adopting similar approaches.

References

Arnason, A. N. 1972. Parameter estimates from mark–recapture experiments on two populations subject to migration and death. Researches on Population Ecology 13:97–113

Brownie, C., J. E. Hines, J. D. Nichols, K. H. Pollock, and J. B. Hestbeck. 1993. Capture–recapture studies for multiple strata including non-Markovian transition probabilities. Biometrics 49:1173–1187

Burnham, K. P. and D. R. Anderson. 2002. Model selection and multi-model inference: a practical information-theoretic approach. Springer, New York

Burnham, K. P., D. R. Anderson, G. C. White, C. Brownie, and K. H. Pollock. 1987. Design and analysis methods for fish survival experiments based on release–recapture. American Fisheries Society Monograph 5:1–437

Carroll, C. and D. G. Miquelle. 2006. Spatial viability analysis of Amur tiger *Panthera tigris altaica* populations in the Russian Far East: the role of protected areas and landscape matrix in population persistence. Journal of Applied Ecology 43:1056–1068

Chapron, G., D. G. Miquelle, A. Lambert, J. M. Goodrich, S. Legendre, and J. Clobert. 2008. The impact on tigers of poaching and prey depletion. Journal of Applied Ecology 45:1667–1774

Cormack, R. M. 1964. Estimates of survival from the sightings of marked animals. Biometrika 51:429–438

Jolly, G. M. 1965. Explicit estimates from capture–recapture data with both death and immigration – stochastic model. Biometrika 52:225–247

Karanth, K. U. and R. S. Chundawat. 2002. Ecology of the tiger: implications for population monitoring. Pages 9–21 *in* K. U. Karanth and J. D. Nichols, editors. Monitoring tigers and their prey: a manual for researchers, managers and conservationists in tropical Asia. Centre for Wildlife Studies, Bangalore, India

Karanth, K. U. and J. D. Nichols. 1998. Estimation of tiger densities in India using photographic captures and recaptures. Ecology 79:2852–2862

Karanth, K. U. and J. D. Nichols, editors. 2002. Monitoring tigers and their prey: a manual for researchers, managers and conservationists in tropical Asia. Centre for Wildlife Studies, Bangalore, India

Karanth, K. U. and B. M. Stith. 1999. Prey depletion as a critical determinant of tiger population viability. Pages 100–113 *in* J. Seidensticker, S. Christie, and P. Jackson, editors. Riding the tiger: tiger conservation in human dominated landscapes. Cambridge University Press, Cambridge, UK

Karanth, K. U. and M. E. Sunquist. 2000. Behavioral correlates of predation by tiger, leopard and dhole in Nagarahole, India. Journal of Zoology (London) 250:255–265

Karanth, K. U., M. E. Sunquist, and K. M. Chinnappa. 1999. Long term monitoring of tigers: lessons from Nagarahole. Pages 114–122 *in* J. Seidensticker, S. Christie, and P. Jackson, editors. Riding the tiger: tiger conservation in human dominated landscapes. Cambridge University Press, Cambridge, UK

Karanth, K. U., J. D. Nichols, N. S. Kumar, W. A. Link, and J. E. Hines. 2004. Tigers and their prey: predicting carnivore densities from prey abundance. Proceedings of the National Academy of Sciences USA 101:4854–4858

Karanth, K. U., J. D. Nichols, N. S. Kumar, and J. E. Hines. 2006. Assessing tiger population dynamics using photographic capture–recapture sampling. Ecology 87:2925–2937

Kendall, W. L., K. H. Pollock, and C. Brownie. 1995. A likelihood-based approach to capture–recapture estimation of demographic parameters under the robust design. Biometrics 51:293–308

Kendall, W. L., J. D. Nichols, and J. E. Hines. 1997. Estimating temporary emigration using capture–recapture data with Pollock's robust design. Ecology 78:563–578

Kenny, J. S., J. L. D. Smith, A. M. Starfield, and C. McDougal. 1995. The long-term effects of tiger poaching on population viability. Conservation Biology 9:1127–1113

Kerley, L. L., J. M. Goodrich, D. G. Miquelle, E. N. Smirnov, I. G. Nikolaev, H. B. Quigley, and M. G. Hornocker. 2003. Reproductive parameters of wild female Amur (Siberian) tigers (*Panthera tigris* altaica). Journal of Mammalogy 84:288–298

Miquelle, D. G., E. N. Smirnov, T. W. Merrill, A. E. Myslenkov, H. B. Quigley, M. G. Hornocker, and B. Schleyer. 1999. Hierarchical spatial analysis of Amur tiger relationships to habitat and prey. Pages 71–99 *in* J. Seidensticker, S. Christie and P. Jackson, editors. Riding the tiger: tiger conservation in human dominated landscapes. Cambridge University Press, Cambridge, UK

Nichols, J. D. 1992. Capture–recapture models: using marked animals to study population dynamics. BioScience 42:94–102

Nichols, J. D. 2005. Modern open-population capture–recapture models. Pages 88–122 *in* S. C. Amstrup, T. L. McDonald, and B. F. J. Manly, editors. Handbook of capture–recapture analysis. Princeton University Press, Princeton

Nichols, J. D. and K. H. Pollock. 1990. Estimation of recruitment from immigration versus in situ reproduction using Pollock's robust design. Ecology 71:21–26

Nichols, J. D., J. E. Hines, J. D. Lebreton, and R. Pradel. 2000. The relative contribution of demographic components to population growth: a direct estimation approach based on reverse-time capture–recapture. Ecology 81:3362–3376

Norris, J. L. and K. H. Pollock. 1996. Nonparametric MLE under two closed capture–recapture models with heterogeneity. Biometrics 52:639–649

Otis, D. L., K. P. Burnham, G. C. White, and D. R. Anderson. 1978. Statistical inference from capture data on closed animal populations. Wildlife Monographs 62:1–135

Pledger, S. 2000. Unified maximum likelihood estimates for closed capture–recapture models using mixtures. Biometrics 56:434–442

Pollock, K. H. 1982. A capture–recapture design robust to unequal probability of capture. Journal of Wildlife Management 46:757–760

Pollock, K. H. and R. Alpizar-Jara. 2005. Classical open–population capture–recapture models. Pages 36–57 *in* S. C. Amstrup, T. L. McDonald and B. F. J. Manly, editors. Handbook of capture–recapture analysis. Princeton University Press, Princeton

Pollock, K. H., J. E. Hines, and J. D. Nichols. 1985. Goodness-of-fit tests for open capture–recapture models. Biometrics 41:399–410

Pollock, K. H., J. D. Nichols, C. Brownie, and J. E. Hines. 1990. Statistical inference for capture–recapture experiments. Wildlife Monographs 107:1–197

Pradel, R., J. E. Hines, J. D. Lebreton, and J. D. Nichols. 1997. Capture–recapture survival models taking account of transients. Biometrics 53:60–72

Rexstad, E. and K. P. Burnham. 1991. Users' guide for interactive program CAPTURE. Abundance estimation of closed animal populations. Colorado State University, Fort Collins, CO

Schwarz, C. J. 2005. Multistate models. Pages 165–195 *in* S. C. Amstrup, T. L. McDonald and B. F. J. Manly, editors. Handbook of capture–recapture analysis. Princeton University Press, Princeton

Schwarz, C. J., J. F. Schweigert, and A. N. Arnason. 1993. Estimating migration-rates using tag-recovery data. Biometrics 49:177–193

Seber, G. A. F. 1965. A note on the multiple-recapture census. Biometrika 52:249–259

Smith J. L. D. 1993. The role of dispersal in structuring the Chitwan tiger population. Behavior 124:165–195

Smith J. L. D., C. McDougal, and M. E. Sunquist. 1987. Female land tenure system in tigers. Pages 97–109 *in* R. L. Tilson and U. S. Seal, editors. Tigers of the world: the biology, biopolitics, management and conservation of an endangered species. Noyes, Park Ridge, NJ

Smith, J. L. D., C. W. McDougal, S. C. Ahearn, A. Joshi, and K. Conforti. 1999. Metapopulation structure of tigers in Nepal. Pages 176–189 *in* J. Seidensticker, S. Christie, and P. Jackson, editors. Riding the tiger: tiger conservation in human dominated landscapes. Cambridge University Press, Cambridge, UK

Stanley, T. R. and K. P Burnham. 1999. A closure test for time-specific capture–recapture data. Environmental and Ecological Statistics 6:197–209

Sunquist, M. E. 1981. Social organization of tigers (*Panthera tigris*) in Royal Chitawan National Park, Nepal. Smithsonian Contributions to Zoology 336:1–98

White, G. C. and K. P. Burnham. 1999. Program MARK: survival rate estimation from both live and dead encounters. Bird Study 46:120–139

White, G. C., K. P. Burnham, and D. R. Anderson. 2001. Advanced features of program MARK. Pages 368–377 *in* R. Field, R. J. Warren, H. Okarma, and P. R. Sievert, editors. Wildlife, land, and people: priorities for the 21st century. The Wildlife Society, Bethesda, MD

Williams, B. K., J. D. Nichols, and M. J. Conroy. 2002. Analysis and management of animal populations. Academic, San Diego, CA

Chapter 10
Hierarchical Spatial Capture–Recapture Models for Estimating Density from Trapping Arrays

J. Andrew Royle and Beth Gardner

10.1 Introduction

Much of the theory and methodology underlying inference about population size is concerned with populations that are well-defined in the sense that one can randomly sample individuals associated with some location or area and, usually, uniquely identify them. However, individuals within populations are spatially organized; they have home ranges or territories, or some sense of "place," within which they live and move about. The juxtaposition of this place with a trap or array of traps has important implications for sampling design, modeling, estimation and interpretation of data that result from trapping data. In particular, this juxtaposition induces two general problems. First, for most populations, the spatial area over which individuals exist (and are exposed to capture) cannot be precisely delineated, and movement of individuals onto and off of a putative sample unit results in a form of non-closure, which has a direct effect on our ability to interpret the estimates of population size, *N*, from closed population models. The second problem is that this juxtaposition induces heterogeneity in capture probability as a result of variable exposure of individuals to capture. Certain individuals, e.g., those with territories on the edge of a trapping array, might experience little exposure to capture, perhaps only coming into contact with one or two traps. Conversely, individuals whose territories are located squarely in the center of a trapping array might come into contact with many traps. As such, these individuals should experience higher probabilities of capture than individuals of the former type.

The use of arrays of camera traps for estimating abundance of large cats is widespread. They have been used in studies of tigers *Panthera tigris* (Karanth 1995; Karanth and Nichols 1998; Karanth et al. 2006), ocelots (Trolle and Kéry 2003, 2005), jaguars (Wallace et al. 2003, Maffei et al. 2004), and other species which are uniquely identifiable by spot or stripe patterns. The conventional approach to the analysis of

J.A. Royle (✉) and B. Gardner
U.S. Geological Survey, Patuxent Wildlife Research Center, Laurel, MD 20708, USA
e-mail: aroyle@usgs.gov

A.F. O'Connell et al. (eds.), *Camera Traps in Animal Ecology: Methods and Analyses*, DOI 10.1007/978-4-431-99495-4_10, © Springer 2011

density from these systems is to apply closed population models, and then attempt to convert those estimates to densities using a wide variety of heuristically motivated but essentially ad hoc methods. For example, ecologists have used various heuristic "adjustments" based on auxiliary location information to estimate the mean or maximum distance moved to adjust the effective sample area (Wilson and Anderson 1985a,b; Karanth and Nichols 1998; Parmenter et al. 2003; Trolle and Kéry 2003). The standard estimator used in camera trapping studies seems to derive from Karanth and Nichols (1998) which places a buffer strip around the trap array (more precisely, around a convex hull containing the array) that is equal to half the mean maximum distance moved by individuals captured in more than one trap. As noted by Williams et al. (2002, p.316), this approach has little theoretical justification, but seemed to perform well in simulation studies of Wilson and Anderson (1985a). Although these procedures appear to work adequately in practice, the model or range of conditions for which they do seems to be poorly understood and difficult to characterize theoretically; thus, there is no basis for their extension. Formalization of the use of auxiliary spatial information requires the precise definition of a model – the linkage of encounter location to some notion of territory or home range (and perhaps movements).

The deficiency with classical closed population models for estimating density from trapping arrays is that "space" has no explicit manifestation in such models. The parameter N is just an integer parameter of the model, and the structure of the model is unchanged regardless of trap spacing, whether it is applied to data from a 10×10 grid of small mammal traps, an irregular array of hair snares for bears, or camera traps that are rotated among blocks of forest habitat. To develop a general framework for inference from trapping arrays requires a radical reformulation of closed population models in a manner that admits space explicitly. The natural way to do that is with the use of point process models (Efford 2004). That is, we entertain the notion that individuals are referenced by points in space, say s_i for $i = 1,2,...,N$. Then, ordinary closed population models (which describe the encounter of individuals) are augmented with a model that describes this point process. The result is a hierarchical model (Royle and Dorazio 2008) – a model which possesses explicit models for both an ecological process (distribution of individuals in space) and the imperfect observations of that process (encounters of individuals in traps).

In this chapter, we develop the hierarchical formulation of spatial capture–recapture (CR) models. Although formulation of hierarchical models for data from trapping arrays is relatively simple, analysis of the models poses a difficult statistical problem because the home ranges, or territories of each individual are unknown. In developing formal analysis techniques for spatial CR models, these are regarded as latent variables (i.e., random effects). Moreover, the number of such activity centers (i.e., the population size N) is also unknown. To attack inference under the hierarchical model, we adopt a Bayesian analysis of the model based on data augmentation (Royle et al. 2007), which has been applied to a number of similar models (e.g., Royle and Dorazio 2008; Royle and Young 2008; Royle 2009; Gardner et al. 2009).

We address three distinct classes of models in this chapter which are distinguished by the specific assumptions imposed on the observation model. The distinct observation models arise as a result of the type of trapping apparatus or the manner by which data are obtained from the traps. In the following sections we present the technical development of the models. We begin with a basic Poisson encounter model, which we will label Model 1. The other models are related to one another by the imposition of specific constraints, which produce a binomial encounter model (Model 2) or a multinomial encounter model (Model 3). Thus, our typology of models consists of:

Model 1 (Poisson observations): an individual can be caught an arbitrary number of times in an arbitrary number of traps during any particular trapping interval.

Model 2 (Binomial observations): an individual can be caught at most one time in any single trap, but in an arbitrary number of traps during any particular trapping interval.

Model 3 (Multinomial observations): an individual can be caught at most one time in at most one trap during any particular trapping interval.

In all cases, the traps can catch multiple individuals, and thus they are variations of the "multi-catch" traps (Efford et al. 2008). Arrays of mist-nets are a common type of multi-catch trap, but these models are relevant conceptually and technically to hair snares, camera traps and other sampling methods. Camera trapping studies will usually be concerned with observation models 1 or 2. We note that a fourth type of observation model occurs in many animal population studies that use conventional traps which capture a single individual. We will not be concerned with models for these "single-catch" situations here. Efford (2004) devised a method for inference under such models using a simulation-based approach known as inverse prediction.

We adopt an approach to developing these models in which we encourage the reader to "think hierarchically." By that we mean instead of solving the most general version of the problem first, we develop a sequence of simpler variations of the model that assume certain quantities are known. In particular, we first develop the basic model under the assumption that N is known, and that we know the home range (centers) of all individuals in the population. Analysis of this simpler model is conceptually and methodologically instructive, and the extension does not introduce much additional technical complexity. We develop two subsequent generalizations that prove to be only incrementally more complex, technically and conceptually. This approach reveals the true simplicity of these models as basic Poisson and binomial generalized linear models (GLMs) with random effects. Indeed, in our adoption of a Bayesian framework for inference using data augmentation (Royle et al. 2007), the models are precisely zero-inflated GLMs with random effects. We develop the basic models for closed populations, and we provide the basis for implementing those models in the freely available software package WinBUGS. We also provide (Sect. 10.8) a spatial CR model for a demographically open system, allowing for both survival and recruitment of individuals.

10.2 Background

We noted in the previous section that a large number of ad hoc approaches have been used to derive an "estimate" of effective sample area for use in the calculation of density. These approaches are ad hoc in the sense that the underlying *models* that justify any particular adjustment are poorly understood. They are distinctly *not* model-based procedures. Conversely, Efford (2004) was the first to formalize a model for spatial CR problems in the context of trapping arrays. He adopted a Poisson point process model to describe the distribution of individuals and a distance sampling formulation of the observation model describing the probability of detection as a function of individual location. While earlier (and contemporary) methods of estimating density from trap arrays lacked a formal description of the spatial model, Efford achieved a formalization of the model, but adopted a more or less ad hoc framework for inference under that spatial model (using a simulation based method known as inverse prediction). By this, we mean that it is not clear in his development how that framework relates to established frameworks for parametric inference (i.e., likelihood or Bayesian inference).

Recently, there has been a flurry of effort devoted to formalizing this model-based framework for the analysis of spatial CR models. Two distinct approaches have been developed: (1) classical inference based on likelihood and (2) that based on Bayesian inference. To motivate the origins and relevance of these approaches, we note that, fundamentally, spatial CR models are analogous to classical "individual covariate" models in CR – within a fully model-based inference framework (i.e., based on the "full likelihood," Borchers et al. 2002; see also Royle 2009). While a model-based treatment of individual covariate models is contrary to the traditional manner in which inference is achieved under those models (i.e., Huggins 1989; Alho 1990), model-based analysis has proven necessary in certain classes of individual covariate models, such as with time-varying individual covariates (Bonner and Schwarz 2006) or covariates with measurement error (e.g., distance sampling; see Royle and Dorazio 2008, Karanth et al., Chap. 7). The model-based formulation is easily adapted to standard individual covariate models as well (Royle 2009).

Because spatial CR models are formulated in terms of a collection of latent variables or random effects (corresponding to individual locations), a natural framework for analysis of the models is based on integrated likelihood (Laird and Ware 1982). That is, while the observation model is conceptualized conditional on the random effects, inference is formally based on the likelihood constructed from the *marginal* probability distribution of the observations (i.e., *unconditional* on the random effect). The random effects are removed from the conditional likelihood by integration (which is accomplished numerically in spatial CR models). This approach to inference has been formalized in the context of trapping array problems by Borchers and Efford (2008), Efford et al. (2008), and Efford et al. (2009), and implemented for some classes of models in the software package DENSITY (Efford et al. 2004).

Bayesian analysis is another natural framework for the analysis of models containing latent variables or random effects. Under this approach, analysis of the model is based on Monte Carlo simulation from the posterior distribution, which is the product of the conditional likelihood, the distribution of random effects and perhaps other distributions. This approach was developed by Royle and Young (2008) and was motivated by work focused on modeling individual effects in CR models. In particular, a convenient reparameterization of individual covariate models can be obtained using a method known as data augmentation (Royle et al. 2007). This was applied to classical individual covariate models in Royle (2009). The close similarity between individual covariate models and spatial CR models, with the activity center of the individual, $\mathbf{s}_i$, representing the individual covariate, led to the application of the data augmentation method described by Royle and Young (2008). This is a fairly distinctive application in that there weren't traps, but rather, a physical area (quadrat) that was searched repeatedly. Thus, individuals could be captured anywhere within the boundary of the quadrat. This model stimulated developments for the analysis of tiger camera trapping data (Royle et al. 2009a) and bear hair snares (Gardner et al. 2009). We note that both of these applications made use of a multinomial (Model 3) observation model instead of the more appropriate binomial model (Model 2).

These two technical formulations (integrated likelihood and Bayesian) both provide rigorous solutions to the inference problems posed by spatial CR data. The technical distinctions are that Borchers and Efford (2008) assume a Poisson point process that is unconditional on N, whereas Royle and Young (2008) and related work assume a binomial point process model which is conditional on N. More importantly, Borchers and Efford develop the analysis in a way that is unconditional on the point process (which is removed from the conditional likelihood by integration). Conversely, the analysis of Royle and Young (2008) is conditional on the underlying point process.

We believe that the hierarchical modeling approach is flexible and accessible to practitioners because, fundamentally, the models are simply binomial or Poisson GLMs with random effects. Thus, they are conceptually accessible to practitioners with some basic statistical understanding and experience. For example, we demonstrate in this chapter how the models are developed in WinBUGS as GLMs with random effects (Royle et al. 2009b). We believe that practitioners will have some flexibility in developing models that fit their specific situation. We provide an example with moving traps that does not pose any additional difficulty in defining the model, and thus can be analyzed directly by MCMC in WinBUGS. More generally, integrated likelihood for complex point process models may prove difficult, and so analysis of the model that is conditional on the underlying point process will prove to be more versatile and generalizable.

10.3 Model Formulation

As we noted in the Introduction, the basic deficiency with applying closed population models to encounter history data from trap arrays is that space and movement have no explicit manifestation in such models, i.e., the models are not "spatial." Under these traditional models, N is just an integer-valued parameter that has no spatial context whatsoever. Thus, we seek to formalize the manner in which spatial organization of individuals is relevant to how they are observed. The natural way to accomplish this is with spatial point process models (Efford 2004). In particular, we suppose that individuals are referenced by *fixed* points in space, say $\mathbf{s}_i$ for $i = 1, 2, \ldots, N$, and we describe properties of these point locations with a point process model. We then develop hierarchical extensions of ordinary closed population models that use the point process model. That is, we augment these standard observation models with a model for the underlying point process describing individual locations. The point locations are assumed to be fixed for the duration of the sampling, which represents a form of closure. Thus, individuals may not be static during sampling, but we suppose there exists a spatial attribute of each individual, the point $\mathbf{s}_i$, which *is* fixed for the duration of the study. Strictly speaking, the definition of these points is purely an abstraction, but we might view them conceptually as home range centers (Efford 2004) or points about which animal movements can be described probabilistically (Royle and Young 2008).

Two approaches for characterizing this point process have been suggested. Efford (2004), Borchers and Efford (2008) and related work adopt a Poisson point process assumption which is not conditional on N. In contrast, Royle and Young (2008) and Gardner et al. (2009) adopt a *binomial point process* assumption which is conditional on N. This is something of a minor distinction between the basic formulations of the model, but it is somewhat more important in the development of inference procedures as we noted in Sect. 10.2.

To begin, we suppose that each individual in the population can be characterized by a fixed point, $\mathbf{s}_i = (s_{1i}, s_{2i})$, which is a two-dimensional coordinate representing a point in space about which the movements of individual i are concentrated (i.e., its "activity center"). Further, we suppose that there exists a population of N centers $\mathbf{s}_i; i = 1, 2, \ldots, N$ distributed over some region, say S, the state-space. In practice, S will be prescribed (e.g., by specifying coordinates of some polygon that contains a trapping array). As an example, consider Fig. 10.1. This figure shows a 10 × 10 array of traps with unit spacing (the black dots) within some hypothetical region bounded by a square of dimension 18 × 18 units, which is shown by the dashed boundary. This large square is S. We do not observe the individual activity centers. Instead, they are latent (unobserved) variables in the model.

Next we need to describe the juxtaposition of individual activity centers with the trapping array. Sampling is carried out by a network of J traps, having locations $\{\mathbf{x}_j ; j = 1,2,..., J\}$. In the subsequent development of the model we will suppose that the probability of an individual being captured in some trap j is a function of the distance from the trap to its activity center, and one or more parameters that will be estimated. In the context of the models that we will consider, density estimation is equivalent to estimating the density of activity centers in S, or some subset of S, such as a national park or wildlife refuge.

The sequence of observations generated from trapping arrays are y_{ijk} for individual $i = 1,2,..., n$, trap $j = 1,2,..., J$, and sample (trapping "occasion" or "interval") $k = 1,2,..., K$. Although these observations can be binary, $y_{ijk} = 1$ if individual i is captured and $y_{ijk} = 0$, they generally may be trap-specific frequencies (i.e., the number

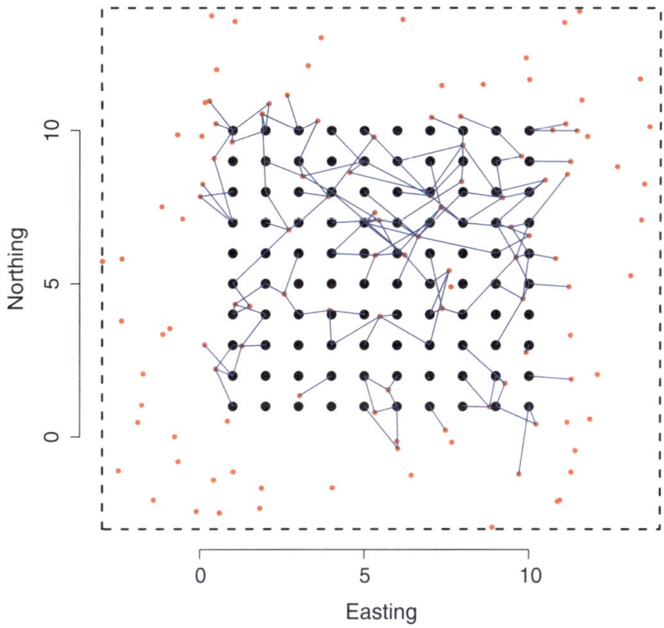

Fig. 10.1 Simulated realization of trapping grid with captured individuals. Simulated captures of individuals (*red dots*) were made by a 10×10 grid of traps (*black dots*). The trap(s) in which each individual was captured are indicated with *blue lines*

of times individual i is captured in trap j during occasion k). Regardless, we will refer to them as encounter or capture histories. Note that this is a three-dimensional array whereas, in classical CR models, we obtain a two-dimensional matrix of encounter observations, usually constructed with rows of the matrix representing individuals and columns representing successive sample periods. In the present case, the extra spatial information provided by trap identity increases the complexity of the encounter history by this third (spatial) dimension. Thus, each *individual* in the population possesses an encounter history *matrix*, which we will represent as a $J \times K$ matrix, where J is the number of traps and K is the number of samples of the population. In certain cases, depending on the specific model under consideration, we will be able to reduce the three-dimensional array of observations, by sufficiency, to a smaller two-dimensional array ($n \times J$) by summing the observations over the replicate samples.

Subsequently, we describe models for the two-dimensional encounter history of each individual that is conditional on that individual's activity center, $\mathbf{s}_i$. For all of the models we describe here, we assume that individuals are independent of one another in terms of their detection in traps. This would not be realistic in some situations, especially in classical trapping grid problems where traps hold or kill individuals (Efford 2004). Moreover, it is probably not a biologically reasonable assumption for territorial carnivores. That said, there have not yet been extensions of spatial CR models that allow for more realistic dependence between individuals.

10.3.1 Observation Models

A number of basic observation models are possible. These derive primarily from the type of, or restrictions on, the physical trap, or the manner in which data are obtained from the trap. We begin by considering a Poisson encounter model, which applies to the situation in which an individual can be caught an arbitrary number of times in an arbitrary number of traps. Strictly speaking, this probably makes sense for camera traps because individuals can visit (and be "observed") an arbitrary number of times during any sampling occasion. However, in camera trapping studies it is often the case that multiple detections of individuals are processed into a single "net detection." Moreover, multiple encounters during a single short occasion (e.g., a night) are not likely to be independent (Royle et al. 2009b) and thus may contribute relatively little information. While we use the Poisson model here as our starting point, we note that other models for detection frequencies in this situation are possible. For example, Efford et al. (2009) consider a negative binomial model. Two other models can be derived directly from the Poisson model. One model that is most relevant to camera trapping studies is the Bernoulli or binomial encounter model. Under this model, an individual can be caught at most one time in any single trap, but in each of an arbitrary number of traps. Thus, encounter (or not) in each trap is a Bernoulli trial. For a study based on K sampling occasions, in the absence of time-variation in parameters, the resulting encounter frequencies have a binomial distribution. Finally, we consider a multinomial encounter model in which individuals can be caught at most a single time and in only a single trap during any particular trapping interval. We do not provide an analysis of this model here because the model is a technical misspecification in most camera trapping situations (see Royle et al. 2009a).

10.3.1.1 Model 1: The Poisson Model

For observations y_{ijk} that are detection frequencies of individual i in trap j during occasion k, then a natural choice for the observation model is the Poisson distribution:

$$y_{ijk} \sim \text{Poisson}(\lambda_0 g_{ij}),$$

where λ_0 is the baseline encounter intensity and g_{ij} is some decreasing function of distance between trap and activity center. Here, we only consider a "half-normal" function of the form:

$$g_{ij} = exp(-d_{ij}^2/\sigma^2) \tag{10.1}$$

where, in this expression, σ^2 is a parameter that will be estimated from data. Thus, for a trap that is located precisely at an individual's activity center, λ_0 is the expected number of captures in that trap.

There are several reasons why we favor the half-normal function. First, it is a common "detection function" used in distance sampling and also in the spatial CR model of Efford (2004) and subsequent developments. Secondly, it can be motivated as arising under an explicit model of movement-induced exposure of individuals to trapping (Royle and Young 2008). For this reason, we sometimes use the term "exposure function" to refer to *g*. Finally, the half-normal exposure function leads to a precise representation of the model as a Poisson regression with random effects. In particular, note that the log-transform of the Poisson mean has the form:

$$\log(E[y_{ijk}]) = \alpha + \beta d_{ij}^2,$$

where $\alpha = \log(\lambda_0)$ and $\beta = -(1/\sigma^2)$. In the present case, d_{ij} is a random effect. Evidently, choice of the function g_{ij} merely affects the functional relationship between the Poisson mean and distance – i.e., the "link function." See Royle and Dorazio (2008, Sect. 4.5.2) for some other instances where link function choice is considered in the context of animal sampling.

10.3.1.2 Model 2: The Binomial Encounter Model

Next we consider the case where the observations are binary. We will view the binary observations conceptually as reductions of the counts that we *could* have observed in the more general case. This might be realistic for bear hair snare studies (and other DNA-based sampling) where an individual can be encountered a number of times during any period, and the biological material (hair, etc.) accumulates but cannot be partitioned into distinct visits after it is collected. To formalize this, suppose that we obtain binary observations y_{ijk} such that

$$\pi_{ijk} = \text{pr}(y_{ijk} = 1) = 1 - \exp(-\lambda_0 g_{ij})$$

which is the probability $y > 0$ under the Poisson encounter frequency model. Thus, the parameters of the model are fundamentally the same, but the observable quantity is a reduced-information summary of what we would prefer to observe (see Royle and Nichols (2003) and Royle (2004) for a similar pairing of models). We realize it could be confusing to use the same variable name (y) for the new data type, but whenever the context might be ambiguous we will be clear about whether this is the Bernoulli or Poisson observation.

This model has a representation as a binomial GLM with a linear effect of distance on a suitable transform of the binomial parameter π_{ijk}. In particular, the complementary log-log transform of π_{ijk} yields:

$$\log(-\log(1 - \pi_{ijk})) = \alpha + \beta d_{ij}^2$$

where α and β are as before. Again, we have a GLM (binomial in this case) with a random effect, d_{ij}. It is worth noting, although we don't develop this issue further here, that the random effects d_{ij}^2 are spatially correlated. The relevance and consequences of this will be developed elsewhere.

10.3.1.3 Model 3: The Multinomial Observation Model

The final model considered here is that which applies to the situation where an individual can only be captured at most a single time during a sample occasion. This is typical of avian mist net studies (Borchers and Efford 2008), although such a model was also used to estimate density in camera trap studies by Royle et al. (2009a). We refer to this model as a multinomial observation model because the "trap of capture" is a multinomial trial. We can think of this as the outcome of rolling a J–sided die. To motivate this model from the basic Poisson model we note that, under the Poisson encounter model with independent traps, the total number of captures for each individual during each period (i.e., when we sum over traps) is also a Poisson random variable (by compound additivity of Poisson random variables):

$$y_{i.k} = \sum_j y_{ijk} \sim \text{Poisson}(\lambda_0 \sum_j g_{ij}). \tag{10.2}$$

Then, the distribution of the individual trap-encounter frequencies *conditional* on $y_{i.k}$ is a multinomial distribution (this is a standard distribution result – independent Poisson random variables when conditioned on their total, have a multinomial distribution). Thus, the trap frequencies y_{ijk} have a multinomial distribution:

$$\{y_{ijk}\}_{j=1}^{J} \mid y_{i.k} \sim \text{Multinom}(y_{i.k}; \left\{\frac{g_{ij}}{\sum_j g_{ij}}\right\}). \tag{10.3}$$

To obtain the relevant observation model, we condition on the event that $y_{i.k} = 1$. That is, an individual is captured only one time. Then, the trap-of-capture is equal to trap j with probabilities as given in (10.3). This component of the model describes the trap-of-capture. But we also need to describe the $\Pr(y_{i.k} = 1)$ – the apparent or "total" probability of capture. We can obtain this by accumulating the positive mass under the Poisson assumption. That is, define:

$$\bar{p}_i = \Pr(y_{i.k} > 0) = 1 - \exp(-\lambda_0 \sum_j g_{ij}) \tag{10.4}$$

Therefore, if $\mathbf{y}_{ik}$ is a multinomial observation for individual i during occasion k, then we can describe the cell probabilities of this $J + 1$ dimensional multinomial distribution by

$$\pi_j = \bar{p}_i \frac{g_{ij}}{\sum_j g_{ij}} \text{ for } j = 1, \ldots, J$$

and, for the last cell corresponding to "not captured", $1 - \bar{p}_i$.

In Royle et al. (2009a) and Gardner et al. (2009), different forms for $\bar{p}_i$ were used, as they did not derive it from the Poisson assumption.

10.4 Analysis of the Models

We proceed with the analysis of these two models (for binomial and Poisson observations) as if we knew precisely the collection of $\mathbf{s}_i$ for all N individuals in the population. In this case the inference problem is to estimate the parameters λ_0 and σ. The aim in doing this is to describe the basic formulation of the inference problem. In addition, this conditional-on-**s** formulation of the model reveals the simplicity of the hierarchical model for camera trap array data. In particular, while the general models are GLMs with random effects, when the $\mathbf{s}_i$ are fixed and known, the models are simple Poisson or binomial GLMs – i.e., with fixed effects only. Extension of the model to allow for **s** to be unknown is technically and conceptually straightforward.

For fixed and known $\mathbf{s}_i$, Bayesian analysis of the model can be carried out easily using the freely available software package WinBUGS (Gilks, et al. 1994). We will not provide an introduction to Bayesian methods here since Bayesianism is a popular inference paradigm and there is a wealth of background material available in the literature, in addition to many accessible texts. In adopting a Bayesian analysis of the model, we require prior distributions for parameters. For that, we use priors that are customarily used to reflect the absence of prior information: for σ, a uniform prior on [0, 5] and, for λ_0, a gamma prior with scale parameters 0.1 and 0.1. Note that the upper bound of the uniform prior for σ must depend on the coordinate system and should be large enough so that the posterior mass for that parameter is not concentrated near the upper bound.

10.4.1 Poisson Detection Frequencies

The WinBUGS model specification for the Poisson model, when $\mathbf{s}_i$ are fixed, is shown in Panel 10.1. We see that this only requires a few lines of WinBUGS model description, and most of that is computing the distances between traps and activity centers. The coordinates of the activity centers are `sx` and `sy` and they are input to WinBUGS as data along with `X`, which is a matrix of the coordinates of the traps, `N`, `J`, and `K` which are the number of individuals, number of traps and number of samples. Finally, the dependent variable is the three-dimensional array `y`. The WinBUGS pseudo-code rendering reveals how remarkably simple the model is. This is the power of hierarchical modeling. When we think about the model *conditional* on the right latent variables, a very simple probability structure emerges. The ***R*** code for simulating data and then fitting this model is available from the authors.

10 4.1.1 Model Extensions and Reductions by Sufficiency

The model specification in Panel 10.1 is more general than may be required in many situations because it specifies the observation model in terms of the finest scale observation – encounter frequencies of every individual, in every trap, and for all *K* occasions. When specified at the level of the most basic observation, the model allows for many extensions. For example, we could add time effects or individual effects to a model for λ_0 by adding only a couple lines to the code in Panel 10.1.

When there are no time effects, we can improve the efficiency of the analysis by recognizing that the total number of captures of each individual in trap *j* is a Poisson random variable with mean $K \times \lambda_0 g_{ij}$. Thus, *K*, the number of replicates, is just an additive offset in the linear model for the mean. One implication of this is that we can choose $K = 1$ without compromising identifiability of the model (see Efford et al. 2009 for additional context). That is, replicate samples are not required.

Due to additivity of Poisson random variables, other simplifications are possible in some cases. We neglect implementation of the possible simplifications and extensions as there are no additional technical considerations.

10.4.2 Model 2: Bernoulli Encounter Process

Under the Bernoulli observation model, an individual may be captured in each trap only once during each sampling occasion. Thus, whether an individual is captured or not in any particular trap can be viewed as a Bernoulli outcome. Implementation of this model, for *fixed* $\mathbf{s}_i$, requires a barely noticeable modification to the WinBUGS model specification, which is shown in Panel 10.2.

```
model {
   sigma2~dunif(0,5)
   lam0~dgamma(.1,.1)
   for(i in 1:N) {
      for(j in 1:J){
         dist2[i,j]<- ( pow(sx[i]-X[j,1],2) + pow(sy[i]-X[j,2],2) )
         mu[i,j]<- lam0*exp(-dist2[i,j]/sigma2)
         for(k in 1:K){
            y[i,j,k]~dpois(mu[i,j])
         }
      }
   }
}
```

Panel 10.1 WinBUGS model specification for the trapping grid model when s_i are known for $i = 1, 2, \ldots, N$. The activity centers, **s**, are input to WinBUGS as data in the form of `sx` and `sy`, vectors of the *x*- and *y*-coordinates of all *N* activity centers. This is an over-simplification of the model that reveals the simplicity of the essential structural component of the model – the relationship between observations, individual activity centers, and trap locations

```
model {
   sigma2~dunif(0,5)
   lam0~dgamma(.1,.1)
   for(i in 1:N){
      for(j in 1:J){
         dist2[i,j]<- ( pow(sx[i]-X[j,1],2) + pow(sy[i]-X[j,2],2) )
         mu[i,j]<- 1-exp( -lam0*exp(-dist2[i,j]/sigma2) )
         for(k in 1:K){
            y[i,j,k] ~ dbern(mu[i,j])
         }
      }
   }
}
```

Panel 10.2 WinBUGS model specification for the Bernoulli encounter model when $\mathbf{s}_i$ are known for $i = 1, 2, \ldots, N$

This specification is more general than may be required in many cases because, as with the Poisson version in Panel 10.1, the observation model is specified in terms of the Bernoulli probabilities for *each* binary observation. While potentially inducing considerable computational burden, this specification allows for simple extensions. For example, we could add time effects or individual effects to a model for λ_0 by adding only a couple lines to the WinBUGS specification. We can improve the efficiency of fitting the model in some cases by recognizing that the total number of captures of each individual in trap j is a binomial random variable based on a sample of size K. Then, the data can be reduced to the $N \times J$ matrix of capture frequencies (number of captures out of K samples). An example of this is given in Panel 10.5 using a slightly more complicated version of the model described in Sect. 10.5.

10.4.3 Analysis of Simulated Data

We simulated data for a hypothetical 10×10 array of camera traps having unit spacing, and then we will fit the model to the simulated data. $N = 120$ individuals were uniformly distributed over an 18×18 square that contains the grid of traps (Fig. 10.1). We used $\lambda_0 = 0.15$ and $\sigma = 1.5$. These individuals were subjected to $K = 6$ survey periods and 61 of them were captured. The total number of captures for all 120 individuals is:

```
capture frequency      0  1  2  3  4 5 6 7 8
number of individuals 59 15 13 12  6 7 3 3 2
```

Since multiple captures are allowed under the Poisson encounter model, there are some capture frequencies >6. For these data, there were 37 instances in which an individual was captured in more than one trap during a sample occasion (2 traps in 28 instances, and 3 traps in 9 instances). We could tinker around with the parameters of this model (λ_0 and σ) in order to achieve the desired structure in the

Table 10.1 Estimates of model parameters for the Poisson and Bernoulli encounter models fitted to the simulated data set shown in Fig. 10.1

Parameter	Mean	SD	MC error	2.5%	Median	97.5%
Poisson						
λ_0	0.143	0.0148	0.00027	0.115	0.142	0.174
σ	1.646	0.1204	0.00235	1.434	1.637	1.901
Bernoulli						
λ_0	0.142	0.0150	0.00031	0.114	0.141	0.173
σ	1.653	0.1183	0.00220	1.438	1.647	1.907

data. When the data are reduced to binary indicators of capture for each individual and each trap, we lose the information that comes from multiple captures in each sample. That is, an individual captured twice in the same trap in the same sample occasion is registered as a single capture. When the simulated data set is reduced in this way, the new (individual) capture frequencies are

```
capture frequency       0  1  2  3  4 5 6 7 8
number of individuals  59 15 16 10 5 9 1 4 1
```

and there were 32 instances where an individual was captured in multiple traps during any particular occasion (twice, 25 times and thrice, 7 times). Since there isn't much loss of information here (subjectively speaking), we would expect both models to yield similar results.

The results of fitting both models to this simulated data set are given in Table 10.1. In general, the posterior standard deviations should be less under the Poisson model, but they are within Monte Carlo error in this case (based on only about 4,500 posterior draws). It might be useful to carry out a full-scale simulation study to see how much information is obtained from multiple captures.

10.5 Model Extension: Unknown *s* and *N*

Suppose we know N, the true number of individuals available for trapping in the region S, but not the location of the activity centers **s**. Conceptually, these can be thought of as *random effects* in the usual sense of the concept as it is used in classical statistics. For analysis of random effects models, we adopt a prior distribution for **s** and proceed with standard methods for analyzing such models. Precisely how we proceed depends in large part on whether we adopt a classical approach to the analysis of random effects or a Bayesian approach.

In the classical treatment of random effects, we would remove them from the likelihood by integration. This was the strategy recently adopted by Borchers and Efford (2008) (see also Efford et al. 2009) for spatial CR models. Alternatively, Bayesian analysis of the random effects model is straightforward and also more accessible to ecologists using the WinBUGS software. The natural prior distribution for **s** is that it is uniformly distributed. In the present context, we assume that points are uniform over some region *S*, which we will denote by

$$\mathbf{s} \sim \text{Uniform}(S).$$

The region *S* is referred to as the state-space of the point process and it is the region over which potentially catchable individuals could be drawn – i.e., a prior distribution for **s**. While *S* could be made arbitrarily large, too large an area will generate additional computational burden. We require specification of *S* whether or not we carry out a Bayesian analysis of the model. If we were to do a non-Bayesian analysis based on integrated likelihood, we would still have to prescribe the limits of integration which is precisely equivalent to prescribing *S*.

To implement this in WinBUGS it is helpful to describe *S* by a regular polygon and that way the uniform assumption on **s** can be described with the expressions `sx[i]~dunif(Xl,Xu)` and `sy[i]~dunif(Yl,Yu)` as illustrated in Panel 10.3. The upper and lower limits `(Xl, Xu, Yl, Yu)` are input into WinBUGS as data. We see the implementation of this model for fixed *N* is straightforward. The interested reader can explore application of this minor embellishment to the simulated data. Royle et al. (2009a) used a discrete representation of *S* so that they could clip out non-habitat. In that case, **s** was assumed uniform on the available points. Borchers and Efford (2008) used a discrete approximation to *S* in order to facilitate a numerical approximation to the integral required to carry-out the integrated likelihood.

```
model {
  sigma2~dunif(0,5)
  lam0~dgamma(.1,.1)

  for(i in 1:N) {
    sx[i]~dunif(Xl,Xu)
    sy[i]~dunif(Yl,Yu)
    for(j in 1:J) {
      dist2[i,j]<- ( pow(sx[i]-X[j,1],2) + pow(sy[i]-X[j,2],2) )
      mu[i,j]<- lam0*exp(-dist2[i,j]/sigma2)
      for(k in 1:K) {
        y[i,j,k] ~ dpois(mu[i,j])
      }
    }
  }
}
```

Panel 10.3 WinBUGS model specification for the trapping grid model when $\mathbf{s}_i$ are unknown but *N* (the number of individuals) is known

We may wish to estimate the number of activity centers located within any arbitrary, polygon. For example, we may be interested in the number of individuals having an activity center located within some polygon *X* (e.g., a national park), say *N(X)*. Or, the density of individuals in *X*, *D(X)* = *N(X)/A(X)* where *A(X)* is the known area of *X*. In order to compute these things we simply tally-up the individual coordinates **s** that are located within the polygon for the simulated draws of $\mathbf{s}_i; i = 1,2,\ldots, N$ from the posterior distribution. WinBUGS properly simulates those summary statistics from the required posterior distribution. Doing this requires a few more lines of WinBUGS model specification. We have to add the following lines to the model specification:

```
tmp1[i]<- step(sx[i]  - xmin)
tmp2[i]<- step(xmax  - sx[i])
tmp3[i]<- step(sy[i]  - ymin)
tmp4[i]<- step(ymax  - sy[i])
incenter[i]<-tmp1[i]*tmp2[i]*tmp3[i]*tmp4[i]
```

and then, outside all the loops, add these lines:

```
Nin<-sum(incenter[1:N])
D<-Nin/81
```

In this case, 81 is the area of the smallest square containing the 10 × 10 array of traps for which we simulated the data in the previous example, and `xmin`, `xmax`, etc., would be the boundaries of that square, which are provided as data to WinBUGS. This is implemented in the WinBUGS specification shown in Panel 10.4. To improve efficiency of the analysis in WinBUGS, this version of the model is described in terms of the $N \times J$ matrix of capture frequencies obtained by summing the *K* frequencies for each individual as described in Sect. 10.4.1.

10.6 Unknown *N*: Data Augmentation

Here we generalize the model one step further, allowing *N* to be unknown. The difficulty in analyzing the more general model is that the dimension of the parameter space (the number of "random effects" – i.e., activity centers) is itself an unknown quantity. It is this problem that motivated the analysis of similar models using the method of data augmentation (Royle et al. 2007). Using data augmentation, Bayesian analysis can be accomplished very directly, in effect allowing us to analyze a version of the "complete data" model – the model with a fixed number of activity centers, which we just accomplished in the previous section. We noted previously that spatial CR models are conceptually and technically similar to a broad class of CR models known as individual covariate models. That context has been developed by Royle and Dorazio (2008, Chap. 7) and Royle (2009).

To introduce data augmentation, we think first about simply developing a Bayesian analysis of the more general model (i.e., *N* unknown). This requires that we describe prior distributions for the parameters N, λ_0, and σ. A natural choice of

```
model {

   sigma2~dunif(0,5)
   lam0~dgamma(.1,.1)

   for(i in 1:N) {

      sx[i]~dunif(Xl,Xu)
      sy[i]~dunif(Yl,Yu)

      tmp1[i]<- step(sx[i] - xmin)
      tmp2[i]<- step(xmax - sx[i])
      tmp3[i]<- step(sy[i] - ymin)
      tmp4[i]<- step(ymax - sy[i])
      incenter[i]<-tmp1[i]*tmp2[i]*tmp3[i]*tmp4[i]

      for(j in 1:J) {
         dist2[i,j]<- ( pow(sx[i]-X[j,1],2) + pow(sy[i]-X[j,2],2) )
         mu[i,j]<- lam0*exp(-dist2[i,j]/sigma2)
         log(lam[i,j])<-log(K)+ log(mu[i,j])
         y[i,j] ~ dpois(lam[i,j])
      }
   }
  Nin<-sum(incenter[1:N])
  D<-Nin/81
}
```

Panel 10.4 WinBUGS model specification for the Poisson version of the trapping grid model (Model 1) when s_i are unknown but N (the number of individuals in S) is known. This specification also computes the derived parameters $N(X)$ (population size) and $D(X)$ (density) for the minimum area rectangle enclosing the trapping grid, X. This version of the model is based on pooled encounter frequencies over all K surveys

priors that reflect the absence of information about these parameters is to assume a discrete uniform prior for N on the integers $0,\ldots,M$ for some large value of M (i.e., $N \sim \mathrm{Du}(0, M)$). Priors for λ_0 and σ are defined as before. We focus our attention on the discrete uniform prior for N. Choice of M is not a critical consideration except that it must be chosen large enough so as to not truncate the posterior distribution of N, which can be checked after a trial analysis.

In principle, the model could be analyzed under this prior specification by conventional MCMC methods for sampling from the posterior distribution. However, we note that, because N is unknown, the dimension of the parameter space – i.e., the number of random effects $\mathbf{s}_i$ – is also unknown. As such, each time that a new draw of N is made from the posterior distribution, the number of random effects (activity centers) changes. Properly updating parameters in this setting has proved to be a challenging problem of some technical complexity. This technical problem motivated an approach to the analysis of such models using the method of data augmentation in Royle et al. (2007).

A heuristic description of data augmentation is that it arises by simply adding excess "all zero" encounter histories to the data set. That is, for M sufficiently large, then we can augment the data set with $M - n$ all-zero encounter histories and recognize that the resulting model for the augmented data is a zero-inflated version of the model for the complete data set (i.e., as if N were known). In models with individual effects, data augmentation is a convenient framework because it allows us to retain a maximal set of random effects in the (augmented) data set, and their values are updated at each iteration of the MCMC algorithm.

Formally, data augmentation is justified as a reparameterization of the model that arises under the discrete uniform prior for N. In particular, note that the discrete uniform prior for N can be constructed by specifying a binomial prior for N: $N \sim \text{Bin}(M, \psi)$, and then placing a uniform (0,1) prior on ψ. When ψ is removed from the binomial component by integration, the result is $N \sim Du(0, M)$. While this may seem only a mathematical curiosity, it suggests a convenient implementation of Bayesian analysis for such models. Namely, we can think of the discrete uniform prior as suggesting a super-population consisting of M individuals, where $M - n$ of them have corresponding "all-zero" encounter histories (because they were not captured). We recognize that some of the M individuals are fixed zeros (i.e., individuals not in the population) whereas some of them are sampling zeros – they correspond to individuals in the population that were not captured. This can be formalized by the introduction of a set of latent indicator variables $z_1, z_2, \ldots, z_M$ such that $z_i = 1$ if individual i is a member of the population and $z_i = 0$ if individual i is a fixed zero. We assume $z_i \sim \text{Bernoulli}(\psi)$. To implement data augmentation, we augment the N observed encounter histories with $M - n$ "all-zero" histories, and then specify the model for the augmented data set in terms of the zero-inflated version of the "known-N" model. As an example, for the Poisson model, the model for the individual and trap-specific encounter frequencies (i.e., summed over K occasions) is

$$y_i \sim \text{Poisson}(\lambda_0 g_{ij}) \text{ if } z_i = 1$$

$$y_i = 0 \text{ if } z_i = 0$$

Under data augmentation, the parameter ψ formally replaces the parameter N, the two being related by the prior specification $N \sim \text{Bin}(M, \psi)$. Because the dimension of the parameter space is fixed, this facilitates a formal analysis by standard methods of MCMC. While developing the MCMC algorithm for analysis of the augmented data is straightforward under this model, we avoid those technical details because the model can also be implemented in WinBUGS.

10.6.1 Implementation

This incremental (but important) extension of the model requires only a little bit of additional WinBUGS model specification. In particular, we must define latent indicator variables, z_i, associated with the data augmentation which we assume to be Bern(ψ) random variables. Then, we note that the counts for the *augmented* data are zero-inflated Poisson or zero-inflated Binomial counts. To implement this, redefine the parameter of the (Poisson or Binomial) distribution to be the product

```
model {

   sigma2~dunif(0,5)
   lam0~dgamma(.1,.1)
   psi ~ dunif(0,1)

   for(i in 1:N) {
      z[i]~dbern(psi)
      sx[i]~dunif(Xl,Xu)
      sy[i]~dunif(Yl,Yu)

      for( j in 1:J){
         dist2[i,j]<- ( pow(sx[i]-X[j,1],2) + pow(sy[i]-X[j,2],2) )
         mu[i,j]<- lam0*exp(-dist2[i,j]/sigma2)
         log(lambda[i,j])<-log(K)+ log(mu[i,j])
         tmp[i,j]<-lambda[i,j]*z[i]
         y[i,j] ~ dpois(tmp[i,j])
      }
   }
   N<-sum(z[1:M])
}
```

Panel 10.5 WinBUGS model specification for the Poisson encounter process model. In this specification, the activity centers **s** and N are unknown

of the indicator z[i] and the parameter of the "known-N" model. For example, expanding on the Poisson model from Panel 10.5, the relevant specification to zero-inflate the Poisson distribution is:

```
z[i] ~ dbern(psi)
log(lambda[i,j])<-log(K)+ log(mu[i,j])
tmp[i,j]<- lambda[i,j]*z[i]
y[i,j] ~ dpois(tmp[i,j])
```

This construction of the zero-inflation process means that, if `z[i]=1`, then the observations are Poisson with mean `lambda[i,j]`; whereas, if `z[i]=0`, then the observations are *fixed zeroes*, which is represented here as a Poisson random variable having mean 0.

This can be seen in context in Panel 10.5. When this model is fit to the sample data, setting $M = N$, i.e., with no additional all-zero encounter histories beyond the actual known number of individuals, then the posterior distribution of ψ concentrates mass at $\psi = 1$, as we might expect.

10.7 Application to Nagarahole Tiger Data

We provide an analysis of data on tigers from the Nagarahole reserve in the state of Karnataka, southwestern India, which has been studied via camera trap methods by Karanth and associates from 1991 until the present (e.g., Karanth 1995;

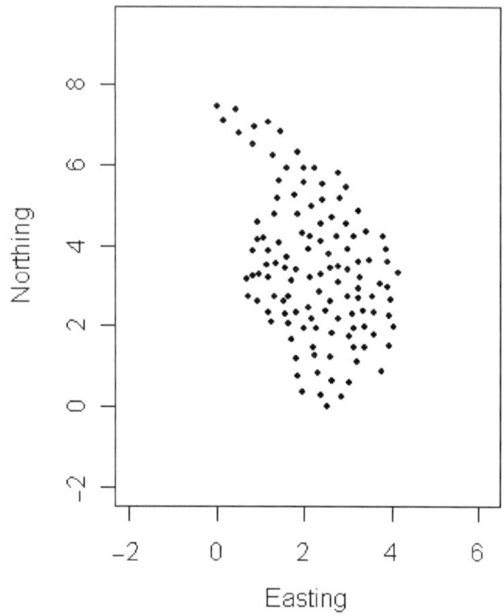

Fig. 10.2 Nagarahole reserve tiger camera trapping array, composed of 120 traps. A unit of distance on this graph is 5 km

Karanth and Nichols 1998, Karanth et al. 2006). The specific data set examined here was obtained in 2006 from sampling at 120 trap stations, each referenced by latitude-longitude and by UTM coordinates (Fig. 10.2). The analysis of the data here follow Royle et al. (2009b) except where noted below. Two camera traps were placed at each location. The sampling took place over 48 nightly intervals between 24 January and 16 March.

Sampling over this period resulted in Bernoulli encounter data on 44 individuals. Thus, each individual encounter history was a $K \times J$ (48×120) matrix where element (k, j) is equal to 1 for each occasion and trap location where a detection occurred, and 0 for each occasion and trap location where the animal was not detected. The Bernoulli encounter model is appropriate for these data. We did not consider a model with time-varying parameters, and so the encounter histories were reduced to individual and trap-specific frequencies by summing over the $K = 48$ occasions. An important feature of the study design is that not all 120 trap stations were operated simultaneously. Instead, the reserve was subdivided into four blocks of approximately 30 trapping stations each, and each block was run for 12 consecutive days. Then, cameras were moved to the next block for another 12 days and the process repeated until all four blocks were sampled. This design follows sample design 4 of Karanth and Nichols (2002, p. 133). Thus, some of the zeros are structural zeros (as opposed to sampling zeros) because if a trap is not operational during a sampling occasion, then an animal cannot be detected.

In the Nagarahole study, traps were moved around such that only 30 of the 120 locations contained a trap on any particular night. This is a standard design for camera trap studies, as well as in hair snares used for obtaining DNA, and other methods of

Table 10.2 Posterior summaries of model parameters for the tiger camera trapping data. *N(X)* is the number of activity centers in *S*, the state-space of the latent point process, and *D* is the density per 100 km^2. The number of unique individuals observed was 44

Parameter	Mean	SD	2.5%	Median	97.5%
λ_0	0.015	0.004	0.008	0.015	0.024
σ	0.338	0.086	0.212	0.325	0.546
ψ	0.567	0.114	0.374	0.557	0.821
D	12.935	2.494	8.750	12.697	18.444
N(X)	301.566	58.134	204.000	296.000	430.000

detecting individuals. This is simple to handle in the analysis merely by defining an indicator of whether trap *j* was in operation at occasion *k*, say m_{jk}, and then:

$$\Pr(y_{ijk} = 1) = 1 - \exp(-\lambda_0 m_{jk} g_{ij})$$

where $m_{jk} = 1$ if trap *j* is operational during occasion *k* and $m_{jk} = 0$, otherwise. Thus, whenever a trap is not operational, $\Pr(y_{ijk} = 1) = 0$, as it should.

For the analysis of these data, we defined *S* to be a rectangle containing the 120 locations shown in Fig. 10.2. Here, *S* was defined to be two units larger than the minimum and maximum coordinate in the four cardinal directions, an area representing approximately 2,331 km^2. Note that the coordinate system was scaled so that a standard unit was 5 km, and thus also are the units of σ. In previous analyses (Royle et al. 2009a, b), we developed an implementation of the model for this discrete state-space situation in the **R** programming language so that non-habitat could be clipped out of *S*. Here, we provide the complete analysis in WinBUGS using the continuous state-space.

Posterior summaries from fitting the Bernoulli encounter model are provided in Table 10.2. Recall that the number of unique individuals observed was 44. The estimate (posterior mean) of *N*(*S*) is approximately 301 individuals and the posterior mean of the density over *S* is 12.935 tigers per 100 km^2. This is slightly larger than reported in Royle et al. (2009b) who reported about 12.2 tigers per 100 km^2 based on area within the rectangle judged to be suitable habitat. In the present case, the 95% posterior interval for density is (8.75, 18.44). The parameter λ_0 is the capture rate in a trap for an individual having s located precisely on a trap location. Thus, $\lambda_0 = 0.015$ indicates a probability of capture for such a situation of $1 - \exp(-0.015) \approx 0.015$.

10.8 Demographically Open Systems

Many camera trap studies take place over multiple years, or are long enough in duration that we do not expect demographic closure to be satisfied. Indeed, there is considerable biological interest in survival and recruitment in the conservation and management of species. Therefore, one useful extension of spatial CR models is the case where populations are open to mortality and recruitment (Karanth et al., Chap. 9).

An important practical motivation for pursuing such models is that, due to the low density of many species studied by camera trap methods, using data from multiple years can greatly improve estimates of density.

The hierarchical formulation of spatial CR models can be extended directly to demographically open systems, and we describe that extension here, based on Gardner et al. (2010). We focus here on the basic conceptual and technical formulation, but do not provide an analysis of data. The interested reader should refer to that paper for analysis and results. The analysis is based on a study of Pampas cats (*Leopardus colocolo*) for $T = 2$ years of data, with multiple samples per year. As with the closed population models, open population models are easily specified for either a binomial or Poisson observation model. Thus, the encounter data are y_{ijkt} for individual i, trap j, sample occasion k, and year t. The observations y_{ijkt} are binary, if an individual can only be captured a single time per trap and occasion, or detection frequencies if individuals can be recorded an arbitrary number of times. Under normal circumstances, when cameras are functioning properly, the type of data obtained may be primarily the result of data processing protocols. We will assume time invariance of all parameters within a year, and define y_{ijt} to be the total number of captures out of K samples (note that K may vary by year but we avoid that generality here in order to be concise). The data structure is consistent with that which arises under the "robust design" (Pollock 1982). Because $T = 2$, there is only one survival parameter and one recruitment parameter. We note that the extension to $T > 2$ years is direct, but we avoid the conceptual generality here in order to achieve clarity.

Gardner et al. (2010) exploited the conceptual linkage between spatial CR and individual covariate models in closed populations to develop a spatial CR model for open populations. An individual effects parameterization of open models was provided by Royle and Dorazio (2008, Chap. 10), and Gardner et al. (2010) then used that formulation to develop spatial CR models. In particular, the underlying model describing survival and recruitment is unchanged. However, the observation model is modified to accommodate a spatial individual covariate in the form of individual activity centers, precisely as in the closed population situation. As we have done previously, we develop the model "conditional-on-N" and then attack inference using data augmentation, by introducing excess $y_{ijt} = 0$ pseudo-observations, producing a data set of size M. In particular, N in this context will be the population of individuals that have ever been alive during a study, i.e., what Schwarz and Arnason (1996) (see also Crosbie and Manly 1985) referred to as the "super-population." To indicate the population size of a particular year, we use N_t.

The state model underlying the Jolly–Seber model has a simple representation in terms of a hidden Markov process model. Define a latent variable $z(i, t)$ which describes the "alive state" of individual i at year t: $z(i, t) = 1$ indicates that the individual is alive and $z(i, t) = 0$ indicates that the individual is not alive. In the context of data augmentation, "not alive" includes individuals that have died, or individuals that have not yet been recruited. The state model is composed of the following two components: First the initial state is described by

$$z(i, 1) \sim \text{Bern}(\psi)$$

and then a model describing the transition of individual states from $t = 1$ to $t = 2$:

$$z(i,2) \sim \text{Bern}(\phi\, z(i, 1) + \gamma\,(1 - z(i, 1))).$$

Thus, if $z(i, 1) = 1$, then the individual may survive with probability ϕ whereas, if $z(i, 1) = 0$, then the "pseudo-individual" may be recruited with probability γ. The observation model, based on K sampling occasions, is

$$y_{ijt}|s_i \sim \text{Poisson}(K\lambda_0 g_{ij}(s_i, x_j) z(i, t)),$$

where g_{ij} is some function of the distance between the activity center i and trap j, just as in the closed model. Note that under this construction of the observation model, y_{ijt} is the outcome of a Poisson random variable if individual i is alive at time t (i.e., $z(i, t) = 1$). Otherwise, if $z(i, t) = 0$, then $y_{ijt} = 0$ with probability 1.

Within the data augmentation framework, the observations are augmented with a large number of $y_{ijt} = 0$ observations to bring the total size of the data set to M individuals. This set of M individuals contains individuals that were observed in the sample, individuals that were not observed in the sample, and also "potential" individuals that may never be recruited. In formulating the model based on data augmentation, the model has a formal equivalence with "multi-season occupancy" models (see Royle and Dorazio 2008, Chaps. 9 and 10). Under this formulation of the model, the recruitment parameter is expressed relative to the pool of available zeros, which does not yield an interpretation that is directly useful. We can define the total number of recruits to be

$$R = \sum_{i=1}^{M} z(i, 2)(1 - z(i, 1))$$

which is just the total number of individuals alive at time $t = 2$ that were not alive at time $t = 1$. Then, the recruitment *rate* is $r = R/N_1$ where $N_1 = \sum_i z(i, 1)$. Similarly, $N_2 = \sum_i z(i, 2)$.

Given the description of the model in terms of individual encounter histories, we now only need to introduce a model describing the individual locations, and then the relationship between the observations and individual locations. In Gardner et al. (2010), we assumed that individual activity centers were static over the two years, i.e., $\mathbf{s}_{i,t} \equiv \mathbf{s}_i \sim \text{Unif}(S)$. There are no further considerations in analysis of this hierarchical model – the model can be described directly (see Panel 10.6) and analyzed using WinBUGS. Alternatively, we could allow the activity center to change across years which seems like it would be easy to implement. One possibility is to assume that $\mathbf{s}(i, t) \sim \text{Normal}(\mathbf{s}(i, t-1), \delta^2\mathbf{I})$ so that individuals home range centers are perturbed randomly from their previous value. Using such models we could conceivably test hypotheses about home range dynamics. We emphasize the conceptual and technical parallels of spatial CR models with individual covariate models. See Bonner and Schwarz (2006), King et al. (2008) and Royle and Young (2008) for some context related to time-varying individual covariates.

```
model {
    sigma2 ~dunif(0, 10)
    lam0~dgamma(.1,.1)
    psi ~dunif(0, 1)
    phi ~dunif(0, 1)
    gamma~dunif(0, 1)
   for (i in 1:M) {
     z[i,1] ~dbern(psi)
     mu[i] <- phi*(z[i,1]) + gamma*(1-z[i,1])
     z[i,2] ~dbern(mu[i])
     SX[i] ~dunif(xl, xu)
     SY[i] ~dunif(yl, yu)
    for(j in 1:J){
       D2[i,j] <- pow(SX[i]-[j,1], 2) + pow(SY[i]-[j,2],2)
       mu[i,j] <- lam0*exp(-D2[i,j]/sigma2)
       log(pmean[i,j])<-log(K[j]) + log(mu[i,j])
    for(t in 1:2){
          tmp[i,j,t]<-pmean[i,j]*z[i,t]*op[j,t]
          y[i,j,t]~dpois(tmp[i,j,t])
  }
  }
  a[i]<-(1-z[i, 1])*z[i,2]
  }
  N1<-sum(z[1:M,1])
  N2<-sum(z[1:M,2])
  R<-sum(a[1:M])
  }
```

Panel 10.6 WinBUGS model specification for the open population model with a Poisson encounter process model. In this specification, the activity centers **s** and N are unknown. Here, `K[j]`, entered as data, represents the number of times that trap j was operational while `op[j,t]` indicates which trap j was operational during year t

10.9 Summary and Discussion

Historically, density estimation in camera trapping studies has been addressed using a number of largely ad hoc or heuristic methods based on closed population estimators of population size applied to individual encounter history data. The conceptual limitation of closed population estimators is that, while the estimate of N may be valid in the sense of estimating the size of a population exposed to sampling, movement of individuals makes it difficult to associate a precise area with the population size estimator. Conventional methods have sought to estimate an effective sample area using informal methods not formally linked (by a statistical model) to the observed encounter history data. Because the underlying models are not specified precisely, they are not sufficiently flexible or extensible. For example, these ad hoc techniques cannot accommodate moving traps, open systems, or multiple captures in a single occasion.

Considerable work has recently been directed at developing a model-based framework for inference in spatial CR models. Efford (2004) presents the first development of such a model based on a Poisson point process governing the distribution of individuals in space. Inference in spatial CR models was formalized in a recent paper by Borchers and Efford (2008) who used an approach based on integrated likelihood – a classical (i.e., non-Bayesian) method for handling random effects. This approach is implemented in the software package DENSITY (Efford et al. 2004). In this chapter, we summarized the hierarchical modeling framework for a broad class of spatial CR models based on a Poisson encounter frequency model. We adopted a Bayesian inference strategy for the hierarchical model based on data augmentation (Royle et al. 2007) which was developed in the context of spatial CR models by Royle and Young (2008) and subsequent efforts.

One of the themes of this chapter is that the different classes of models (Poisson, binomial and multinomial) are closely related to one another by formal reductions or constraints on the observations. A second theme is that fundamentally these models are just GLMs (binomial or Poisson) with random effects (indeed, GLMs in which the random effects are spatially correlated), a class of models referred to as generalized linear mixed models (GLMMs). For example, the Bernoulli model has, for $\pi_{ijk} = \Pr(y_{ijk} = 1)$,

$$\log(-\log(1-\pi_{ijk})) = \alpha + \beta d_{ij}^2,$$

where d_{ij} is a random effect (i.e., it is unobserved). More precisely, d_{ij} depends on the latent activity centers $\mathbf{s}_i$. Note that the half-Gaussian "detection function" leads to the quadratic distance term, but we could use a simple exponential function instead, creating a more standard generalized linear mixed model. Regardless of the precise form of the detection function used, we see that the resulting model, for binary observations, is a standard "logistic regression," although with an alternative link function. This particular link function is usually referred to as the complementary-log-log link.

With this representation as a generalized linear mixed model, we see clearly the linkage between spatial CR models and more traditional "individual covariate" models (Royle 2009). Moreover, this representation also raises the issue elucidated by Link (2003), who noted that the population size estimator is strongly affected by the choice of the distribution for the individual random effect (indeed, N may not even be identifiable). In the present context, choice of the point process model may induce the same ambiguity, as it will largely determine the distribution of d_{ij}. However, as we noted above, the random effect is spatially correlated which may partially mitigate the problem that Link (2003) identified. That the random effect is spatially correlated is interesting for a number of reasons that will be taken up elsewhere.

For analysis of the models, we developed a progression beginning first with the case where activity centers $\mathbf{s}$ are fixed, and N (the number of such centers) is known.

This establishes the fundamental simplicity of these models as GLMs. Extending the model slightly to allow for **s** to be unknown, but N fixed, produces a random effects model that is naturally analyzed by putting a prior distribution on **s**. In the case of spatial CR models, the natural choice is the uniform distribution. Models with unknown **s** are formally related to individual covariate models, where the individual covariate is measured with error (Royle and Dorazio 2008, Chap. 7). The third case considered in the conceptual progression is that where N is unknown. Analysis of this most general case is aided by the technique of data augmentation (Royle et al. 2007) in which we augment the data set with a large number of all-zero encounter histories, corresponding to hypothetical, uncaptured individuals. We admit that the augmented data set is a zero-inflated version of the "known-N" data set. Bayesian analysis of this model under data augmentation is straightforward, as is specification of the model in WinBUGS.

We believe the hierarchical formulation will prove flexible in the development of extensions of spatial CR models. One area that we have developed in this chapter is the extension of these models to demographically open systems. Within the hierarchical modeling framework, inclusion of an individual covariate (activity center) is not difficult. It is an extension of the basic individual effects parameterization of Jolly-Seber type models described by Royle and Dorazio (2008, Chap. 10). The resulting model represents a hierarchical extension of a simple space–time point process with imperfect observation of points. Development of more complex space–time point processes is a natural consideration (e.g., Rathbun and Cressie 1994). Generalizing the point process to allow for non-independence between points also seems like an important direction for extension of spatial CR models. For example, it is natural to consider point process models that exhibit interaction (e.g., inhibition models, Markov point processes, and processes that allow for clustering) which can be used to model the interaction of individuals defending territories, or sex differences, or even interactions among species. The filling-up of available habitat by individuals (especially rare species) is an ecological process of some importance in conservation biology and management (e.g., Fretwell 1972). Spatial CR models allow us to make explicit inferences about the underlying point process. We believe that as the underlying point process model is made more complex (e.g., containing interactions and conditional dependencies), the integration required to compute the integrated likelihood (Borchers and Efford 2008) may become computationally inhibiting. However, Bayesian analysis of the hierarchical formulation only requires (in principle) the capability to carry-out conditional simulation of the activity center locations. Finally, camera traps can collect data in "real time," and so extension of the models to continuous time measurements might yield efficiencies. While camera traps can obtain multiple captures within a sample interval, the data are often not rendered at their finest resolution for operational reasons. Moreover, observations are likely to be highly dependent at some time scale and thus some consideration should be given to models of temporal clustering of observations.

References

Alho, J. M. 1990. Logistic regression in capture–recapture models. Biometrics 46:623–635

Bonner, S. J. and C. J. Schwarz. 2006. An extension of the Cormack–Jolly–Seber model for continuous covariates with application to *Microtus pennsylvanicus*. Biometrics 62:142–149

Borchers, D. L. and M. G. Efford. 2008. Spatially explicit maximum likelihood methods for capture–recapture studies. Biometrics 64:377–385

Borchers, D. L., S. T. Buckland, and W. Zucchini. 2002. Estimating animal abundance: closed populations. Springer, London

Crosbie, S. F. and B. F. J. Manly. 1985. Parsimonious modelling of capture–mark–recapture studies. Biometrics 41:385–398

Efford, M. 2004. Density estimation in live-trapping studies. Oikos 106:598–610

Efford, M. G., D. K. Dawson, and C. S. Robbins. 2004. DENSITY: software for analysing capture–recapture data from passive detector arrays. Animal Biodiversity and Conservation 27:217–228

Efford, M. G., D. L. Borchers, A. E. Byrom. 2008. Density estimation by spatially explicit capture-recapture: likelihood-based methods. Pages 255–269 *in* D. L. Thomson, E. G. Cooch, and M. J. Conroy, editors. Modeling demographic processes in marked populations. Springer, New York.

Efford, M. G., D. K. Dawson, and D. L. Borchers. 2009. Population density estimated from locations of individuals on a passive detector array. Ecology 90:2676–2682

Fretwell, S. D. 1972. Populations in a seasonal environment. Monographs in Population Biology 5:1–217

Gardner, B., J. A. Royle, and M. T. Wegan. 2009. Hierarchical models for estimating density from DNA mark-recapture studies. Ecology 90:1106–1115

Gardner, B., J. Reppucci, M. Lucherini, and J. A. Royle. 2010. Spatially-explicit inference for open populations: estimating demographic parameters from camera-trap studies. Ecology (in press.)

Gilks, W. R., A. Thomas, and D. J. Spiegelhalter. 1994. A language and program for complex Bayesian modelling. The Statistician 43:169–178

Huggins, R. M. 1989. On the statistical analysis of capture experiments. Biometrika 76:133–140

Karanth, K. U. 1995. Estimating tiger *Panthera tigris* populations from camera-trap data using capture–recapture models. Biological Conservation 71:333–338

Karanth, K. U. and J. D. Nichols. 1998. Estimation of tiger densities in India using photographic captures and recaptures. Ecology 79:2852–2862

Karanth, K. U. and J. D. Nichols, editors. 2002. Monitoring tigers and their prey: a manual for researchers, managers and conservationists in Tropical Asia. Centre for Wildlife Studies, Bangalore

Karanth, K. U., J. D. Nichols, N. S. Kumar, and J. Hines. 2006. Assessing tiger population dynamics using photographic capture–recapture sampling. Ecology 87:2925–2937

King, R., S. P. Brooks, and T. Coulson. 2008. Analysing complex capture–recapture data: incorporating time-varying covariate information. Biometrics 64:1187–1195

Laird, N. M. and J. H. Ware. 1982. Random-effects models for longitudinal data. Biometrics 38: 963–974

Link, W. A. 2003. Nonidentifiability of population size from capture–recapture data with heterogeneous capture probabilities. Biometrics 59:1123–1130

Maffei, L., E. Cuellar, and A. Noss. 2004. One thousand jaguars (*Panthera onca*) in Bolivia's Chaco? Camera trapping in the Kaa-Iya National Park. Journal of Zoology 262:295–304

Parmenter, R., T. Yates, D. Anderson, K. Burnham, J. Dunnum, A. Franklin, M. Friggens, B. Lubow, M. Miller, G. Olson, et al. 2003. Small-mammal density estimation: a field comparison of grid-based vs. web-based density estimators. Ecological Monographs 73:1–26

Pollock, K. H. 1982. A capture–recapture design robust to unequal probability of capture. Journal of Wildlife Management 46:752–756

Rathbun, S. L. and N. A. C. Cressie. 1994. A space–time survival point process for a longleaf pine forest in Southern Georgia. Journal of the American Statistical Association 89:1164–1174

Royle, J. A. 2004. N-mixture models for estimating population size from spatially replicated counts. Biometrics 60:108–115

Royle, J. A. 2009. Analysis of capture–recapture models with individual covariates using data augmentation. Biometrics 65:267–274

Royle, J. A. and R. M. Dorazio. 2008. Hierarchical modeling and inference in ecology: the analysis of data from populations, metapopulations, and communities. Academic, San Diego, CA

Royle, J. A. and J. D. Nichols. 2003. Estimating abundance from repeated presence-absence data or point counts. Ecology 84:777–790

Royle, J. A. and K.V. Young. 2008. A hierarchical model for spatial capture–recapture data. Ecology 89:2281–2289

Royle, J. A., R. M. Dorazio, and W. A. Link. 2007. Analysis of multinomial models with unknown index using data augmentation. Journal of Computational and Graphical Statistics 16:67–85

Royle, J. A., J. D. Nichols, K. U. Karanth, and A. Gopalaswamy. 2009a. A hierarchical model for estimating density in camera trap studies. Journal of Applied Ecology 46:118–127

Royle, J. A., K. U. Karanth, A. M. Gopalaswamy, and N. S. Kumar. 2009b. Bayesian inference in camera trapping studies for a class of spatial capture–recapture models. Ecology 90: 3233–3244

Schwarz, C. J. and A. N. Arnason. 1996. A general methodology for the analysis of capture–recapture experiments in open populations. Biometrics 52:860–873

Trolle, M. and M. Kéry. 2003. Estimation of ocelot density in the Pantanal using capture–recapture analysis of camera-trapping data. Journal of Mammalogy 84:607–614

Trolle, M. and M. Kéry. 2005. Camera-trap study of ocelot and other secretive mammals in the northern Pantanal. Mammalia 69:409–416

Wallace, R., H. Gomez, G. Ayala, and F. Espinoza 2003. Camera trapping for jaguar (*Panthera onca*) in the Tuichi Valley, Bolivia. Journal of Neotropical Mammalogy 10:133–139

Williams, B. K., J. D. Nichols, and M. J. Conroy. 2002. Analysis and management of animal populations. Academic, San Diego

Wilson, K. R. and D. R. Anderson. 1985a. Evaluation of two density estimators of small mammal population size. Journal of Mammalogy 66:13–21

Wilson, K. R. and D. R. Anderson 1985b. Evaluation of a nested grid approach for estimating density. Journal of Wildlife Management 49:675–678

Chapter 11
Inference for Occupancy and Occupancy Dynamics

Allan F. O'Connell and Larissa L. Bailey

11.1 Introduction

This chapter deals with the estimation of occupancy as a state variable to assess the status of, and track changes in, species distributions when sampling with camera traps. Much of the recent interest in occupancy estimation and modeling originated from the models developed by MacKenzie et al. (2002, 2003), although similar methods were developed independently (Azuma et al. 1990; Bayley and Petersen 2001; Nichols and Karanth, 2002; Tyre et al. 2003), all of which deal with species occurrence information and imperfect detection. Less than a decade after these publications, the modeling and estimation of species occurrence and occupancy dynamics have increased significantly. Special features of scientific journals have explored innovative uses of detection–nondetection data with occupancy models (Vojta 2005), and an entire volume has synthesized the use and application of occupancy estimation methods (MacKenzie et al. 2006). Reviews of the topical concepts, philosophical considerations, and various sampling designs that can be used for occupancy estimation are now readily available for a range of audiences (MacKenzie and Royle 2005; MacKenzie et al. 2006; Bailey et al. 2007; Royle and Dorazio 2008; Conroy and Carroll 2009; Kendall and White 2009; Hines et al. 2010; Link and Barker 2010). As a result, it would be pointless here to recast all that these publications have so eloquently articulated, but that said, a review of any scientific topic requires sufficient context and relevant background information, especially when relatively new methodologies and techniques such as occupancy estimation and camera traps are involved. This is especially critical in a digital age where new information is published at warp speed, making it increasingly difficult to stay abreast of theoretical advances and research developments.

A.F. O'Connell (✉)
U.S. Geological Survey, Patuxent Wildlife Research Center, Beltsville, MD 20705, USA
e-mail: aoconnell@usgs.gov

L.L. Bailey
Department of Fish, Wildlife, and Conservation Biology, Colorado State University, Fort Collins, CO 80523, USA

A.F. O'Connell et al. (eds.), *Camera Traps in Animal Ecology: Methods and Analyses*,
DOI 10.1007/978-4-431-99495-4_11,

First, we review the principles that guide occupancy estimation (e.g., sampling design considerations) and its role in ecological studies. We provide the basic model framework for single and multi-season occupancy models (MacKenzie et al. 2002, 2003), discuss the underlying assumptions of occupancy models and options for analyzing occupancy data, and present proactive methods of evaluating different sampling designs prior to study initiation (Bailey et al. 2007). Case studies are then used to illustrate some extensions of the basic occupancy models that can provide greater flexibility and utility with respect to making inferences about species occurrence and related system dynamics. Specifically, we discuss a multi-method approach (that includes camera traps) used to provide inferences about occupancy at multiple scales (Nichols et al. 2008), abundance-induced heterogeneity models (Royle and Nichols 2003; Wenger and Freeman 2008), and a species co-occurrence model (MacKenzie et al. 2004). We also summarize recent advances that will likely be utilized in camera trapping studies in the future, including multistate models (Nichols et al. 2007; MacKenzie et al. 2009) and a one dimensional spatial autocorrelation model (Hines et al. 2010).

11.2 Occupancy in Animal Ecology

Estimating population parameters such as abundance or density has long been the focus of many animal population studies (see O'Brien, Chap. 6; Karanth et al., Chap. 7; Maffei et al., Chap. 8). Information on population size and associated vital rates is often necessary to evaluate the status of and monitor changes in populations (see Karanth et al., Chap. 9), but requires relatively large populations where individuals can be uniquely identified or physically captured and marked [i.e., capture–recapture (CR) methods, Chaps. 6–9]. In cases where capture and marking of animals is very difficult, or when densities are expected to be low, occupancy information can be a useful alternative, providing estimates of species distributions and the processes driving distributional patterns. In addition, the type and amount of data necessary to estimate occupancy (i.e., species occurrence information) are usually less expensive and time-consuming to collect than abundance or density information.

Fundamentally, occupancy is a function of abundance (i.e., number of individual animals) and parameters that govern the dynamic process of how animals are distributed in the environment (Royle and Dorazio 2008). For some species, occupancy can even be viewed as a surrogate for abundance (MacKenzie and Nichols 2004). Occupancy has been widely used in a variety of ecological investigations to address basic ecological questions related to geographic range, metapopulation dynamics, habitat relationships, resource selection, and species interactions. In general, occupancy is defined as the probability that a site or patch is occupied by a target species, but several pseudonyms can be found in the literature including patch or site occupancy or proportion of area occupied (PAO) (see MacKenzie et al. 2006 for more details on the origin and interpretation of these pseudonyms). The collection of occupancy data in animal ecology is relatively simple, and sampling can take many forms, from observations of the animals themselves (e.g., photographs from camera traps) to some indicators that the species is present (e.g., scat or tracks).

Making valid inferences about changes in the occurrence of a species requires addressing two important sources of variation common to most animal sampling programs: spatial variation and detectability (Lancia et al. 1994; Thompson, 2002). Many programs cannot sample the entire area of interest so a probability-based method for selecting sample units is recommended (e.g., simple or stratified random sample). The sampled units are then used to draw inference to the entire area of interest. The second issue of concern is detectability, a near-universal situation in animal population sampling where survey methods do not detect all the animals or species present in the sampled area (Pollock et al. 2002). Historically, species occurrence information has long been referred to as presence–absence information whereas detectability, or the probability of detecting the target species at an occupied site, was largely ignored. Recent advances in ecological sampling now permit efficient use of detection–nondetection information to draw inferences about whether a species is present or absent, given the imperfect observation process that involves detecting the species at occupied locations (MacKenzie 2005; Royle and Dorazio 2008). In the case of occupancy data, there is no uncertainty about the occupancy state at locations where the species is detected, but at locations where the species is not detected the occupancy state is ambiguous (i.e., the species could be present or absent). Detectability has long been considered a key element in CR methods because of the need to assess uncertainty in the recapture process (i.e., not all marked individuals are recaptured). When a detection probability is not incorporated into the modeling of detection–nondetection information, the reliability of the inference breaks down because the relationship between the count (the number of locations where the species was detected) and the parameter of interest (occupancy) is not known. Instead, the naive estimates of occupancy (proportion of sites at which the species is detected) are negatively biased by some unknown but variable amount (Bailey et al. 2004). All occupancy estimation methods discussed in this chapter deal explicitly with this issue of non-detection and attempt to provide unbiased estimates of occupancy and related dynamic parameters (MacKenzie et al. 2006).

11.3 Model Framework, Assumptions and Analytical Options

11.3.1 Standard Models for Occupancy Estimation and Modeling Occupancy Dynamics

The occupancy estimation technique developed by MacKenzie et al. (2002, 2003) simultaneously estimates probabilities for detectability and occupancy in a maximum likelihood framework (but see Royle and Dorazio 2008 for a Bayesian treatment). The basic sampling scheme involves multiple visits to a randomly selected subset of sites within an area of interest. Detection–nondetection information is collected during each visit, and the visits occur during a short time interval when sites are closed to changes in the occupancy state (i.e., sites are either occupied or not during the sampling interval). We use ψ_i to denote the probability that site *i* is occupied by the target species, and p_{ij} is the probability of detecting the species at site *i* during the *j*th independent visit to the site. Dropping the *i* notation for each site, the probability of detecting the species at least once during *T* visits can be written as $p^* = 1 - \prod_{j=1}^{T}(1 - p_j)$. Accordingly, $(1-p^*)$ is the probability of failing to detect the species at an occupied site.

Detection–nondetection data are compiled for each sampled site, and model parameters are used to describe the stochastic processes that could have produced the data. For example, a site sampled on five occasions (visits) may have a detection history of '*00101*'. The probability of observing this history would be written as:

$$\Pr(00101) = \psi(1-p_1)(1-p_2)p_3(1-p_4)p_5 \tag{11.1}$$

A detection history consisting of all zeros, 00000, would have two possible explanations: either the site was occupied, but the species was not detected during any visit, or the site was unoccupied. Written as a mathematical expression we have the following:

$$\Pr(00000) = \psi(1-p_1)(1-p_2)(1-p_3)(1-p_4)(1-p_5) + (1-\psi) \tag{11.2}$$

The original 'single-season' model (MacKenzie et al. 2002) assumes the occupancy state of sampled locations is closed during the sampling 'season'; however, the occupancy state can change over time (i.e., between 'seasons') as one would expect due to significant events such as habitat enhancement or human disturbance, resulting in local colonization and extinction events.

MacKenzie et al. (2003) extended the 'single season' occupancy models to include two dynamic parameters: ε_t = the probability that an occupied site in season *t* becomes unoccupied in season $t+1$ (i.e., local extinction) and γ_t = the probability that an unoccupied site in season *t* is occupied by the target species in season $t+1$ (i.e., colonization). These 'multi-season' models still assume that sites are visited multiple times within a season, over a period during which the occupancy state at each site is static [notice this design resembles Pollock's (1982) robust design in CR studies]. Under this model, a site sampled three times during each of two seasons might yield a detection history of '*100 000*'. The target species was clearly present at the site during the first season, but two possibilities may explain the nondetection during the second season, specifically that the species is either present but not detected or the species went locally extinct:

$$\Pr(100000) = \psi p_{1,1}(1-p_{1,2})(1-p_{1,3})\left[(1-\varepsilon_1)\prod_{j=1}^{3}(1-p_{2,j}) + \varepsilon_1\right] \tag{11.3}$$

Here, detection probability may vary among seasons *t* or visits within seasons *j*, p_{tj}.

In addition to the closure assumption within seasons (no change in occupancy state within each season), the original occupancy models require that: (1) sites and detections are independent, (2) species are not misidentified, and (3) probabilities of occupancy and detection are constant across all sites within a stratum or can be modeled using covariates (MacKenzie et al. 2006). Modeling occupancy, extinction, and/or colonization as a function of covariates is often a primary focus of occupancy studies. Models incorporating different combinations of covariates represent competing hypotheses about factors believed to influence species occurrence and vital rates (extinction and colonization probabilities). These models are then fit to the data and evaluated via model selection procedures (e.g., Burnham and Anderson 2002). An alternative approach to estimation and modeling is to view the above models hierarchically under a Bayesian framework, and statistical inference is achieved using Markov chain Monte Carlo (MCMC) methods (e.g., Royle and Dorazio 2008). Using either procedure, it is possible to model occupancy, dynamic parameters, and detection probability as functions of measured covariates.

Several software programs are available for analyzing occupancy data including program PRESENCE (http://www.mbr-pwrc.usgs.gov/software/presence.html) and program MARK (White and Burnham 1999) for maximum likelihood estimates. R or WinBUGS code can be written for Bayesian estimation, and Royle and Dorazio (2008) provide sample WinBUGS code for occupancy modeling.

11.3.2 Abundance-Induced Detection Heterogeneity

As noted previously, occupancy estimation operates under the basic premise that heterogeneity in detection probability is adequately modeled via time- or site-specific covariates. In many systems, however, variation in local abundance can create heterogeneity in a species' detection probability that is difficult to model with conventional covariates (e.g., habitat features, Royle 2005; MacKenzie et al. 2006). Logically, the probability of detecting at least one member of a species will tend to increase with the species' local abundance. In study systems where local populations vary among sites, this phenomenon may lead to considerable heterogeneity in species detection probabilities, and failing to account for this heterogeneity will yield to underestimation of occupancy.

Royle and Nichols (2003) addressed abundance-induced heterogeneity directly by exploiting the relationship between abundance, detection and occupancy. Specifically, they noted that site-specific detection probability could be written as:

$$p_i = 1 - (1 - r)^{N_i} \tag{11.4}$$

where p_i is the probability of detecting at least one individual of the target species at site i, r is the individual detection probability, and N_i is the abundance at site i. If local abundance is unknown, but is likely closed (i.e., no change) during the sample period, it may be modeled with an appropriate parametric distribution.

Often, site-specific abundance is estimated using a Poisson distribution with mean λ, where λ may represent the average number of individuals per site. Under the assumption of the Poisson distribution, occupancy is a derived parameter, $\psi = \Pr(N > 0) = 1 - e^{-\lambda}$. If a species is absent for a large proportion of sites, a zero-inflated Poisson distribution may be more appropriate (Wenger and Freeman 2008), and the negative binomial distribution has also been considered, though this distribution is often difficult to fit (Royle and Nichols 2003; MacKenzie et al. 2006).

Abundance-induced heterogeneity may be important when animal populations exist at low densities that vary over the area of interest (e.g., N_i <10, MacKenzie et al. 2006). Such conditions existed for a camera trap study involving a suite of five ungulate species in Peru (Tobler et al. 2009, also see Sect. 11.6.2). Preliminary analyses confirmed high levels of heterogeneity in detection probability for all species among camera trap locations; thus, the authors used the Poisson model implemented in program PRESENCE to explore how habitat influenced mean abundance, λ, among different study areas and sampling seasons. Occupancy estimates were then derived for each habitat, study area, and time period (season) using the best supported model.

11.4 Study Designs for Occupancy Models

All occupancy models described in this chapter require some type of replication at sampled units (sites) over a time period where the occupancy state is closed. Investigators need to carefully consider what constitutes a 'site' and 'season' in their given biological systems (MacKenzie et al. 2006). For example, occupancy-based research on a single species warrants a different sampling approach compared to large-scale efforts to monitor species richness of entire communities. Although design of occupancy studies using camera traps will vary greatly depending on study objectives (see also Sect. 11.8 below), some of the more obvious questions to consider revolve around the camera system itself and the ecology of the target species. Definition of a site (or patch) selection, and allocation of sites across the area of interest (e.g., randomization, stratification) should be evaluated with respect to the study objectives and the home range size and habits of the target species. In camera trapping studies, sites may be naturally occurring, discrete patches of habitat (e.g., islands of habitat or salt licks, Terborgh et al. 2001; Tobler et al. 2009). More often, sites are arbitrarily defined as a camera trap location within quadrats of a specified size (e.g., 1 ha quadrats). The definition, and spacing, of sites relate to the model assumptions of closure and independence. If sites are defined as quadrats of habitat that are smaller than an individual's home range, or spacing among sites allows for a single individual to be detected at multiple sites within a season, then both assumptions are violated. Ironically, good camera trap design for estimation of abundance or density (as discussed in Chaps. 6–9) may violate key assumptions of occupancy estimation.

Design considerations should also include some thought about the number of visits or sampling occasions (e.g., days) needed to detect rare species (i.e., the length of a season), what species can be expected to be photographed (e.g., terrestrial versus arboreal), type of camera trap (e.g., active versus passive camera systems, see Swann et al., Chap. 3), and the environmental covariates an investigator deems important to the detection process. Investigators increase their chance of success if they think about these issues before sampling begins. Such issues are at the heart of the Wildlife Picture Index (WPI), an indicator effort recently proposed to monitor trends in tropical biodiversity (O'Brien et al. 2010). The WPI is a composite index based on the geometric mean of relative occupancy estimates scaled over time (Buckland et al. 2005). We suggest that practitioners of camera trap surveys interested in conducting occupancy studies consider a priori the issues above, otherwise use of indices such as the WPI are applied with unrealistic expectations.

Once definitions of site, season, and occasions per season have been established, there are various programs to aid investigators in determining how to best allocate effort among sites to achieve desired scientific or conservation objectives (e.g., Bailey et al. 2007). These programs are also useful at evaluating the impact of potential assumption violations (e.g., closure). Program GENPRES (http://www.mbr-pwrc.usgs.gov/software.html), a companion to program PRESENCE, and program MARK (White and Burnham 1999) both have flexible simulation capabilities to allow investigators to consider a wide variety of sampling designs utilizing any and all of the occupancy models described above. Well-defined scientific or conservation objectives are easily translated into models representing competing hypotheses about the status and dynamics of the study system. Information from pilot studies or other systems can be used to provide initial parameter estimates for exploring sampling design trade-offs (e.g., O'Connell et al. 2006). Results from study design evaluations can be extremely useful during the critical planning phase, prior to the initiation of field work. Additionally, such exercises can provide realistic expectations for managers and funding agencies and help evaluate monitoring programs (e.g., Mattfeldt et al. 2009).

11.5 Suggestions for Presenting Results of Occupancy Analysis

In this chapter we have focused on various likelihood-based models that can be fit to camera trap data yielding maximum likelihood estimates (see Royle and Dorazio 2008 for use of Bayesian estimation methods). Most studies will fit multiple models, representing competing hypotheses, and evaluate these hypotheses via model selection methods (e.g., Burnham and Anderson 2002). Several concise papers have provided practitioners with important guidance for presenting results of these types of analyses (e.g., Anderson et al. 2001; Anderson and Burnham 2002), but we emphasize a few basic points here in hopes of improving the information conveyed by camera trap studies that utilize occupancy estimation methods.

We expect that most surveys or studies using occupancy analysis will focus on how covariates (e.g., habitat type) influence occupancy, dynamic parameters, or detection probabilities. It is imperative that if such models are supported by the data, investigators report the effect size and a measure of its associated precision, or graph the relationship between the parameter of interest and the covariate(s). For example, if a species' occurrence is influenced by distance to the nearest road, stream, or amount of forested habitat, authors should graph the relationship or report the estimated effect size and direction (positive or negative) to adequately communicate results to readers and provide comparisons to other published studies.

It is well known that relationships between naïve estimates of occupancy probabilities and relevant covariates can be biased when species detection probabilities are <1 (Gu and Swihart 2004; MacKenzie et al. 2006). For example, parameters specifying occupancy–habitat relationships show the largest bias when detection probabilities are also related to the habitat variables of interest (Gu and Swihart 2004; MacKenzie et al. 2006: Fig 2.3). Assuming that detection probabilities are equal over sampled sites (e.g., $p(.)$ models) has the potential of producing biased results if species detection probabilities vary among covariates.

Finally, caution should be taken when reporting and interpreting an 'overall occupancy probability' (often denoted as $\hat{\bar{\psi}}$). If this estimate is obtained by averaging over site-specific occupancy estimates for each sampled site (as was the case in early versions of program PRESENCE), the estimate may not represent the occupancy probability of a randomly chosen site in the area of interest. The two quantities are equivalent only if sample sites are chosen via a simple random sample, thus assuring that the distribution of covariates in the sample is representative of the covariate distribution in the area of interest. If this is not the case (e.g., if sites are chosen via a stratified random sample), then the overall occupancy estimate should be calculated as a weighted average of stratum-specific estimates and the delta method can be used to provide a variance estimate (MacKenzie et al. 2006, p.121–122).

11.6 Occupancy Estimation with Camera Trap Data: Model Extensions

Camera traps provide detection–nondetection information for any species that can be photographed, along with a variety of ancillary information (e.g., behavior, age, and physical condition). Cameras can be deployed as the sole detection method or in combination with other methods in a manner that is consistent with the sampling framework previously described to yield data that are well suited for occupancy models. Estimates of occupancy are especially appealing when species cannot be uniquely identified, and these techniques have given new life to simple detection–nondetection data. Indeed, recent camera trapping studies have used the standard single-season occupancy model detailed above to establish baseline species' distributions in previously unsampled regions (Johnson et al. 2009), to explore the importance of forest structure or forest retention on the occupancy probability of various mammal species (Linkie et al. 2007; Baldwin and Bender 2008; McShea et al. 2009), and to evaluate the influence of human activities such as recreation on species occurrence and behavior (Zielinski et al. 2008).

To date no study has attempted to use camera trap data to estimate occupancy dynamics (vital rates) over time, but given that declines in geographic range are often used to evaluate a species status and assign risk categories (e.g., IUCN criteria), we suspect that occupancy dynamic models will be utilized in the future.

Next, we use several camera trap studies to illustrate extensions to the standard occupancy models that relax model assumptions or estimate additional parameters.

11.6.1 Multiple Methods and Multiple Scales

Camera traps are commonly used to detect multiple species, but no sampling device can detect all species with equal efficiency (Gompper et al. 2006; O'Connell et al. 2006; Long et al. 2007; Tobler et al. 2009). Increasingly, monitoring programs are being developed that target multiple species requiring the use of multiple methods (Manley et al. 2004; Mattfeldt et al. 2009). Under such designs, data from detection methods may simply be combined to indicate whether the target species was detected by at least one method, but researchers often want to test the effectiveness of the different detection methods or evaluate the influence that alternative detection methods can have on occupancy estimates (Bailey et al. 2004; O'Connell et al. 2006; Long et al. 2007).

Camera traps, cubby boxes (i.e., enclosed track plates) and hair snares were all included in sampling arrays used to estimate occupancy and detection probabilities for a variety of mammal species on Cape Cod, Massachusetts and seven other U.S. National Park Service lands in the northeastern USA (O'Connell et al. 2006; Nichols et al. 2008, e.g., Fig. 11.1). Nichols et al. (2008) capitalized on the use of multiple detection methods and extended the standard 'single-season' model to estimate occupancy probabilities at two spatial scales and compare detection probabilities associated with each of the detection methods. The approach permits occupancy estimation at a large spatial scale, corresponding to species' presence within a sample unit (1 ha) and, conditionally, at a smaller scale that corresponds to the species availability at the array or site (local occurrence or use, Fig 11.1). In addition, the authors found relatively strong evidence of temporal variation in detection probabilities for several species (e.g., striped skunks *Mephitis mephitis*) among different detection methods, including camera traps.

We believe that this model may be particularly useful for low-density, highly mobile species where the species may occupy a relatively large sample unit, but may not be near camera locations during a given sampling occasion. The use of additional detection methods, such as animal sign (e.g., hair, tracks or scat), may enhance camera trapping studies in the future allowing investigators to estimate species occurence at large-scale sample units and also local availability or use at trapping locations within occupied units. It should be noted that sign must be correctly assigned to species and a time period corresponding to camera trapping

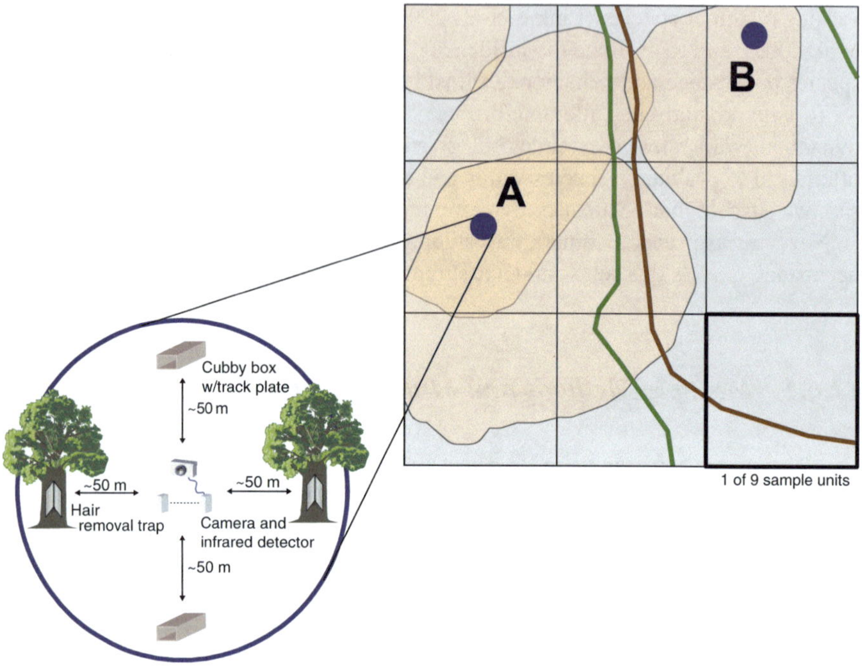

Fig. 11.1 Diagram representing hypothetical ranges for four striped skunks distributed among nine sample units (each grid cell is a sample unit). Two of the sample units (A and B) were randomly selected for sampling, and a detection array (sample station site) was randomly placed within each of these units. Each detection array consisted of a single camera trap at the centre and two track plates and two hair removal traps placed equidistant (~50 m) from the camera in the four cardinal directions (Reprinted with permission from John Wiley and Sons)

effort. In other words, if both camera traps and animal sign are surveyed, detections from each method should indicate a species occurrence (locally) over the same visit or survey occasion (Rhodes et al. 2010).

11.6.2 Species Co-Occurrence and Resource Partitioning

Tobler et al. (2009) used camera traps to explore patterns of spatial partitioning among ungulate species using a species co-occurrence model that allows for imperfect detection (MacKenzie et al. 2004). Expanding the standard single-season occupancy model to include two species, the true occupancy state at any given sample site now includes four mutually exclusive possibilities: occupied by both species ψ^{AB}, occupied by species A only $\psi^{A}-\psi^{AB}$, occupied by species B only $\psi^{B}-\psi^{AB}$, or unoccupied $1-\psi^{A}-\psi^{B}+\psi^{AB}$. If the two species are distributed independently one would expect $\psi^{AB}=\psi^{A}\times\psi^{B}$; however, the co-occurrence model (MacKenzie et al. 2004) includes a species interaction parameter γ where

$\psi^{AB} = \psi^{A} \times \psi^{B} \times \gamma$. Values of $\hat{\gamma} < 1$ suggest that the two species co-occur less often than expected, suggesting possible avoidance or competitive exclusion, whereas $\hat{\gamma} > 1$ would indicate that the species tend to occur more often than expected under the assumption of independence. A similar interaction parameter is also included to explore interaction in the detection process, or to test whether the detection probability of one species is influenced by the occurrence or detection of the other species (MacKenzie et al. 2004).

The five ungulate species studied by Tobler et al. (2009) exhibited considerable spatial overlap and $\hat{\gamma} \approx 1$ for most species comparisons indicating little spatial resource partitioning. These authors did not suspect interactions in the detection process and assumed that detection probabilities were not influenced by the presence of a conspecific, but this may not be true in many carnivore studies that involve highly territorial species. The two-species occupancy model has been used to investigate such relationships in systems where species interactions in both occurrence and the detection process are likely (e.g., Bailey et al. 2009).

11.7 Recent Advances

To our knowledge these advances have not yet been used in camera trapping studies, but we discuss them to serve as an impetus for the future.

11.7.1 Multistate Occupancy Models

The standard single and multi-season models have been extended to include different categories of occupancy (Royle 2004; Royle and Link 2005; Nichols et al. 2007; MacKenzie et al. 2009). Often these models are used to classify occupied sites by whether reproduction occurred at the sites (i.e., occupied sites either supported reproduction or not), but the definition of occupancy categories, or states, is very general and can include abundance categories (low or high relative abundance) or animal behaviors such as territoriality or seasonal migration. The model can even be used to model occupancy and habitat dynamics simultaneously (Martin et al. 2010).

11.7.2 Occupancy Models with Spatially Clustered Subunits

Although the original occupancy models were developed assuming temporal replication (multiple, independent visits to each site), a number of additional options have been proposed to achieve 'repeat surveys' including: (1) conduct multiple surveys within a single visit, but separate detections into multiple time periods (e.g. Nichols et al. 2008, p.1323) (2) use multiple independent observers during a single visit, or (3) conduct surveys at multiple spatial subunits within a larger sample unit. This latter method requires that the target species has a nonzero probability of being detected at all sampled subunits within occupied sites (this can be insured by sampling subunits with replacement): criteria that are not met in many study systems (Kendall and White 2009). Hines et al. (2010) developed a model to accommodate a sampling design based on spatial replicates (subunits) that are sampled only once and are expected to exhibit spatial dependence. The model was motivated by large-scale tiger sampling that relies on detection of sign along trails and roads, where it is likely that tiger sign is present on successive sections (subunits) of sampled trails or roads (Hines et al. 2010). Using successive sections of a trail as replicates likely violates the independence assumption of standard occupancy models, but the Hines et al. (2010) model decomposes the detection process into two components: the probability that the species is present at the subunit and the probability of detection, given the subunit is occupied. The species presence at, or use of, subunit t may be modeled as a function of the species occurrence in the previous subunit $t-1$ (i.e., as a first order Markov process) and relevant covariates. The model also provides a way of formally testing for this type of spatial dependence in the data.

To date this model has not been applied to studies involving camera traps, but since cameras are often placed at trail intersections (e.g., Tobler et al. 2009), these models may be useful in the future.

11.8 Concluding Remarks

Camera traps and occupancy estimation or modeling have each become well-established tools for sampling and making inferences about animal populations. We believe that they complement each other, allowing for time-efficient, cost-effective collection of quality data, often in the form of indisputable documentation that allows for strong inferences. The combination of using camera traps and occupancy models to generate unbiased estimates of animal occurrence represents significant progress in how we conduct ecological investigations. Although much of the recent work with camera traps and occupancy estimation has focused on mammals, the recent application of these techniques to other taxa further demonstrates the flexibility and power of modeling of detection–nondetection data with photographic images (e.g., Nomani et al. 2008; Winarni et al. 2009).

With this in mind, we believe it is important to end with a cautionary note about using these techniques. Investigators frequently overlook the steps necessary to conduct effective research or implement useful monitoring programs. First and foremost, we consider it essential that investigators give considerable thought to: (1) why they are sampling – either to advance science or inform conservation management decisions, (2) what it is they hope achieve (i.e., objectives), and (3) how they can best design a research or monitoring program to achieve the desired results. All too often, investigators rush into the field and begin sampling, bypassing these questions and thereby limiting the value of their work before they even begin. These concepts have been discussed in detail elsewhere (Nichols et al., Chap. 4, Nichols and Williams 2006), but we point them out here because they are essential to the design and implementation of useful sampling programs that target and collect animal occurrence information.

In an era where the risk of extinction for animal species is increasing and biodiversity assessments are sorely needed, quantitative methods such as occupancy estimation can be used to strengthen scientific inference that can benefit the management of animal populations. Unfortunately, the use of estimation methods to analyze camera trap survey data has been slow to develop (Conroy and Carroll 2009). Because occupancy estimation methods utilize data that are relatively easy to collect and which generate unbiased estimates, these methods have a variety of current applications, and we anticipate that occupancy estimation may soon become a preferred method for analysis of camera trap surveys in the future.

Acknowledgements We would like to thank Jim Nichols and Ullas Karanth for comments that improved earlier drafts of this chapter, and Mathias Kobler for allowing us to examine and discuss his datasets.

References

Anderson, D. R. and K. P. Burnham. 2002. Avoiding pitfalls when using information-theoretic methods. Journal of Wildlife Management 66:912–918

Anderson, D. R., W. A. Link, D. H. Johnson, and K. P. Burnham. 2001. Suggestions for presenting the results of data analyses. Journal of Wildlife Management 65:373–378

Azuma, D. L., J. A. Baldwin, and B. R. Noon. 1990. Estimating the occupancy of spotted owl habitat by sampling and adjusting for bias. USDA Technical Report PSW-124. Berkeley, CA

Bailey, L. L., T. R. Simons, and K. H. Pollock. 2004. Estimating site occupancy and detection probability parameters for terrestrial salamanders. Ecological Applications 14:692–702

Bailey, L. L., J. E. Hines, J. D. Nichols, and D. I. MacKenzie. 2007. Sampling design trade-offs in occupancy studies with imperfect detection: examples and software. Ecological Applications 17:281–290

Bailey, L. L., J. A. Reid, E. D. Forsman, and J. D. Nichols. 2009. Modeling co-occurrence of northern spotted and barred owls: accounting for detection probability differences. Biological Conservation 142:2983–2989

Baldwin, R. A. and L. C. Bender. 2008. Distribution, occupancy, and habitat correlates of American marten (*Martes americana*) in Rocky Mountain National Park, Colorado. Journal of Mammalogy 89:419–427

Bayley, P. B. and Peterson, J. T. 2001. An approach to estimate presence and richness of fish species. Transactions of the American Fisheries Society 130:620–633

Buckland, S. T., A. E. Magurran, R. E. Green, and R. M. Fewster. 2005. Monitoring change in biodiversity through composite indices. Philosophical Transactions of the Royal Society B 360:243–254

Burnham, K. P. and D. R. Anderson. 2002. Model selection and multimodel inference: a practical information-theoretic approach, Second edition. Springer, New York, NY, USA

Conroy, M. J. and J. P. Carroll. 2009. Quantitative conservation of vertebrates. Blackwell, West Sussex, UK

Gompper, M. E., R. W. Kays, J. C. Ray, S. D. LaPoint, D. A. Bogan, and R. J. Cryan. 2006. A comparison of noninvasive techniques to survey carnivore communities in Northeastern North America. Wildlife Society Bulletin 34:1142–1151

Gu, W. and R. K. Swihart. 2004. Absent or undetected? Effects of non-detection of species occurrence on wildlife-habitat models. Biological Conservation 116:195–203

Hines, J., J. D. Nichols, J. A. Royle, D. I. MacKenzie, A. M. Gopalaswamy, N. S. Kumar, and K. U. Karanth. 2010. Tigers on trails: occupancy modeling for cluster sampling. Ecological Applications 20:1456–1466

Johnson, A., C. Vongkhamheng, and T. Saithongdam. 2009. The diversity, status and conservation of small carnivores in a montane tropical forest in Northern Laos. Oryx 43:626–633

Kendall, W. L. and G. C. White. 2009. A cautionary note on trading spatial for temporal sampling in studies of site occupancy. Journal of Applied Ecology 46:1182–1188

Lancia, R. A., J. D. Nichols, and K. H. Pollock. 1994. Estimating the number of animals in wildlife populations. Pages 215–253 *in* T. A. Bookhout, editor. Research and management techniques for wildlife and habitats, Fifth edition. The Wildlife Society, Bethesda, MD

Link, W. A. and R. J. Barker. 2010. Bayesian inference: with ecological applications. Academic, San Diego, CA

Linkie, M., Y. Dinata, A. Nugroho, and I. A. Haidir. 2007. Estimating occupancy of a data deficient mammalian species living in tropical rainforests: sun bears in the Kerinci Seblat region, Sumatra. Biological Conservation 137:20–27

Long, R. A., T. M. Donovan, P. Mackay, W. J. Zielinski, and J. S. Buzas. 2007. Comparing scat detection dogs, cameras, and hair snares for surveying carnivores. Journal of Wildlife Management 71:2018–2025

MacKenzie, D. I. 2005. What are the issues with presence–absence data for wildlife managers? Journal of Wildlife Management 69:849–860

MacKenzie, D. I. and J. D. Nichols. 2004. Occupancy as a surrogate for abundance estimation. Animal Biodiversity and Conservation 27:461–467

MacKenzie, D. I. and J. A. Royle. 2005. Designing efficient occupancy studies: general advice and tips on allocation of survey effort. Journal of Applied Ecology 42:1105–1114

MacKenzie, D. I., J. D. Nichols, G. B. Lachman, S. Droege, R. A. Royle, and C. A. Langtimm. 2002. Estimating site occupancy rates when detection probabilities are less than one. Ecology 83:2248–2255

MacKenzie, D. I., J. D. Nichols, J. E. Hines, M. G. Knutson, and A. B. Franklin. 2003. Estimating site occupancy, colonization, and extinction when a species is detected imperfectly. Ecology 84:2200–2207

MacKenzie, D. I., L. L. Bailey, and J. D. Nichols. 2004. Investigating species co-occurrence patterns when species are detected imperfectly. Journal of Animal Ecology 73:546–555

MacKenzie, D. I., J. D. Nichols, J. A. Royle, K. H. Pollock, L. L. Bailey, and J. E. Hines. 2006. Occupancy estimation and modeling: inferring patterns and dynamics of species occurrence. Academic, New York, NY

MacKenzie, D. I., J. D. Nichols, M. E. Seamans, and R. J Gutierrez. 2009. Modeling species occurrence dynamics with multiple states and imperfect detection. Ecology 90:823–835

Manley, P. N., W. J. Zielinski, M. D. Schlesinger, and S. R. Mori. 2004. Evaluation of a multiple-species approach to monitoring species at the ecoregional scale. Ecological Applications 14:296–310

Martin, J. S., S. Chahamaillé-Jaames, J. D. Nichols, H. Fritz, J. E. Hines, C. J. Fonnesbeck, D. I. MacKenzie, and L. L. Bailey. 2010. Simultaneous modeling of habitat suitability, occupancy, and relative abundance: African elephants in Zimbabwe. Ecological Applications 20:1173–1182

Mattfeldt, S. D., L. L. Bailey, and E. H. Campbell Grant. 2009. Monitoring multiple species: estimating state variables and exploring the efficacy of a monitoring program. Biological Conservation 142:720–737

McShea, W. J., C. Stewart, L. Peterson, P. Erb, R. Stuebing, and B. Giman. 2009. The importance of secondary forest blocks for terrestrial mammals within an acacia/secondary forest matrix in Sarawak, Malaysia. Biological Conservation 142:3108–3119

Nichols, J. D. and K. U. Karanth. 2002. Statistical concepts; assessing spatial distribution. Pages 29–38 *in* K. U. Karanth and J. D. Nichols, editors. Monitoring tigers and their prey. A manual for managers, researchers, and conservationists. Centre for Wildlife Studies, Bangalore, India

Nichols, J. D. and B. K. Williams. 2006. Monitoring for conservation. Trends in Ecology and Evolution 21:668–673

Nichols, J. D., J. E. Hines, D. I. MacKenzie, M. E. Seamans, and R. J. Gutiérrez. 2007. Occupancy estimation and modeling with multiple states and state uncertainty. Ecology 88:1395–1400

Nichols, J. D., L. L. Bailey, A. F. O'Connell, Jr., N. W. Talancy, E. H. Campbell, E. H. C. Grant, A. T. Gilbert, E. M. Annand, T. P. Husband, and J. E. Hines. 2008. Multi-scale occupancy estimation and modeling using multiple detection methods. Journal of Applied Ecology 45:1321–1329

Nomani, S. Z., R. R. Carthy, and M. K. Oli. 2008. Comparison of methods for estimating abundance of gopher tortoises. Applied Herpetology 5:13–31

O'Brien, T. G., J. E. M. Baille, and M. Cuke. 2010. The wildlife picture index: monitoring top trophic levels. Animal Conservation. doi:10.1111/j.1469-1795.201000357.x

O'Connell, A. F. Jr., N. W. Talancy, L. L. Bailey, J. R. Sauer, R. Cook, and A. T. Gilbert. 2006. Estimating site occupancy and detection probability parameters for mammals in a coastal ecosystem. Journal of Wildlife Management 70:1625–1633

Pollock, K. H. 1982. A capture–recapture design robust to unequal probability of capture. Journal of Wildlife Management 46:752–757

Pollock, K. H., J. D. Nichols, T. R. Simons, G. L. Farnsworth, L. L. Bailey, and J. R. Sauer. 2002. Large scale wildlife monitoring studies: statistical methods for design and analysis. Environmetrics 13:105–119

Rhodes, J. R., D. Lunney, C. Moon, A. Matthews, and C. A. McAlpine. 2010. The consequence of using indirect sign that decay to determine species' occupancy. Ecography (in press)

Royle, J. A. 2004. Modeling abundance index data from anuran calling surveys. Conservation Biology 18:1378–1385

Royle, J. A. 2005. Site occupancy models with heterogeneous detection probabilities. Biometrics 62:97–102

Royle, J. A. and R. M. Dorazio. 2008. Hierarchical modeling and inference in ecology. The analysis of data form from populations, metapopulations, and communities. Academic, San Diego, CA

Royle, J. A. and W. A. Link. 2005. A general class of multinomial mixture models for anuran calling survey data. Ecology 86:2505–2512

Royle, J. A. and J. D. Nichols. 2003. Estimating abundance from repeated presence–absence data or point counts. Ecology 84:777–790

Terborgh, J., L. Lopez, P. Nuñez, M. Rao, G. Shahabuddin, G. Orihuela, M. Riveros, R. Ascanio, G. H. Adler, T. D. Lambert, and L. Balbas. 2001. Ecological meltdown in predator-free forest fragments. Science 294:1923–1926

Tobler, M. W., S. E. Carrillo-Percastegui, and G. Powell. 2009. Habitat use, activity patterns and use of mineral licks by five species of ungulate in south-eastern Peru. Journal of Tropical Ecology 25:261–270

Thompson, S. K. 2002. Sampling, Second edition. Wiley, New York, NY

Tyre, A. J., B. Tenhumberg, S. A. Field, D. Niejalke, K. Parris, and H. P. Possingham. 2003. Improving precision and reducing bias in biological surveys by estimating false negative error rates in presence–absence data. Ecological Applications 13:1790–1801

Vojta, C. 2005. Old dog new tricks: innovations with presence–absence information. Journal of Wildlife Management 69:845–848

Wenger, S. J. and M. C. Freeman. 2008. Estimating species occurrence, abundance, and detection probability using zero-inflated distributions. Ecology 89:2953–2959

White, G. C. and K. P. Burnham. 1999. Program MARK: survival estimation from populations of marked animals. Bird Study 46:120–139

Winarni, N. L., T. G. O'Brien, J. P. Carroll, and M. F. Kinnaird. 2009. Movements, distribution, and abundance of Great Argus Pheasants (*Argusianus Argus*) in a Sumatran rainforest. Auk 126(2):341–350

Zielinski, W. J., K. M. Slauson, and A. E. Bowles. 2008. Effects of off-highway vehicle use on the American marten. Journal of Wildlife Management 72:1558–1571

Chapter 12
Species Richness and Community Dynamics: A Conceptual Framework

Marc Kéry

12.1 Introduction

The study of animal communities has a long history in many branches of ecology, especially community ecology, biogeography and conservation biology. Furthermore, characterizing the size, composition and dynamics of animal communities is also important from a management perspective. For instance, community characteristics such as total size (i.e., species richness) or the size of certain subsets (e.g., number of rare or Red listed species) are often used to direct conservation efforts or to monitor their effectiveness. Camera traps can be used to study the size, composition and dynamics of animal communities, especially for large and medium-sized mammals and birds, terrestrial animals and particularly for nocturnal species. Although camera trap data can be treated in much the same way as data from other methods of sampling animal communities, it is particularly suited for capture–recapture (CR)-type analyses, given the ease with which discrete capture periods are defined. One important feature of camera trap data as used for community inference is that the surveyed communities are typically not very large. Hence, the inferential challenges caused by the possible presence of a very large number of very rare or elusive species (Mao and Colwell 2005) are presumably greatly alleviated.

Regardless of the method used to record species in animal communities, observations are usually not perfect, i.e., not every species present in a community will be detected and not every species occurring will be detected at each site and point in time where it occurs (Kéry 2002; Schmidt 2005). This holds even when studying only segments of communities, such as certain guilds or taxonomic groups. Treating species counts as if they were the genuine numbers of species present is equivalent to making the assumption that every species present in a study area is indeed detected, or to put more strongly, it means to assume that *each individual* of each species present is detected perfectly. Under binomial sampling, the probability of detecting a species at a site (P) is a function of the probability p to detect each

M. Kéry (✉)
Swiss Ornithological Institute, 6204, Sempach, Switzerland
e-mail: marc.kery@vogelwarte.ch

A.F. O'Connell et al. (eds.), *Camera Traps in Animal Ecology: Methods and Analyses*, DOI 10.1007/978-4-431-99495-4_12, © Springer 2011

single one among N individuals, $P = 1 - (1 - p)^N$ (Royle and Nichols 2003), and P can only be 1 when $p = 1$.

On the other hand, treating species counts just as indices of relative species richness is equivalent to assuming that the proportion of species detected is the same, on average, over all dimensions of comparison, i.e., that the expectation of detection probability, $E(P)$, is constant. When comparing species richness across space, habitat types, along an elevational gradient or over time, the use of raw species counts assumes that there is no correlation between the proportion of detected species any of these variables, respectively. "On average" does not mean that the proportion of species detected must remain exactly constant along these dimensions. Rather, this proportion may vary, but it must not do so in relation to the factors that are the focus of interest with respect to species richness.

Both assumptions (i.e., $P = 1$ and $E(P) = 1$) are hardly ever stated explicitly and much less tested, although the first assumption must clearly be wrong in the field in all but exceptional cases. Even the second assumption is at least questionable and should be tested if possible. Surprisingly though, most managers as well as scientists in community ecology, biogeography, conservation biology and related disciplines have been remarkably resistant to an explicit recognition of either assumption and of the consequences of violating them.

Arguably, a better approach than making untested or unstated assumptions is to explicitly account for imperfect detection when making inference about animal communities. In spite of only relatively few applications in ecology and management, the formal estimation of species richness and related quantities, corrected for imperfect detection, has a fairly long history in statistical ecology and has been subject to a number of reviews, including Bunge and Fitzpatrick (1993), Colwell and Coddington (1994), Nichols and Conroy (1996), Williams et al. (2002, pages 555–573), Chao (2005), also see O'Brien et al., Chap. 13.

At least four broad classes of approaches have been used to account for unseen species in community studies (Dorazio et al. 2006): (1) extrapolation of species accumulation curves (Soberón and Llorente 1993; Gotelli and Colwell 2001), (2) parametric models for the apparent species abundance distribution of those species that were detected (Pielou 1977, pages 269-290), (3) nonparametric models based on sampling theory (Bunge and Fitzpatrick 1993) and (4) community analogs of closed population CR models for estimation of abundance, where species take the place of individuals (Otis et al. 1978; Burnham and Overton 1979).

Here, I summarize one line of research on the estimation of the size, composition and dynamics of communities belonging to category 4 along with some associated software, recognizing that other approaches (e.g., Chao 1987, 2005; Chao and Lee 1992; Chao et al. 2006; Mao and Colwell 2005) and software (e.g., EstimateS; http://viceroy.eeb.uconn.edu/estimates) exist. The reason for my choice is that CR models have the attractive property of being based on an explicit recognition of the observation process that relates the true system state, e.g., species richness, to our observations of the system, e.g., a species count. Thus, arguably, the CR framework provides the most mechanistic representation of the system that produced that species count. Further, it is perhaps the most flexible and general approach, since by a mere redefinition of the unit of analysis (e.g., individuals, occupied areas or species), one moves from the inference about a population to that about species distributions and finally to an animal community or a metacommunity. This generality lets one take useful concepts and models from a fairly large branch of research in statistical ecology, represented for instance by treatises such as Seber (1982), Williams et al. (2002) and Royle and Dorazio (2008), and simply carry them over to the community case.

Capture–recapture methods, in the widest sense, are based on the observations of a system (here, one or more communities) at repeated points in time (often termed occasions), and the observation or non-observation of its individually recognizable units (here, species) provides the information required to infer both features of the focal system (e.g., size, composition and dynamics of the community) *and* features of the observation process, represented by detection probability and possibly other nuisance parameters, such as temporary emigration probabilities. Note that as defined here, CR models include site occupancy models (MacKenzie et al. 2002).

There are only two places where the analogy between individuals in a population and species in a community break down, reincarnation and individual heterogeneity (Nichols et al. 1998a). First, when a species goes extinct in a community, recolonization is possible, but the same is not possible when an individual dies in a population. Consider a detection history *101*, indicating detection at occasion 1 and 3 and nondetection at occasion 2. For an individual, the internal zero may only be associated with the events "alive but nondetected" or perhaps "alive and temporarily emigrated." However, if this is the detection history for a species at a particular site, it may also mean that the species became extinct and then recolonized the site. In order to resolve this ambiguity the modeling of open communities requires data in the format of a robust design (Pollock 1982; Williams et al. 2002, pages 523-554). In the robust design, samples are taken at two temporal scales and changes may occur among primary sampling occasions but not among the secondary sampling occasions that are nested within the primary occasions. The subsampling at the second level of this design provides the information about detection probability which is required to distinguish temporary emigration, or extinction followed by recolonization, from nondetection.

Second, species are much more different than are individuals that belong to the same species. The detectability P of a species at a site is related to its local abundance N by the well-known relationship $P = 1-(1-p)^N$ for the case where individuals are detected individually (Royle and Nichols 2003). Since species' abundances in a community may differ by orders of magnitude, different species are likely to be detected with different probabilities. In addition, there is variation among individuals of different species in coloration, behavior, social status, sex, age, habitat use, physiological or many other state variables, and these will all contribute further to variation in detection probability. Heterogeneity among species in detection probability has been demonstrated many times, including for avian (Boulinier et al. 1998a, b; Kéry and Royle 2008a, b; Kéry and Schmidt 2008) and butterfly communities (Dorazio et al. 2006; Kéry and Plattner 2007; Kéry et al. 2009).

A classic result in CR, in the community context, is that unmodeled species-specific heterogeneity will lead to a negative bias of N, i.e., of community size (Dorazio and Royle 2003), and this will partly be propagated into the estimates of other quantities such as colonization and extinction. Depending on the magnitude of that heterogeneity, estimates of species richness and other community descriptors can be severely biased. Therefore, in a community context, by default any estimation method should account for heterogeneous species detectability.

In addition, species heterogeneity in detection has a further intriguing consequence: it makes it very difficult or impossible to obtain a representative sample of a community. The sample of species observed will virtually always be biased towards the more detectable species and in many cases seriously so. This is an added reason for why it is useful to base inference about communities on an explicit accounting of the detection process. Otherwise, the inference will be biased almost certainly towards the common and more visible members of the community, and these may not represent the entire community well (Kéry et al. 2008). For instance, owing to their smaller population size, rare species likely experience higher extinction rates. Since they are probably also the ones that are most often missed in the sample, the sample extinction rate will be a biased estimate with respect to the entire community about which inference is sought (Nichols et al. 1998a; Alpizar-Jara et al. 2004).

In principle, any robust design CR-type of model that allows for individual detection heterogeneity is a candidate for inference about animal communities. However, the choice of how to model heterogeneity is not trivial and there are many types of models to account for it, e.g., non-parametric methods (Burnham and Overton 1979; Chao 1987), finite-mixture distributions (Norris and Pollock 1996; Pledger 2000) or continuous mixture distributions such as the beta-binomial and the logistic-normal (Coull and Agresti 1999; Dorazio and Royle 2003). Unfortunately, resulting inferences may not always concur across models. Furthermore, data-based criteria such as the deviance or Akaike's Information Criterion (AIC) cannot be trusted as an aid to that choice (Dorazio and Royle 2003). Indeed, Link (2003) has shown that N, here defined as community size, is an unidentified parameter across classes of such heterogeneity models.

What does this mean in practice? It means that if we use two different ways of specifying the heterogeneity among species in detection probability and they give a different answer, we have no way to let the data tell us which one is the better answer (for some examples of this, see Dorazio and Royle 2003 and Link 2003). However, the conclusion that we should not use heterogeneity models at all would mean "throwing the baby out with the bathwater": both in principle and from empirical studies we *know* that species differ widely. And we also know that unmodeled heterogeneity can severely bias low inference about community size N and related parameters. Hence, not to use a heterogeneity model just because we don't know which one is the most appropriate and instead use a model with constant detection probability (across species) will most probably result in inference of much of poorer quality than using the "wrong" heterogeneity specification. In addition, heterogeneity models have performed well in several studies where population size was known (e.g., Greenwood et al. 1985; Manning et al. 1995; Pledger 2000; Conn et al. 2006).

However, it is certainly advisable to eliminate as much individual heterogeneity as possible at the design stage or by using explicit covariates that can explain differences between species in the analysis stage. It has also been argued that there are so many factors that might induce heterogeneity among species that the result might be best represented by a smooth distribution for detection and that continuous mixtures might be more natural than finite-mixtures and therefore be preferred (Dorazio and Royle 2003). However, this view has been challenged (Pledger 2005), and it is true that in the name of parsimony an "unnatural" model specification may indeed be sometimes useful.

So at present, it seems wise to tentatively use the class of mixtures that appears best in a particular case, as well as to eliminate as much heterogeneity as possible in both the design and the analysis stages of a study. Furthermore, high detection probabilities alleviate the problem of divergent inferences (Link 2003), so a high search effort always pays – see later section entitled "*Design Considerations.*"

A useful taxonomy of the inferential situations encountered in the study of animal communities is provided by combining the two dichotomies, single-site *vs.* multiple sites and static (also, single "season") *vs.* dynamic (also, multiple "seasons"). Table 12.1 shows some of the quantities that may be of interest in each of the four resulting situations. Note that the quantities for the dynamic situations (cells 2 and 4, right column) include those for the associated static cells (1 and 3 (left column), respectively). By season is meant a time period that is sufficiently short that the studied community does not change or hardly so, i.e., can be assumed closed (Kendall 1999). Evidently, what is "short" will depend completely on the dynamics of the studied community. For instance, the duration over which a community can be considered closed will be much longer for mammals or birds than for insects. As

Table 12.1 A taxonomy of inferential situations in animal communities showing some quantities of interest. All quantities shown for the static situations may also be of interest in the associated dynamic situation for each point in time. A season is defined as a time period within which the community does not change, i.e., is closed

	Static (single season)	Dynamic (multiple seasons)
Single site	(1) Occurrence (z_i)	(2) Survival rate (ϕ_t)
	Species richness ($N = \sum z_i$)	Extinction rate ($\varepsilon_t = 1 \;\; \phi_t$)
	Community integrity	Colonization rate (γ_t)
		Trend ($T_t = N_{t+1} / N_t$)
		Turnover rate
Multiple sites	(3) Occurrence (z_{ij})	(4) Survival rate (ϕ_{jt})
	Occupancy rate (ψ_i)	Extinction rate ($1 - \phi_{jt}$)
	Finite-population occupancy rate (ψ_i^{fp})	Colonization rate (γ_{jt})
	Local species richness (N_j)	Local trend ($T_{jt} = N_{j,t+1} / N_{j,t}$)
	Global species richness (N)	Global trend (T_t)
	Similarity among species	Turnover rate
	Similarity among sites	
	Species accumulation	

Notation: N species richness, *z* presence/absence indicator, *i* index for seen and unseen species, *j* index for space, *t* index for time (seasons)

noted elsewhere (Royle and Kéry 2007), "dynamic" is a more precise term than "multi-season" since multi-season data could be analyzed simply by treating season as a group and not specifying any relationships between the estimated quantities across seasons. Furthermore, the former emphasizes the model whereas the latter emphasizes the design.

The following overview is organized around Table 12.1. I will proceed from single-site to multi-site inferential situations and, within that, from the static to the dynamic case. Interestingly, historically, the methodological development in the field has progressed in the same way. Following that, I will consider survey design issues and finally offer some outlook. Throughout, I will avoid most technical details and rather mention the main idea of a model and then point the reader to the relevant primary literature.

12.2 Inference About Single Sites

12.2.1 Static Community at a Single Site

The simplest possible inferential situation is a single static community (Table 12.1, cell 1) in which we have an observed species list of length *C* (for count) and at least two surveys providing replicate observations for each species, being zero for a species not detected on that particular occasion (replicate survey) and 1 for a species that was detected. The main inferential quantity of interest in this situation is species richness, or alternatively, the number of species present in the community that did not make it onto the list. On the usual assumption that those species that were detected are informative about those that were not, one may estimate the number of unseen species, and therefore, total species richness using any reasonable CR estimator.

The customary CR framework as applied to this inferential situation requires temporal replicate observations of a community. However, the framework has also been applied for spatial replicates (Nichols et al. 1998a, b; Boulinier et al. 1998a, b; Cam et al. 2002a, b; Doherty et al. 2003; Dorazio and Royle 2003). While this is a creative idea, it is important to recognize that the approach confounds detection heterogeneity with small-scale heterogeneity in the occurrence of species. Finally, note that spatial sampling must be done with replacement in order to provide the best chance of obtaining an unbiased estimate (Kendall and White 2009).

Key to inference in closed population CR is getting an adequate model for the patterns in species-specific detection probability, p. A useful taxonomy of patterns in detection probability in the context of models for closed populations is due to Otis et al. (1978). They distinguish three kinds of effects on p: individual heterogeneity h, behavioral response b and time t. The associated models are denoted M_h, M_b and M_t, respectively, and the first specifies unstructured differences among individuals, the second differences in detection according to whether an individual is captured for the first time or was captured previously, and the last differences among capture occasions. Pairwise combinations and one three-way combination also exist, as does the case where p is constant; this is called model M_0. One or several estimators for most of these models were implemented in the program CAPTURE (Otis et al. 1978; also see http://www.mbr-pwrc.usgs.gov/software/capture.html), though not, for instance, for the most complex model M_{tbh}. CAPTURE can be run most easily nowadays by a call from within program MARK (White and Burnham 1999; http://welcome.warnercnr.colostate.edu/~gwhite/mark/mark.htm).

As argued before, a model for species richness by default should include individual heterogeneity. Another effect that may potentially be important to include is behavior (*b*), that is, the tendency for detection probability to depend on whether a species has been detected previously. This may occur because of the species or because of the observer. If a species is very elusive and represented by only a few individuals, then these may learn from the negative effects of being captured and behave more elusively afterwards, leading to a negative trap response; that is, recapture rates that are lower than first capture rates. It is possible to model temporal change in this capture effect also, for instance, the effect of a first capture on later capture probabilities may last for just one capture occasion (immediate trap response) or for all subsequent occasions (permanent trap response). Other patterns are possible, though the closed models in CAPTURE implement a permanent trap response. Positive trap responses are also possible, mostly mediated by a change in the knowledge of an investigator: for instance, once he knows where to watch and how to search, the probability to detect some species may increase. Behavioral response may be an important effect to consider for community modeling, but probably much less so for the "blind" and relatively unobtrusive camera traps. One exception might be if traps are moved around in a study area until certain species are detected leading to a situation where a likely positive trap response should be accounted for in order to obtain adequate inference about the community. Variation in detection probability among occasions (effect *t*) is always likely to occur to some degree and should be accounted for if it is important.

One of the most widely used estimators for species richness has been the jackknife estimator for model M_h (Burnham and Overton 1979; Boulinier et al. 1998a; also see Chap. 13). It is based on capture frequencies f_i, i.e., the number of species captured i times, and the aim is to extrapolate to f_0, the number of species never detected, and has worked reasonably well in many applications (see examples cited in Boulinier et al. 1998a). One drawback is that this application requires completely balanced data in the number of replicate surveys (i.e., the number of actual detections does not need to be the same for each species). Furthermore, no effects of additional covariates can be modeled. Also, its performance with small samples (e.g., 2–3 occasions) has not been studied well, although it has yielded apparent sensible results also in such situations (Kéry and Schmid 2004, 2006; Kéry and Plattner 2007). There is also a limiting form of the jackknife that can be applied to apparent species abundance data from a single occasion (Burnham and Overton 1979), and that is implemented in the repository of software maintained by J.E. Hines at the Patuxent Wildlife Research Center (see program SPECRICH, http://www.mbr-pwrc.usgs.gov/software/specrich.html).

One of the drawbacks of the estimators contained in the program CAPTURE is that not all of them are likelihood-based and, therefore, they cannot all be compared in a common currency, for instance using the deviance or using AIC (Burnham and Anderson 2002). This problem has been addressed by Norris and Pollock (1996) and Pledger (2000), who described a modeling framework for closed populations based on finite-mixtures. That is, to account for individual heterogeneity, species are assumed to belong to one of a few groups (typically just two or three) with a constant detection probability for individuals within each group. The detection parameters to be estimated are the constant detection probability for each group and the mixture proportions (i.e., the proportion of species in that group) for all but one group. The proportion of that one group is given by 1 minus the mixture proportions of all the others.

Individual heterogeneity in detection may be a consequence of many different factors and so is almost surely continuous rather than discrete. Hence, the idea to represent it by a finite-mixture distribution may strike one as unnatural. Nevertheless, finite-mixture models have compared quite well in comparisons with populations of known size (Pledger 2000). Therefore, it seems that the "trick" of attributing species in one of a small number of virtual groups may be useful in eliminating a fairly large part of the individual heterogeneity. In addition to individual heterogeneity, it is possible to specify time and behavior effects, hence, maximum likelihood estimates for all combinations of effects in the taxonomy of Otis et al. (1978) can be obtained, including for model M_{tbh}, making this framework potentially very useful. These models can all be implemented in program MARK. Alternatively, they could be implemented as individual covariate models in a Bayesian mode of inference and using program WinBUGS (Spiegelhalter et al. 2003; Royle 2009).

Another way to specify individual heterogeneity is by a continuous rather than by a finite mixture distribution. The beta-binomial distribution is a continuous distribution on the interval (0, 1) and thus appears a natural candidate to specify heterogeneity in a probability. Another one is the logistic-normal distribution (Coull and Agresti 1999), which specifies normally distributed species-specific effects on the logit-transform of detection probability. Dorazio and Royle (2003) compare three classes of heterogeneity specifications, finite-mixtures, and beta-binomial and logistic-normal continuous mixtures, and find them all useful for estimating species richness. (However, they note the existence of non-identifiability problems across these classes.) Continuous mixture distributions can be fitted using maximum likelihood as do Dorazio and Royle (2003) or in a Bayesian mode of inference using program WinBUGS (Spiegelhalter et al. 2003). For code examples see the volume by Royle and Dorazio (2008).

For a static community at a single site, a comparison of species richness (N) among different sites or times may be achieved by first using a regional pool of species that could potentially occur in that particular community (Karr 1990; Cam et al. 2002a, b). If such information is available, constructing a ratio of the number of species occurring and the number of potentially occurring species may be used as a measure of ecological integrity for that community (Table 12.1). This is a metric of obvious use for management applications (Cam et al. 2002a, b).

This idea provides one of the leads to the use of site occupancy models for inference about species richness. Site occupancy models are a fairly recent class of models (MacKenzie et al. 2002, 2003, 2006; also see O'Connell and Bailey, Chap. 11) that estimate, in their original form, the proportion of sites, e.g., sample quadrats or potential territories of territorial species, that are occupied, corrected for imperfect detection, based on replicate surveys of at least some of these sites. Actually, site occupancy models are a very general class of models and can be viewed as providing a unification of many different models in the CR literature (see Royle and Dorazio 2008).

Interestingly, there is a duality between models used to estimate the size of a closed population, where some individuals are overlooked, and site occupancy models, where all "potential individuals" (sites) are seen, but some are just not recognized to be "occupied." Indeed, asymptotically, both models are equivalent (Although note that individual covariates are usually observed in the site occupancy context which creates great inferential benefits.). This recognition has given rise to the important notion that classical closed population CR models can be reparameterized by zero-inflation and applied to a zero-augmented version of the original data set. That is, one can add to the data set an arbitrary number of all-zero detection histories and fit a site occupancy model, where the occupancy parameter takes the place of community size. This data augmentation yields great computational advantages and permits extension of the idea of the site occupancy model to a great many other CR models, both open and closed (Royle et al. 2007; also see Royle and Dorazio 2008).

In addition, by redefining a site to be a species among a list of species that could potentially occur in a region, one can then directly use a site occupancy model to estimate "relative species richness" or community integrity (Karr 1990; Cam et al. 2002a,b), corrected for imperfect detection, i.e., species richness relative to some baseline list of species that may be thought to represent a regional pool of species present.

Site occupancy models can be fit using programs MARK and PRESENCE 2 (http://www.mbr-pwrc.usgs.gov/software/doc/presence/presence.html). In both programs, heterogeneity in detection probability among species (or sites) can be incorporated by finite mixtures (Pledger 2000). My own limited experience with finite mixtures in a previous version of PRESENCE suggests that the numerical optimization for heterogeneity in occupancy models may not be sufficient (i.e., that the algorithm often does not converge). In that case, one is stuck with an M_0 version of an occupancy model whereby inference will typically underestimate species richness, although estimates of extinction probability may be less affected (Alpizar-Jara et al. 2004).

Royle (2006) describes site occupancy models with heterogeneity, specifically, using finite mixtures and beta-binomial and logistic-normal continuous mixtures as well as the Royle–Nichols formulation of abundance-induced heterogeneity (Royle and Nichols 2003). He fits these models using maximum likelihood, and Royle and Dorazio (2008, book and website) provide some example code in program R (R Development Core Team 2008) and also for Bayesian implementations in WinBUGS.

Perhaps the most flexible, general, and to a biologist, accessible way of estimating species richness in the single site static case is by fitting a logistic-normal site occupancy model in program WinBUGS (Spiegelhalter et al. 2003) to a zero-inflated version of the species detection data. Thus, in what is essentially a logistic regression, species detections can be modeled as coming from a Bernoulli distribution, and the logit-transform of the Bernoulli parameter (detection probability) can be expressed as the sum of both fixed and random effects. The former could be known species attributes such as approximate population size or body size, and the latter are zero-mean normal deviates to take account of unstructured heterogeneity in detection among species. The other effects in the catalog of Otis et al. (1978), time and behavior, or their additive combinations or interactions, can equally be specified in this linear predictor in the flexible fashion of a linear model. Example code to fit site occupancy models in WinBUGS can be found in Royle and Kéry (2007), Royle and Dorazio (2008), and Kéry (2010), as well as in the volume of ecological examples distributed with the program OpenBUGS (A. Thomas, St. Andrews, pers. comm.).

Finally, although the CR framework allows one to obtain a rigorous estimate of the likely number of species that were not detected, it does not identify those species. This could only be done in an informal way. For instance, when one estimates that four species were present that were never observed, and experienced naturalist may then draw up a list of species that might be expected in the studied community and, by comparison with the actually detected species, guess at what species might have been the four. Alternatively, in a site occupancy estimation context with a regional species list, the value of the occurrence indicators z_i (see Table 12.1) could be estimated and may give some additional insight into which species were most likely missed.

12.2.2 Dynamic Community at a Single Site

Next we consider extensions to more than one season, that is, to a single, dynamic community (Table 12.1, cell 2). In addition to the quantities shown in cell 1 of the table, several other relevant metrics may now be defined and become estimable provided adequate data are available. Foremost among these are the community analogs of vital rates, survival and fecundity rate, in population dynamics. That is, species survival rates ϕ_t, or alternatively, species extinction rates $\varepsilon_t = 1 - \phi_t$, and species colonization rates γ_t, where t indexes the intervals between successive seasons, t and $t+1$. Furthermore, turnover is a function of both survival and colonization and is useful to characterize the stability of a community. Finally, the community analog of the population growth rate λ is called T_t in Table 12.1.

Nichols et al. (1998a) appear to be the first to develop an estimation framework for dynamic ecological communities with imperfect detection. Their sampling situation is across two seasons and is assumed to be that of the so-called robust design (Williams et al. 2002, pages 523-554), where samples are replicated at two temporal scales. Between the primary sampling occasions (i.e., seasons) the community is open but among the nested secondary (i.e., within-season) sampling occasions, the community is assumed closed. Importantly, the subsampling within each primary occasion is required to distinguish nondetection from extinction followed by recolonization. Based on the Jackknife estimator of species richness for each primary sampling period, Nichols et al. developed estimators for species richness in each primary period and for survival, colonization and turnover rates as well as for the trend in species richness between years. They define turnover rate as the probability that a randomly selected species in year $t + 1$ is a species that was not part of the community in year t, i.e., is a new species. Hines et al. (1999) developed program COMDYN that implements this scheme for two seasons.

This seminal framework has allowed powerful insights into animal communities while fully correcting for imperfect detection (e.g., Boulinier et al. 1998b, 2001; Lekve et al. 2002; Doherty et al. 2003; Kéry and Schmid 2004) and has been adapted to the spatial case (Nichols et al. 1998b) for comparing two communities at a single point in time; see later. However, there are also obvious limitations at least in the COMDYN implementation, for instance the restriction to a single estimator for community size (sometimes another model than M_h may be adequate), the restriction to fully balanced data (in some communities, the number of replicate observations may not be the same for all species, see later), the limitation to two seasons (often, data are available from more, and sometimes many more, seasons). Also, species identity is not retained in the analysis leading to a loss of efficiency, because detections of species A in year 2 will probably be informative also about its detectability in year 1, yet this information is not exploited. Furthermore, COMDYN is often used in a two-step fashion when comparisons across multiple sites are desired (e.g., Doherty et al. 2003). This is not ideal, but the estimation framework of COMDYN would not easily allow a generalization to multiple sites with possible spatial correlation.

MacKenzie et al. (2006) suggested that dynamic site occupancy models could be used for inference about dynamic communities provided that a comprehensive species list, one which would include all species that would possibly occur at the study site, could be defined. Indeed, a site occupancy model applied to robust design data from samples of a single community over time provides a very useful framework to analyze such data (see also Chap. 13). However, it would seem that the tremendous differences among species in a community in detection probability would have to be accounted for in that model in order for it to provide useful estimates of species richness; otherwise, at least species richness is likely to be greatly underestimated. The bias engendered in the other parameters (colonization, extinction/survival, and turnover) is not entirely obvious, though it may be acceptable for extinction/survival (Alpizar-Jara et al. 2004).

Dynamic site occupancy models can be fitted fairly easily using the Bayesian software WinBUGS (Spiegelhalter et al. 2003; for WinBUGS code see upcoming volume of ecological examples; A. Thomas, St. Andrews, pers. comm.) where the inclusion of species-specific random effects in detection probability is fairly straightforward (Royle and Kéry 2007). The hierarchical, or state-space, formulation of the model in WinBUGS appears to be very flexible and can easily be extended to contain more effects for any of its primary biological parameters (first-year occurrence, and extinction and colonization rates for each time interval), for instance, to account for the Otis et al. (1978) taxonomy of effects or for further covariates. In addition, finite-population inference is trivial in a Bayesian implementation, for instance, an estimate of species richness at a studied site along with its uncertainty (SE, CI) is easy to obtain (Royle and Kéry 2007; Link and Barker 2010).

12.3 Inference About Multiple Sites

12.3.1 Static Metacommunity

A metacommunity is composed of a collection of communities. Even in the static case, there is a multitude of potentially interesting quantities to characterize a metacommunity, see Table 12.1, cell 3. The fundamental quantity from which all others can be derived is the species- (i) and site- (j) specific occurrence indicator z_{ij} for every species in the metacommunity, be it observed or not. These z_{ij} can be collected in a matrix, Z, with rows denoting potentially occurring species and columns the sites. This Z matrix is also called the presence–absence matrix and has been termed the fundamental unit of analysis in biogeography and community ecology (McCoy and Heck 1987).

The Bernoulli parameter governing the random variable z_{ij} equals the population occupancy rate ψ_i for species i, while finite-population occupancy rate ψ_i^{fp} for species i is estimated by the proportion among all studied sites at which z_{ij} equals 1 (indicating the occurrence of species i at site j). Local species richness at site j is simply the sum of z_{ij} over all species at site j, and global species richness can be similarly derived. Furthermore, the similarity of, or conversely, the difference between, species and sites, respectively, can be expressed by the proportion of sites where two species co-occur or the proportion of species that two sites co-host, respectively. Species accumulation is the relationship between the cumulative number of distinct species and the number of spatial, and sometimes, temporal, sampling units.

To account for the imperfect observation process in inference about communities at multiple sites, several approaches are possible, and I will present three: (1) site-by-site analysis followed by a second-step analysis, (2) the spatial analog of the robust design approach by Nichols et al. described in the previous section and (3) the novel multi-species site occupancy model of Dorazio and Royle (2005).

First, the simplest inference about a metacommunity is obtained by site-wise application of any method described for the single-site static case. For instance, numerous authors have applied the jackknife estimator independently at a collection of sites in a first step, and then in a second step, analyzed resulting estimates of species richness, $\hat{N}$, or other quantities such as mean species detection probability ($\hat{P}$, computed by the ratio $C/\hat{N}$) in a linear model to study the relationships between these quantities and various explanatory variables. Examples include Boulinier et al. (1998a), Doherty et al. (2003), Kéry and Schmid (2004, 2006) and Jiguet et al. (2005). This approach is not wrong, but it is likely to be inefficient. For instance, detections of species A at site 2 may be informative about its detection probability also at site 1, but a piecewise approach does not take species identity into account. Furthermore, when treating estimates from the first analysis step as data in the second step, their uncertainties are hard to take account of. If they aren't, this will lead to too short confidence intervals overall and to tests that are too liberal. On the other hand, a second-step analysis that properly takes into account the full variance–covariance matrix of all estimates from the first step would compromise most of the attractive simplicity of the two-stage approach (Link 1999).

As an aside, Cam et al. (2002a) showed how typical species accumulation data can be used in conjunction with a sampling-based model, the removal model, to estimate species accumulation curves free of the distorting effects of imperfect detection. Cam et al. (2002b) show that detection error leads to a bias in the observed species accumulation curves: their slope is exaggerated when imperfect detection is not accounted for.

Second, for the special case of two sites compared, Nichols et al. (1998b) developed a spatial analog of their framework for modeling the temporal dynamics of a single community, described in Nichols et al. (1998a) and in the section above. Using robust design data, they develop, among others, estimators for species richness at each site, relative species richness (the spatial analog to trend between two successive years), the number of species occurring at one but not at the other site and *vice versa* and the number of shared species. Program COMDYN (Hines et al. 1999) can be used for the computations. Again, although this rigorous estimation framework was a great advance compared with the then existing methods, it suffers from the same drawbacks as were described for the temporal dynamic case: limitation to two populations compared, use of the Jackknife estimator and the requirement of balanced data. The extension of the framework as implemented in COMDYN to more than two sites is not obvious.

Third, Dorazio and Royle (2005; see also Dorazio et al. 2006 and Royle et al. 2007) developed a multi-species extension of the site occupancy model (MacKenzie et al. 2002). Based on species detection data that are replicated both spatially (i.e., at multiple sites) and temporally at each site (i.e., observations are made on ≥ 2 occasions for at least some of the sites), this model essentially provides an *estimate of the true presence–absence matrix* Z. The model thus corrects for the fact that some species present in the metacommunity, of which the actually sampled sites form a sample, do not occur on the sampled sites, and also for species that do occur on the actually sampled sites but were never detected. Having an estimate of z_{ij}, i.e., for the occurrence of every species i, seen or unseen, at each of the sampled sites j, then allows one to obtain detectability-corrected estimates of all the metrics shown in Table 12.1 (cell 3): occupancy for each species, local species richness (for instance at each sampled site or for regions containing several sites) or global species richness, similarity among sites in terms of the occurring species or among species in terms of coinhabited sites, as well as species accumulation curves that are corrected for detection bias (Cam et al. 2002a, b) and that do not depend on the order in which samples are added up (Dorazio et al. 2006). This model provides an extremely powerful framework for analyzing animal communities, both in terms of aggregate traits (such as species richness) and in terms of individual traits (such as occurrence of individual species). For example applications, see Kéry and Royle (2008, 2009) and Kéry et al. (2008). Interestingly, Gelfand et al. (2005) independently developed a similar model. One of the main differences is that they do not extrapolate inferences to species that were never seen at all.

To illustrate this modeling framework, I have summarized the study by Kéry and Royle (2008). We had data from a subset of 26 quadrats in a national breeding bird monitoring program, where each quadrat was surveyed six times in a single breeding season thought to contain a closed community. In fact, there were two observers that each conducted three surveys, but I will ignore some details and present only those features of the study that are essential for illustration of the model. A total of 103 avian species were detected. One of the questions was about the proportion of species that are detected on average.

The observed data are the binary indicators x_{ijk} that are equal to 1 for a species i (i = 1...103) that was detected on occasion k (k = 1...6) in quadrat j (j = 1...26) and equal to 0 for a species that was not detected. We fitted a minimal version of the Dorazio and Royle model, where we assumed that the only effect on detection probability was species identity; thus, we fitted the equivalent of model M_h of Otis et al. (1978), but in a site occupancy context. Since we were not interested in any effects that varied by occasion k, we collapsed the data over that dimension and modeled a two-dimensional data array: x_{ij}, the number of occasions (out of 6) that species i was detected in quadrat j. A conceptualization of the model as applied to our case is shown in Table 12.2, where the fully observed data are shown in dark gray.

The basic idea is to formulate the model in terms of a latent binary process which indicates the presence (z_{ij} = 1) or absence (z_{ij} = 0) of species i in quadrat j. The realizations of this process z_{ij} are only partially observed, i.e., latent, since we do not know whether the value of z_{ij} is 0 or 1 at sites where a species was not observed, that is, where $x_{ij} = 0$. Hence, the values of z_{ij} that correspond to observations with $x_{ij} = 0$ can be viewed as missing observations (NAs) and the aim of the analysis is to impute them (Table 12.2; light gray multi-column rectangle).

The model described by Royle et al. (2007) and also by Royle and Dorazio (2008, pages 379-400) is a hierarchical or state-space model. That is, it is composed of a nested sequence of distributional assumptions that describe the true, but partially latent biological process and the imperfect observation process, conditional upon the realization of the biological process. The observation model for the observed detection frequencies x_{ij} in Kéry and Royle (2008a) is a simple logistic regression,

$$x_{ij} \sim Bin(6, z_{ij} * p_{ij}),$$

where detection frequencies are binomially distributed with sample size equal to the number of occasions, here 6, and where success probability is equal to the product of the realization of the latent occurrence process, z_{ij}, and detection probability p_{ij}. Our model assumes no false positive errors, i.e., species can only be overlooked but not misidentified (an important assumption, see later, "Design considerations"). Hence, when species i occurs at site j and z_{ij} = 1, x_{ij} is binomial with success probability equal to p_{ij}, and when the species does not occur and z_{ij} = 0, x_{ij} is a structural zero.

The model for the latent biological (occurrence) state, z_{ij}, is described by another logistic regression:

$$z_{ij} \sim Bin(1, \psi_{ij}).$$

Table 12.2 Concept of the multi-species site occupancy model of Dorazio and Royle (2005) in the data augmentation parameterization described by Royle et al. (2007) as applied to the data analyzed by Kéry and Royle (2008)

		Observed: x_{ij}					Only partially observed: z_{ij} and w_i					
	Quadrat j	1	2	3	…	26	1	2	3	…	26	w_i
Species i	1	6	3	0	…	4	1	1	NA	…	1	1
	2	0	0	1	…	2	NA	NA	1	…	1	1
	3	3	0	2	…	0	1	NA	1	…	NA	1
	…	…	…	…	…	…	…	…	…	…	…	…
n	103	0	0	1	…	0	NA	NA	1	…	NA	1
n+1	104	0	0	0	…	0	NA	NA	NA	…	NA	NA
	…	…	…	…	…	…	…	…	…	…	…	…
N	?	0	0	0	…	0	NA	NA	NA	…	NA	NA
N+1	?+1	0	0	0	…	0	0	0	0	…	0	0
	…	…	…	…	…	…	…	…	…	…	…	…
	…	…	…	…	…	…	…	…	…	…	…	…
M	203	0	0	0	…	0	0	0	0	…	0	0

The fully observed data (*dark gray shaded rectangle*) contain x_{ij}, the detection frequencies of 103 observed species at 26 quadrats. To these were added 100 detection histories containing only zeroes (the data augmentation part, represented by the *medium gray shaded rectangle below the dark gray rectangle*). The model enables inference about two latent structures, the partially observed true presence–absence matrix Z (i.e., the N-by-26 matrix containing z_{ij}) and the "super-population indicators" w_i (both represented in *light gray shading*). Essentially, the aim of the modeling is to impute the missing values (NA).

Thus, the occurrence of species *i* at site *j* is modeled as a Bernoulli process (a binomial with sample size 1, corresponding to a single coin flip) with occurrence (or occupancy) probability ψ_{ij}. When we assume that occurrence probability is constant across sites and only varies by species, as we did (Kéry and Royle 2008), this becomes $z_{ij} \sim Bin(1, \psi_i)$.

This model can be fitted in a fairly complicated way, as represented by the original model formulation in Dorazio and Royle (2005). It turns out, however, that the fitting of the model is greatly simplified by data augmentation (Royle et al. 2007): to the observed data (dark gray in Table 12.2) are added an arbitrary number of all zero detection histories (medium gray in Table 12.2). In our actual analysis we added 100 "pseudo" or "potential" species, bringing the total dimension of the observed data matrix to $M = 203$ by 26. To these augmented data we fit a reparameterization of the original model, where an additional hierarchical layer was added. This additional layer can be thought of as modeling the availability of species *i* in a larger pool of species (or a super-population of species) and the realizations from this process determine whether a species is part of the sampled community represented by the 26 quadrats or not. Thus, whether or not a species occurs in the sampled metacommunity is another Bernoulli random variable denoted by w_i and governed by the "inclusion" probability Ω: when $w_i = 1$, species *i* occurs in the metacommunity associated with our 26 quadrats, while when $w_i = 0$ it does not. The problem of estimating metacommunity size N is then converted to the equivalent problem of estimating Ω. (Note that the expected value of N is equal to $M*\Omega$.)

Consequently, the hierarchical model now has three levels (note the slight change of the state process compared to above) and can be written in three conditional, that is, dependent, probability statements:

1. Superpopulation process: $w_i \sim Bern(\Omega)$
2. State process (occurrence): $z_{ij} \sim Bern(w_i * \psi_i)$
3. Observation process (detection): $x_{ij} \sim Bin(6, z_{ij} * p_{ij})$

This model could also be described as a three-level, hierarchical, non-standard generalized linear mixed model (GLMM) or alternatively, as a non-standard random effects logistic regression (Kéry 2010).

However, the model as currently described has too many parameters to be useful and some constraints need to be introduced. Assuming that the occupancy and the detection probabilities of the species in the metacommunity are not independent, but stochastically dependent in the form of belonging to some common probability distribution, reduces the complexity of this model and makes it useful for inference. This means that we are making a random-effects assumption about the species-specific effects in occupancy and detection. In our case, we made the customary assumption that the logit transforms of the two parameters are random draws from a normal distribution whose mean and variance we estimate. Specifically, for occupancy we assumed $\text{logit}(\psi_i) = \alpha_i$ and $\alpha_i \sim \text{Normal}(\mu_\alpha, \sigma_\alpha^2)$ and for detection $\text{logit}(p_i) = \beta_i$ and $\beta_i \sim \text{Normal}(\mu_\beta, \sigma_\beta^2)$. (For problems of convergence induced frequently by this logistic-normal specification of heterogeneity in WinBUGS and a workaround see Kéry and Royle 2009). Importantly, it is precisely this random-effects assumption about the effects on occurrence and detection of all species in the metacommunity, i.e., seen or unseen, which constitutes the extrapolation from the seen species to the unseen ones.

We note furthermore that this is a remarkably parsimonious description of a metacommunity and the associated detection process: the entire variation in occurrence and detection of all species, detected and undetected, is described by just five structural parameters in the hierarchical model: the super-population inclusion probability Ω (serving the role of community size) and the two normal means μ_α and μ_β and their associated variances σ_α^2 and σ_β^2. See Kéry and Royle (2008a) for further details and results. One of the results of this study was to point out that a great advantage of the Dorazio-Royle community model is its spatial integration. Thus, compared to a site-wise application of Burnham's jackknife, the integrated model provides much more precise estimates (in terms of their SE) and in addition, the integrated model provided many fewer unreasonable estimates.

As is customary for a random-effects GLM, this model can be extended in a modular way as needed and as warranted by the available data. For instance, it is likely that there exists a correlation between ψ_i and p_i (see Dorazio et al. 2006 for an explanation and an example) and this can be easily accommodated by assuming a bivariate normal distribution for the logit transforms of these parameters, see Kéry and Royle (2009). Further, the linear predictors for either occupancy or detection could be extended to include other effects, either random or fixed. Fixed effects could include habitat effects on occurrence or detection-relevant effects such as season or time of day on detection, see Kéry and Royle (2009) or Kéry et al. (2008) for examples. Importantly, when covariates that vary by occasion are included, the observed data must be modeled in their original, uncollapsed format; that is, we will then model the three-dimensional array as x_{ijk}, and the observation process will be described by a Bernoulli rather than by a binomial distribution with sample size > 1, as above (see Kéry and Royle 2009 for WinBUGS code examples). Further, random site effects could be included on occupancy and these could be made spatially dependent by imposing some distance-related correlation function to obtain a geostatistical multi-species site occupancy model for a metacommunity. It is one of the key advantages of hierarchical models that such extensions are conceptually easy and can be added in a fully modular way (Royle and Dorazio 2008). R and WinBUGS code examples to fit some of these models can be found in Dorazio et al. (2006), Kéry and Royle (2009), Royle and Dorazio (2008), and Zipkin et al. (2009).

One interesting aspect of the modeling of a metacommunity concerns the main community metric, *N*. Much of biogeography deals with explaining large-scale variation in species richness. So how would one model relationships between *N* and covariates such as habitat descriptors in the framework of a multi-species site occupancy model? It appears that this is awkward, since *N* is not a primary parameter but a derived quantity, the sum of z_i over all species. An obvious *ad hoc* tactic alternative would be to estimate *N* for each study site and then model these estimates in a second analysis as a function of the covariates of interest. However, doing this while properly accounting for the variance–covariance of the estimates would be extremely difficult and normally, such analyses are conducted in a way that ignores the uncertainty in the estimates from the first analysis. Another, more integrated, approach would be to apply the hierarchical N-mixture model (Royle 2004) for estimation of population, or here, community size, from spatially and temporally replicated counts of species. This approach would integrate the information across the spatial samples but would not retain species identity, so might be slightly less efficient that a multi-species site occupancy model.

So the direct modeling of species richness under the multi-species site occupancy model is not intuitive. Perhaps, the important thing is this: the organic construction of the model as a representation of an animal metacommunity and its observation focuses one's thinking on the actual process that generates the observed patterns of species richness, that is, the Bernoulli random process governed by occupancy probability of each species, including its relationship with covariates. At the basis of any observed association between species richness and, say, a habitat covariate is exactly the collection of all species-specific relationships between occupancy probability, ψ_i, and that covariate. This is perhaps a good example for one of the great heuristic benefits of hierarchical models; that they force us into a more mechanistic way of thinking.

Camera traps are typically used to study medium and large carnivores and carnivore communities seldom contain many species, even at locations with relatively high biodiversity. In any case, community size *N* is unknown and can be estimated under the multi-species site occupancy model. The main advantage of using this modeling framework may reside in the fact that species "borrow strength" from each other. Estimates for species with very sparse data may be greatly improved in this way (see Nichols et al. 2008 and Zipkin et al. 2009).

12.3.2 Dynamic Metacommunity

The final inferential situation in the classification scheme depicted in Table 12.1 concerns the modeling of spatial and temporal variation in an animal community. This contains some of the most recent methodological developments and so this section will be less well developed than the previous ones. I will mention two approaches, *ad hoc* two-step analyses and dynamic multi-species site occupancy models for robust design data.

First and obviously, some static community quantities such as richness or dynamic ones such as the vital rates and turnover could be estimated at each of a collection of sites using any of the methods previously described. Resulting estimates could then be plugged into a second analysis and modeled as a function of spatial or other covariates that describe each single community. In spite of the drawbacks mentioned earlier, this would seem to be almost the only approach that is accessible to a biologist at present.

Second, the multi-species site occupancy model described in the previous section is currently being extended to the dynamic case. There are at least two ways in which the modeling of the dynamics of metacommunities could be approached: temporal covariate and Markovian, or temporal dependence, models (see Royle and Dorazio 2008, Sect. 12.4.). For both cases, data are required in the robust design format, i.e., there must be at least two replicate observations during a period of closure within each of multiple seasons.

The simpler and "cheap" way to model the dynamics of a metacommunity is by using some temporal covariate that is informative about the occurrence of species. We chose this approach in the context of a Swiss butterfly metacommunity inhabiting 13 sites that was observed over an entire season (April–September; see Kéry et al. 2009). There were 4–7 primary sampling occasions and two nested secondary sampling occasions. Butterflies, like many other animal or plant species in a seasonal environment, have rather well defined flying periods that can be approximated by some simple function of time. To model the occurrence of each species in the dynamic metacommunity over the entire season, we fitted the multi-species site occupancy model to the observations from all primary sampling occasions and related the occurrence of each species in each primary occasion to season (e.g., expressed as the number of days since April 1) in a species-specific quadratic function. This allowed us to obtain estimates of the total size of the butterfly community and especially also estimates of the number of species that occurred in each community over the entire season. In addition, we obtained detection-corrected phenology curves for each species as well as estimates of detection probability for each species.

Since strong temporal correlations of occurrence are typical for many species in seasonal environments, we expect this modeling framework to be useful in many situations. Code examples for running these models in WinBUGS are given in Kéry et al. (2009). Obviously, depending on the biological questions and the amount of data available, all the typical extensions could be considered again, such as additional covariates for occurrence or detection, random-effects modeling or correlated random effects.

The temporal covariate model does not explicitly model the temporal changes in the metacommunity in a mechanistic way, that is, as a function of species appearances (colonization, recruitment) and disappearances (extinctions). However, this can be done, and the most general description of a dynamic metacommunity is the multi-species extension of the dynamic, single-species site occupancy model described by MacKenzie et al. (2003) of which Royle and Kéry (2007) described a hierarchical version implemented in WinBUGS. Again, species-specific parameters can be collected together using random-effects assumptions; this is, what represents the extrapolation from the seen to the unseen part of the communities. A model of this type has been recently developed for the seasonal butterfly metacommunity previously described and can be implemented in WinBUGS (Dorazio et al. 2010; see also Russell et al. 2009).

This concludes our overview of methods of inference about communities and metacommunities under imperfect observations of some of their members. It is likely that especially for the dynamic metacommunity case, much progress is to be expected in the coming years.

12.4 Design Considerations

To obtain data that can be analyzed with methods described in this chapter, many points need to be considered and some of them are very general. As initial advice, I like the forceful simplicity of the prescription given by Yoccoz et al. (2001) for monitoring studies, which applies equally to any study of a wild population or community: first, take an adequate spatial sample, and second, take detection probability into account.

Regarding the former, without some sort of probabilistic selection of sites, formal inference to a broader statistical "population" cannot be made using the laws of probability (Thompson 2002; Thompson 2004; Kéry and Schmidt 2008). Unfortunately, the commandment of a random or other probabilistic spatial sample is violated in a large majority of ecological field studies as well as in many monitoring programs (Pollock et al. 2002). With regard to the latter (detection), application of an estimation framework as described in this chapter takes this issue into account.

In the following, we briefly discuss issues that are relevant for the proper application of these models. Since several methods described are based on site occupancy models, the papers by MacKenzie and Royle (2005) and Bailey et al. (2007) are certainly very relevant. The latter paper also describes a software program, GENPRES, which is useful for studying questions about the design of occupancy studies.

Almost all models described assume a closed community at some point in time, and repeated observations of that community provide the information required to model false negative errors in the observation process, i.e., detection probability. Closure can be attained by limiting the temporal spacing of secondary samples in a robust design. Sometimes it may have to be achieved by discarding some information. For instance, bird communities that are sampled repeatedly over an entire breeding season (mid-April to mid-July) may have some late-arriving migrants that were not yet available during the first one or two replicate visits. Hence, the resulting zeroes do not contain information about the binomial detection process but rather constitute structural zeroes. One of the easiest ways to deal with that is by simply deleting the associated observations (but only for those species) or equivalently, turning them into missing values (Kéry and Royle 2009). This is possible in the more modern estimation frameworks based for instance on site occupancy models, but it wasn't for the jackknife estimator which requires balanced data. An alternative would be to model the seasonal availability of these species along the lines of the butterfly work described in the section on dynamic metacommunities (Kéry et al. 2009).

Rota et al. (2009) have developed a formal hypothesis test of the closure assumption used in occupancy models and show how to compute analytically the power of detecting violations of closure as a function of sample size, and extinction and colonization probabilities. They present results for both the standard sampling protocol and for the "removal" sampling protocol wherein each site is only sampled until first detection of the species.

Population estimation in the presence of individual detection heterogeneity is difficult (Link 2003) and yet heterogeneity is a law of life. Reducing detection heterogeneity between species is useful and ultimately allows for more robust inferences. This can be done by standardization at the design stage or by use of some covariates that can "explain away" some of the differences between species. Also, the discrepancies in inferences under different heterogeneity formulations described by Link (2003) are reduced for higher detection probability, so a high survey effort pays dividends.

All of the CR-based models described in this chapter assume that the observation process contains only false negative errors, and this assumption is of course always violated to some degree: in most situations, misidentifications, representing false positive errors, can occur and lead to the detection of a species that is, in reality, not "there". False positive errors in a site occupancy context have been dealt with by Royle and Link (2006). Their results generate concern since they show that even small false positive error rates can contaminate the data and lead to remarkable bias in estimates of occupancy. As is intuitive, this bias is even increased with increasing number of observations. Incorporating false positives (misclassification) in CR models is an active area of research, but a difficult one. At present, it appears important to try and eliminate false positives to the extent possible, for instance by using well-trained personnel. And most importantly, it is highly beneficial to discard any doubtful records. This will tend to increase the number of false negative errors and hence lower detection probability, but of course, our models deal with this very well. However, discarding doubtful cases will also tend to clean up our data from false-positives, with which our models cannot really deal in a satisfactory way at present.

CR consists of representing in a model the main features in detection probability. In general, the more structure there is, i.e., the more parameters are needed for describing the observation process, the less precise are our inferences about the community. Therefore, it is advantageous to eliminate as many effects on detection as possible. For instance, if possible, try to avoid behavioral effects by using an appropriate design such as truly independent replicate observations. Adding behavioral effects into a model, even though it can be done technically, usually greatly inflates the uncertainty about parameter estimates. Similarly, reducing variation in detection over time pays.

Several choices have to be made regarding sample size: number of sites (communities), number of species (if subsets of a community are studied) and number of temporal replicates within seasons (over which closure is assumed). In an occupancy setting, some of these questions can be valuably informed by the simulation facilities provided in the programs PRESENCE and GENPRES. In general, simulation is the best tool to study required sample sizes to attain a desired level of precision of the estimates or to gauge the effects of the violation of model assumptions.

I have stressed the importance of accounting for species-specific heterogeneity in detection probability when applying an estimation framework in a community context. Typically, only few (if any) temporal replicates are available, e.g., 2 or 3, and it may be asked rightly whether this is enough to model species heterogeneity in detection probability. Otis et al. (1978) always recommend at least five replicates in order to discriminate and fit even single-effect models. More research on minimal required sample sizes would be welcome. However, it is likely that in integrated models, where integration means combined analysis of similar data over many spatial or temporal replicates such as in the Dorazio and Royle (2005) model, the combination of the information about detection across sites (in the static case) and primary occasions (in the dynamic case, cf. butterfly study) is likely to pay great dividends in this respect.

In a practical application, not only should model assumptions be met as well as possible, but a design should be used that achieves a certain precision weighed against some maximum cost. For example, the removal sampling protocol could be useful in reducing the number of temporal replicates at each site. Much remains to be learned about achieving optimal designs, and the aforementioned papers by MacKenzie and Royle and Bailey et al. and their associated recommendations and software are expecially relevant, as is the speculation made by Dorazio et al. (2006), about the importance of sequential sampling designs. Sampling design in community studies is another area of research where we can expect much progress in the future, either from theoretical investigations or from the comparison of empirical studies under different designs.

Acknowledgements I thank Jim Nichols as well as Bob Dorazio, Andy Royle, Allan O'Connell and Elise Zipkin for very valuable comments to this chapter.

References

Alpizar-Jara, R., J. D. Nichols, J. E. Hines, J. R. Sauer, K. H. Pollock, and C. S. Rosenberry. 2004. The relationship between species detection probability and local extinction probability. Oecologia 141:652–660

Bailey, L. L., J. E. Hines, J. D. Nichols, and D. I. MacKenzie. 2007. Sampling design trade-offs in occupancy studies with imperfect detection: examples and software. Ecological Applications 17:281–290

Boulinier, T., J. D. Nichols, J. R. Sauer, J. E. Hines, and K. P. Pollock. 1998a. Estimating species richness: the importance of heterogeneity in species detectability. Ecology 79:1018–1028

Boulinier, T., J. D. Nichols, J. E. Hines, J. R. Sauer, C. H. Flather, and K. P. Pollock. 1998b. Higher temporal variability of forest breeding bird communities in fragmented landscapes. Proceedings of National Academy of Sciences USA 95:7497–7501

Boulinier, T., J. D. Nichols, J. E. Hines, J. R. Sauer, C. H. Flather, and K. P. Pollock. 2001. Forest fragmentation and forest bird dynamics: inference at regional scales. Ecology 82:1159–1169

Bunge, J. and M. Fitzpatrick. 1993. Estimating the number of species: a review. Journal of the American Statistical Association 88:364–373

Burnham, K. P. and D. R. Anderson. 2002. Model selection and multimodel inference: a practical information theoretic approach, Second edition. Springer, New York

Burnham, K. P. and W. S. Overton. 1979. Robust estimation of population size when capture probabilities vary among animals. Ecology 60:927–936

Cam, E., J. D. Nichols, J. R. Sauer, and J. E. Hines. 2002a. On the estimation of species richness based on the accumulation of previously unrecorded species. Ecography 25:102–108

Cam, E., J. D. Nichols, J. E. Hines, J. R. Sauer, R. Alpizar-Jara, and C. H. Flather. 2002b. Disentangling sampling and ecological explanations underlying species-area relationships. Ecology 83:1118–1130

Chao, A. 1987. Estimating the population size for capture–recapture data with unequal catchability. Biometrics 43:783–791

Chao, A. 2005. Species estimation and applications. Pages 7907–7916 *in* N. Balakrishnan, C.B. Read, and B. Vidakovic, editors. Encyclopedia of statistical sciences, Vol. 12, Second edition. Wiley, New York

Chao, A. and S.-M. Lee. 1992. Estimating the number of classes via sample coverage. Journal of the American Statistical Association 87:210–217

Chao, A., R. L. Chazdon, R. K. Colwell, and T.-J. Shen. 2006. Abundance-based similarity indices and their estimation when there are unseen species in samples. Biometrics 62:361–371

Colwell, R. K. and J. A. Coddington. 1994. Estimating terrestrial biodiversity through extrapolation. Philosophical Transactions of the Royal Society B: Biological Sciences 345:101–118

Conn, P. B., A. D. Arthur, L. L. Bailey, and G. R. Singleton. 2006. Estimating the abundance of mouse populations of known size: promises and pitfalls of new methods. Ecological Applications 16:829–837

Coull, B. A. and A. Agresti. 1999. The use of mixed logit models to reflect heterogeneity in capture–recapture studies. Biometrics 55:294–301

Doherty, P. F., Jr., G. Sorci, J. A. Royle, J. E. Hines, J. D. Nichols, and T. Boulinier. 2003. Sexual selection affects local extinction and turnover in bird communities. Proceedings of National Academy of Sciences USA 100:5858–5862

Dorazio, R. M. and J. A. Royle. 2003. Mixture models for estimating the size of a closed population when capture rates vary among individuals. Biometrics 59:351–364

Dorazio, R. M. and J. A. Royle. 2005. Estimating size and composition of biological communities by modeling the occurrence of species. Journal of the American Statistical Association 100:389–398

Dorazio, R. M., J. A. Royle, B. Söderström, and A. Glimskär. 2006. Estimating species richness and accumulation by modeling species occurrence and detectability. Ecology 87:842–854

Dorazio, R. M., M. Kéry, J. A. Royle, and M. Plattner. 2010. Models for inference in dynamic metacommunity systems. Ecology (in press)

Gelfand, A. E., A. M. Schmidt, S. Wu, J. A. Silander, Jr., A. Latimer, and A. G. Rebelo. 2005. Modelling species diversity through species level hierarchical modelling. Applied Statistics 54:1–20

Gotelli, N. J. and R. K. Colwell. 2001. Quantifying biodiversity: procedures and pitfalls in the measurement and comparison of species richness. Ecology Letters 4:379–391

Greenwood, R. J., A. B. Sargeant, and D. H. Johnson. 1985. Evaluation of mark-recapture for estimating striped skunk abundance. Journal of Wildlife Management 49:332–340

Hines, J. E., T. Boulinier, J. D. Nichols, J. R. Sauer, and K. P. Pollock. 1999. COMDYN: software to study the dynamics of animal communities using a capture–recapture approach. Bird Study 46(suppl.):S209–S217

Jiguet, F., O. Renault, and A. Petiau. 2005. Estimating species richness with capture–recapture models: choice of models when sampling in heterogeneous conditions. Bird Study 52:180–187

Karr, J. R. 1990. Biological integrity and the goal of environmental legislation: lessons for conservation biology. Conservation Biology 4:244–250

Kendall, W. L. 1999. Robustness of closed capture–recapture methods to violation of the closure assumption. Ecology 80:2517–2525

Kendall, W. L., and G. C. White. 2009. A cautionary note on substituting spatial subunits for repeated temporal sampling in studies of site occupancy. Journal of Applied Ecology 46:1182–1188

Kéry, M. 2002. Inferring the absence of a species – A case study of snakes. Journal of Wildlife Management 66:330–338

Kéry, M. 2010. Introduction to WinBUGS for ecologists: a Bayesian approach to regression, ANOVA, mixed models and related analyses. Academic Press, Burlington, MA

Kéry, M. and M. Plattner. 2007. Species richness estimation and determinants of species detectability in butterfly monitoring programs. Ecological Entomology 32:53–61

Kéry, M. and J. A. Royle. 2008. Hierarchical Bayes estimation of species richness and occupancy in spatially replicated surveys. Journal of Applied Ecology 45:589–598

Kéry, M. and J. A. Royle. 2009. Inference about species richness and community structure using species-specific occupancy models in the national Swiss breeding bird survey MHB. Pages 639–656 *in* D. L. Thomson, E. G. Cooch, and M. J. Conroy, editors. Modeling demographic processes in marked populations, series: environmental and ecological statistics, Vol. 3. Springer, Berlin

Kéry, M. and H. Schmid. 2004. Monitoring programs need to take into account imperfect species detectability. Basic and Applied Ecology 5:65–73

Kéry, M. and H. Schmid. 2006. Estimating species richness. Calibrating a large avian monitoring program. Journal of Applied Ecology 43:101–110

Kéry, M. and B. R. Schmidt. 2008. Imperfect detection and its consequences for monitoring for conservation. Community Ecology 9:207–216

Kéry, M., J. A. Royle, and H. Schmid. 2008. Importance of sampling design and analysis in animal population studies: a comment on Sergio et al. Journal of Applied Ecology 45:981–986

Kéry, M., J. A. Royle, M. Plattner, and R. M. Dorazio. 2009. Species richness and occupancy estimation in communities subject to temporary emigration. Ecology 90:1279–1290

Lekve, K., T. Boulinier, N. C. Stenseth, J. Gjosaeter, J.-M. Fromentin, J. E. Hines, and J. D. Nichols. 2002. Spatio-temporal dynamics of species richness in coastal fish communities. Proceedings of the Royal Society B: Biological Sciences 269:1781–1789

Link, W. A. 1999. Modeling pattern in collections of parameters. Journal of Wildlife Management 63:1017–1027

Link, W. A. 2003. Nonidentifiability of population size from capture–recapture data with heterogeneous detection probabilities. Biometrics 59:1123–1130

Link, W. A. and R. J. Barker. 2010. Bayesian inference with ecological applications. Academic Press, London

MacKenzie, D. I. and J. A. Royle. 2005. Designing occupancy studies: general advice and allocating survey effort. Journal of Applied Ecology 42:1105–1114

MacKenzie, D. I., J. D. Nichols, G. B. Lachman, S. Droege, J. A. Royle, and C. A. Langtimm. 2002. Estimating site occupancy rates when detection probability rates are less than one. Ecology 83:2248–2255

MacKenzie, D. I., J. D. Nichols, J. E. Hines, M. G. Knutson, and A. B. Franklin. 2003. Estimating site occupancy, colonization, and local extinction when a species is detected imperfectly. Ecology 84:2200–2207

MacKenzie, D. I., J. D. Nichols, J. A. Royle, K. P. Pollock, L. L. Bailey, and J. E. Hines. 2006. Occupancy estimation and modeling: inferring patterns and dynamics of species occurrence. Academic, New York

Manning, T., W. D. Edge, and J. O. Wolf. 1995. Evaluating population-size estimators: an empirical approach. Journal of Mammalogy 76:1149–1158

Mao, C. X. and R. K. Colwell. 2005. Estimation of species richness: mixture models, the role of rare species, and inferential challenges. Ecology 86:1143–1153

McCoy, E. D. and K. L. Heck, Jr., 1987. Some observations on the use of taxonomic similarity in large-scale biogeography. Journal of Biogeography 14:79–87

Nichols, J. D. and Conroy, M. J. 1996. Estimation of species richness. Pages 226–234 *in* D. E. Wilson, F. R. Cole, J. D. Nichols, R. Rudran, and M. Foster, editors. Measuring and monitoring biological diversity. Standard methods for mammals. Smithsonian Institution Press, Washington, DC

Nichols, J. D., T. Boulinier, J. A. Hines, K. P. Pollock, and J. R. Sauer. 1998a. Estimating rates of local species extinction, colonization, and turnover in animal communities. Ecological Applications 8:1213–1225

Nichols, J. D., T. Boulinier, J. A. Hines, K. P. Pollock, and J. R. Sauer. 1998b. Inference methods for spatial variation in species richness and community composition when not all species are detected. Conservation Biology 12:1390–1398

Nichols, J. D., L. L. Bailey, A. F. O'Connell, Jr., N. W. Talancy, E. H. C. Grant, A. T. Gilbert, E. M. Annand, T. P. Husband, and J. E. Hines. 2008. Multi-scale occupancy estimation and modeling using multiple detection methods. Journal of Applied Ecology 45:1321–1329

Norris, J. L. and K. H. Pollock. 1996. Nonparametric MLE under two closed-capture models with heterogeneity. Biometrics 52:639–649

Otis, D. L., K. P. Burnham, G. C. White, and D. R. Anderson. 1978. Statistical inference from capture data on closed animal populations. Wildlife Monographs 62:1–135

Pielou, E. C. 1977. Mathematical ecology. Wiley, New York

Pledger, S. 2000. Unified maximum likelihood estimates for closed capture–recapture models using mixtures. Biometrics 56:434–442

Pledger, S. 2005. The performance of mixture models in heterogeneous closed population capture–recapture. Biometrics 61:868–876

Pollock, K. H. 1982. A capture–recapture sampling design robust to unequal catchability. Journal of Wildlife Management 46:752–757

Pollock, K. H., J. D. Nichols, T. R. Simon, G. L. Farnsowrth, L. L. Bailey, and J. R. Sauer. 2002. Large scale wildlife monitoring studies: statistical methods for design and analysis. Environmetrics 13:105–119

R Development Core Team. 2008. R: a language and environment for statistical computing. R Foundation for Statistical Computing, Vienna, Austria. ISBN 3-900051-07-0, URL: http://www.R-project.org

Rota, C. J., R. J. Fletcher Jr, R. M. Dorazio, and M. G. Betts. 2009. Occupancy estimation and the closure assumption. Journal of Applied Ecology 46:1173–1181

Royle, J. A. 2004. N-mixture models for estimating population size from spatially replicated counts. Biometrics 60:108–115

Royle, J. A. 2006. Site occupancy models with heterogeneous detection probabilities. Biometrics 62:97–102

Royle, J. A. 2009. Analysis of capture–recapture models with individual covariates using data augmentation. Biometrics 65:267–274

Royle, J. A. and R. M. Dorazio. 2008. Hierarchical modeling and inference in ecology. Academic, Amsterdam

Royle, J. A. and M. Kéry. 2007. A Bayesian state-space formulation of dynamic occupancy models. Ecology 88:1813–1823

Royle, J. A. and W. A. Link. 2006. Generalized occupancy models allowing false positive and false negative errors. Ecology 87:835–841

Royle, J. A. and J. D. Nichols. 2003. Estimating abundance from repeated presence–absence data or point counts. Ecology 84:777–790

Royle, J. A., R. M. Dorazio, and W. A. Link. 2007. Analysis of multinomial models with unknown index using data augmentation. Journal of Computational and Graphical Statistics 16:67–85

Russell, R. E., J. A. Royle, V. A. Saab, J. F. Lemkuhl, W. M. Block, and J. R. Sauer. 2009. Modeling the effects of environmental disturbance on wildlife communities: avian responses to prescribed fire treatments in a coniferous forest in Washington. Ecological Applications 19:1253–1263

Schmidt, B. R. 2005. Monitoring the distribution of pond-breeding amphibians when species are detected imperfectly. Aquatic Conservation: Marine and Freshwater Ecosystems 15:681–692

Seber, G. A. F. 1982. The estimation of animal abundance and related parameters. Charles Griffin, London

Soberón, M. J. and B. J. Llorente. 1993. The use of species accumulation functions for the prediction of species richness. Conservation Biology 7:480–488

Spiegelhalter, D., A. Thomas, and N. G. Best. 2003. WinBUGS user manual, version 1.4. MCR Biostatistics Unit, Cambridge

Thompson, S. K. 2002. Sampling. Wiley, New York

Thompson, W. L. 2004. Sampling rare and elusive species. Island, Washington, DC

White, G. C. and K. B. Burnham. 1999. Program MARK: survival estimation from populations of marked animals. Bird Study 46 (suppl.):S120–S138

Williams, B. K., J. D. Nichols, and M. J. Conroy. 2002. Analysis and management of animal populations. Academic, San Diego, CA

Yoccoz, N. G., J. D. Nichols, and T. Boulinier. 2001. Monitoring of biological diversity in space and time. Trends in Ecology and Evolution 16:446–453

Zipkin, E. F., A. DeWan, and J. A. Royle. 2009. Impacts of forest fragmentation on species richness: a hierarchical approach to community modeling. Journal of Applied Ecology 46: 815–822

Chapter 13
Estimation of Species Richness of Large Vertebrates Using Camera Traps: An Example from an Indonesian Rainforest

Timothy G. O'Brien, Margaret F. Kinnaird, and Hariyo T. Wibisono

13.1 Introduction

The number of biological species that occurs at a particular geographic unit, whether that be global, a biogeographic region, country, or national park, is of great relevance to the management and conservation of biodiversity. Major policy initiatives at the international, national, and regional levels have committed entire government programs to attaining measurable targets of this variable in the conservation of biodiversity (Danielsen et al. 2005). The Convention on Biological Diversity lists a reduction in the rate of loss of biodiversity as a goal for 2010 and many of the indicators proposed to measure achievement of that goal are indices that hopefully track changes in species richness (United Nations Environment Programme 2002). It is unlikely that governments will be able to judge their progress without monitoring systems and indicators in place to assess the effectiveness of their interventions (Balmford et al. 2005). Species diversity usually refers to the number of species in a location or "species richness" (Schluter and Ricklefs 1993; Lande 1996). Species richness is often used as a state variable in evaluating the impact of management interventions and anthropogenic disturbance on biodiversity. One of the greatest hindrances to understanding and conserving biodiversity, however, is our inability to determine how many species we have and how fast that number is changing (Balmford et al. 2005; May 1988).

Species richness can be broken down into three, scale-dependent components. Total diversity on a regional scale is called γ-diversity and is composed of average

T.G. O'Brien (✉)
Wildlife Conservation Society, 2300 Southern Blvd, Bronx, NY 10460, USA
and
Mpala Research Centre, P.O. Box 555, Nanyuki, Kenya
e-mail: tobrien@wcs.org

M.F. Kinnaird
Mpala Research Centre, P.O. Box 555, Nanyuki, Kenya
and
Wildlife Conservation Society, 2300 Southern Blvd, Bronx, NY 10460, USA

H.T. Wibisono
Wildlife Conservation Society – Indonesia Program, Jl. Ciremai No. 8, Bogor, Indonesia

A.F. O'Connell et al. (eds.), *Camera Traps in Animal Ecology: Methods and Analyses*,
DOI 10.1007/978-4-431-99495-4_13, © Springer 2011

site-level diversity in a region or α-diversity, and the turnover of species among sites or β-diversity. Typically, β-diversity is calculated from α- and γ-diversity. Whittaker (1972) defined the relationship between measures as $\gamma = \alpha\beta$, a multiplicative relationship. Whittaker's approach leads to β-diversity expressed in dimensionless units of species turnover. Lande (1996) notes that Whittaker's β-diversity is actually the inverse of community similarity in species composition rather than a diversity metric. Lande (1996) defined the relationship between measures additively, $\gamma = \beta + \alpha$, an approach that leads to β-diversity expressed in units of species richness. β-diversity can be further divided into the variety of habitats and the habitat breadth of species (the inverse of the average number of habitats each species occupies: Schluter and Ricklefs 1993).

Ideally, if we knew the abundance of every species, a species abundance distribution, composed of the number of individuals of each species in the community, would tell us everything we need to know about species richness and evenness (a measure of the equitability in distribution of abundance or biomass among species in a community). Unfortunately, we are rarely fortunate to have complete knowledge, and usually some species escape the sampling effort. With rare exception, it is typical that after a sample of n individuals, many species are represented by only one or two individuals recorded in only a few samples. It is well known that, as sampling effort increases, the likelihood of detecting new individuals of rare species increases, but also the detection of new, rare species increases. Mao and Colwell (2005) refer to this as "Preston's demon" or the sampling-dependent movement of the veil line between detected and undetected species (Preston 1948). Estimating the number of undetected species in a sample is the major challenge to understanding species richness and the processes that underlie the dynamics of species richness (Nichols et al. 1998a, b).

There are many approaches to estimating species richness including the extrapolation of species-area or species-effort curves, the use of parametric models of species abundance based on count statistics, use of taxon ratios, and estimation of species richness based on sampling (Bunge and Fitzpatrick 1993; Colwell and Coddington 1994; Magurran 1988). The use of species-area or species-effort curves is based on sampling relationships. It is well known that the number of species S_n in a sample increases as sampling effort n increases (n = number of individuals sampled, the size of the area sampled, the number of habitats sampled) up to the limit of total species, S_{max}, in the sample area. Species-area/-effort curves exploit the relationship between S_n and n by fitting curves to the pattern of species accumulation and estimating the asymptote. Colwell and Coddington (1994) review asymptotic and non-asymptotic methods of calculating species richness based on species accumulation curves. They point out that the basic problem with extrapolation is that different models give different S_{max} for the same data (a given sample effort n produces different S_n). Because the appropriate curve depends on the rate of species accumulation, which in turn depends on the distribution of species abundance, as well as on other sources of variation in animal detection probabilities (individuals of different species are not likely to be equally detectable), no single model is expected to work across a range of accumulation patterns. This is problematic because the answer depends on the method and the choice of method is subjective.

Cam et al. (2002a) present a probabilistic, non-parametric estimator of species richness for use with species accumulation data. They make the connection between species richness estimation and abundance estimation using capture-removal models in which the detection probability changes after the first detection (Model M_b: Otis et al. 1978). Removal models are appropriate for species accumulation data because a species is removed from the population after its first detection so the only statistics are the number of new species detected at each sample period. Cam et al. (2002a) recommend the Otis et al. (1978) model M_{bh} in which each species may have a different detection probability. They suggest that the jackknife estimator of Pollock and Otto (1983) may be the most appropriate for species accumulation data. This approach has the advantage of being less dependent on assumptions about the underlying distribution of species richness and is based on the sampling process rather than curve-fitting.

Colwell and Coddington (1994) also review the utility of species abundance distributions to estimate species richness based on count data. These models include the classic lognormal distribution (Preston 1948) and log-series distribution of abundance (Williams 1964). Approaches based on such distributions typically assume abundances to be known. The major problem with these methods relates to the general problem of using count data addressed in several chapters in this book: true abundance is not known and counts are unlikely to be proportional to true abundance for all species. The use of counts that are biased in many different directions depending on the species set, results in species richness estimates that are biased in unknown directions. In addition, for the case of the parametric models, the choice of interval size for abundance categories affects the final answer, and this choice is subjective. The lognormal distribution is very sensitive to the number of rare species represented by only one individual. Finally, there is no measure of reliability for estimates of species richness resulting from the lognormal distribution. The log-series model has no theoretical limit on the number of species in a distribution. If there is a measure of effort or number of individuals, however, a log-series fit may give reasonable estimates of species richness. Log-series estimates, however, still depend on sample effort and relative abundance, producing estimates of S_{max} that are difficult to interpret because the degree of bias is unknown. In addition, proper interpretation of species count data requires estimation of the detection probabilities.

Good (1953) first suggested the use of probabilistic models to estimate the number of species in a sample as an alternative to fitting species frequencies to particular distributions of data (e.g., Chambers and Yule 1942; Corbett et al. 1943; Preston 1948). Burnham and Overton (1979) suggested the use of models that incorporate heterogeneity in species-specific detection probabilities into estimates of species richness. The various-order jackknife estimators for the detection heterogeneity model of Burnham and Overton (1978, 1979) is the same as the M_h estimator in closed population estimation permitting heterogeneous capture probabilities among individuals (Otis et al. 1978; Williams et al. 2002). For example, species richness $\hat{S}$ can be estimated by the first-order jackknife:

$$\hat{S} = S_{obs} + (t-1)f_1 / t, \tag{13.1}$$

where S_{obs} is the number of species observed in t samples and f_1 is the number of species found in only one sample. The general form of the jackknife estimator for species richness is

$$\hat{S}_l = S_{obs} + \sum_{i=1}^{l} \alpha_{il} f_i, \quad (13.2)$$

where the α_{il} are constants corresponding to the jackknife estimators of order l (Burnham and Overton 1978, 1979) and f_i are the number of species seen at exactly i of l samples. Thus, f_1 is the number of species observed at only one sample, f_2 is the number of species observed at two samples and so on. A different kind of estimator was proposed by Chao et al. (1992) that relaxes the assumption of equal catchability within each species across samples and is analogous to the M_{th} model described by Otis et al. (1978). Nichols and Pollock (1983) and Nichols et al. (1986) applied Burnham and Overton's model to the problem of estimating taxonomic diversity and extinction rates in the fossil record. Boulinier et al. (1998) and Nichols et al. (1998a, b) developed these ideas into spatial and temporal models of species richness that allow estimation of extinction, colonization and turnover, all important statistics in community dynamics. Using Pollock's (1982) robust design, these models incorporate species detection probabilities into the estimation process (see O'Brien, Chap. 6 and Kéry, Chap. 12). The program COMDYN (Hines et al. 1999) was developed to estimate species richness and associated dynamics parameters using a robust design framework.

Because species richness can be estimated from detection/nondetection data, recent developments have focused on the use of occupancy models to estimate species richness at a site or at a collection of sites (MacKenzie et al. 2006). Occupancy models are especially useful for estimation involving site level species richness when a list of potential species occurring at the site or in the region is available. Species richness at a particular site will be determined by local environmental conditions (i.e., habitat) and by the regional species pool (the list) that contains all possible species for the area. Cam et al. (2000, 2002b) refer to relative species richness, $\hat{\varphi}$, as the ratio of species at a single site (α-diversity) to the number of species in the regional species pool (γ-diversity) or the proportion of total species at a given site. Cam et al. (2000) use the estimated relative species richness, $\hat{\varphi}$, as an indicator: sites with favorable conditions should have greater $\hat{\varphi}$ than sites that are degraded. $\hat{\varphi}$ may also be used as a temporal indicator to measure changes in species richness at a site over time. The estimators of Cam et al. (2000) are based on the nonparametric estimator of Burnham and Overton (1979), developed to deal with heterogeneous detection probabilities not associated with species-specific covariates.

When the regional species pool is known, each species may serve as a "site" in the context of occupancy sampling. Replicated sampling is carried out (e.g., at a single location) and a detection history is constructed for each species in the regional pool. The detection history matrix will often contain regional pool species that are never detected in the study area. The proportional occupancy of the species, $\hat{\psi}$, is then interpreted as the probability that a member of the regional pool is present in the sample, which is the same as $\hat{\varphi}$ described by Cam et al. (2000). Occupancy modeling allows tracking of changes in species richness over time and the modeling of covariates that might affect detectability, extinction or colonization

(MacKenzie et al. 2003, 2006). The use of covariates is not possible in the capture –recapture (CR) models of Cam et al. (2000) because undetected species are not used in the estimation, providing no ability to use covariate information of such species.

13.2 Camera Traps and Species Lists

Typically, ecologists and conservation biologists are concerned with a component of biodiversity, rather than total biodiversity. Even studies of biodiversity components, however, tend to be complex. Studies attempting to document species richness of only terrestrial vertebrate taxa must use a wide range of sampling methods appropriate for different taxa. This might, for example, involve line transects for arboreal mammals, terrestrial and arboreal trapping for rodents, point counts for birds, drift fence trapping for reptiles and amphibians, harp traps for bats, acoustic recording for birds, frogs and bats, spoor and sign count surveys, and camera trapping. Camera trapping is especially useful for medium-sized to large, terrestrial and semi-terrestrial mammals, and for large ground birds. While these birds and mammals may represent a relatively small proportion of the overall species richness at a site or region, it is frequently an especially important threatened subset of vertebrate species. Thus, many management activities specifically target this group.

Camera trapping offers some distinct advantages over other methods for surveying terrestrial mammals and birds. The method is non-invasive in the sense that taking a photograph does not represent a form of harassment of wildlife. Because camera traps operate in the absence of humans, there is no observer bias, or response of animals to humans, factors that need to be considered in line transects and point count surveys. There is a possibility that use of a flash may affect the behavior of nocturnal animals, but there is little evidence to support the idea that animals alter movement behavior in response to being photographed. Camera traps can operate for extended periods of time in remote locations, which makes them especially useful as a sampling tool. They can operate 24 h per day which means that the same technique may be used for diurnal and nocturnal sampling. Finally, they provide an unambiguous record of the species, date and time of detection.

Camera trapping may be especially useful for estimating species richness at a particular location, especially when the opportunities for other forms of sampling are limited. Although no new species have yet been discovered using camera traps, a number of significant re-discoveries and new location records for mammals and birds have been possible using remote camera detection methods. On the island of Sumatra in Indonesia, camera trapping led to the re-discovery of the giant pitta *Pitta caerulea*, Sunda Ground Cuckoo (M. Linkie, pers. commun) and Sumatran striped rabbit (*Nesolagus netscheri*) in two national parks. In Lao PDR, camera traps documented the Annamites Mountains striped rabbit *Nesolagus timinsi* and the Saola *Pseudoryx nghetinhenis*. In the United States, Zielinski et al. (2005), used camera traps to document changes in diversity of the forest carnivore community of the Sierra Nevada mountains.

Table 13.1 Mammal Surveys (excluding bats, small rodents) using camera trapping and transect-based sampling and old species lists, based on combined, previously published species lists. The regional species pool is the expected occurrence of medium and large mammals in the area and relative S_{obs} is the proportion of the regional species pool detected by camera trapping and by transect-based methods, not corrected for detectability

	Pantanal	Amazon	Thailand[a,b]	Indonesia[a,b]	Tanzania[b]
Camera trap	16	13	30	39	25
Transects	26	23	31	19	41
Old lists	–	31	57	22	45
Regional species pool	43	50	60	55	55
Relative S_{obs} (cameras)	0.37	0.26	0.50	0.71	0.46
Relative S_{obs} (transects)	0.60	0.46	0.52	0.34	0.91
Source	Trolle (2003a)	Trolle (2003b)	Lynam et al. (2006)	O'Brien et al. (unpubl. data)	Foley et al. (unpubl. data), J. Kingdon (pers. comm.)

[a] Camera trap surveys conducted over > 1 time period
[b] Transect surveys conducted over > 1 time period

Table 13.1 illustrates the usefulness of camera trapping in mammal surveys. Where camera trap survey efforts have exceeded 1,000 trapdays (e.g., Thailand and Indonesia) the cameras have documented the occurrence of 50% or more of the total medium and large mammal fauna expected to occur at the site, based on a regional species list. The relatively poor performance of camera traps in the Pantanal and the Amazon samples probably results from inadequate sampling efforts (<500 trapdays, Trolle 2003a). In all survey efforts, however, camera traps recorded species undocumented by other survey techniques. In Bukit Barisan Selatan National Park, Indonesia, in addition to the giant pitta and striped rabbit, the use of camera traps revealed two previously unrecorded pheasant species, a second pitta, otter civet *Cynogale bennetti*, as well as several new records for other civets. Camera traps are especially useful for nocturnal terrestrial mammals that may hide during the day and for forest mammals. Camera traps are less useful detecting the smaller terrestrial species (i.e., rats) or arboreal species such as squirrels and medium-sized primates, but even these mammals occasionally appear in camera trap photos.

Although species lists are informative about the potential richness of a site, there are limitations. Regional species lists or pooled species lists represent the accumulated knowledge of a site rather than a snapshot estimate of current species richness. Historical presence does not equate to presence today so the regional lists only give an idea of the potential species richness. Species lists resulting from a sample based on transects, point counts, or camera trapping, represent count data and are subject to all of the limitations of count data due to differences in detectability among species.

13.3 Estimating and Monitoring Species Richness: An Example from Indonesia

In 1998, we established a camera trapping program in Bukit Barisan Selatan National Park (BBSNP: Fig. 13.1) on the island of Sumatra, Indonesia to monitor changes over time in the abundance of tigers and their prey species (O'Brien et al. 2003), to document edge avoidance by the Sumatran tiger *Panthera tigris sumatrae*, elephant *Elephas maximus* and rhinoceros *Dicerorhinus sumatrensis* (Kinnaird et al. 2003), and to assess the effects of hunting on wildlife species (Wibisono 2006). The camera trapping design consists of 10 sampling blocks, each 20 km^2 in size (2 km × 10 km), oriented from the park boundary inward and spaced at 10–15 km intervals along the 150 km length of the park (Fig. 13.1). Camera traps were deployed at randomized UTM coordinates within each square kilometer of a sampling block (one camera per km^2) and operated for approximately 30 days or until all film was exposed. Blocks were surveyed sequentially until all 10 blocks were completed. Parkwide surveys were conducted in 1998/1999, 2000/2001, 2002/2003, 2003/2004 and 2005/2006 (hereafter referred to by the first year).

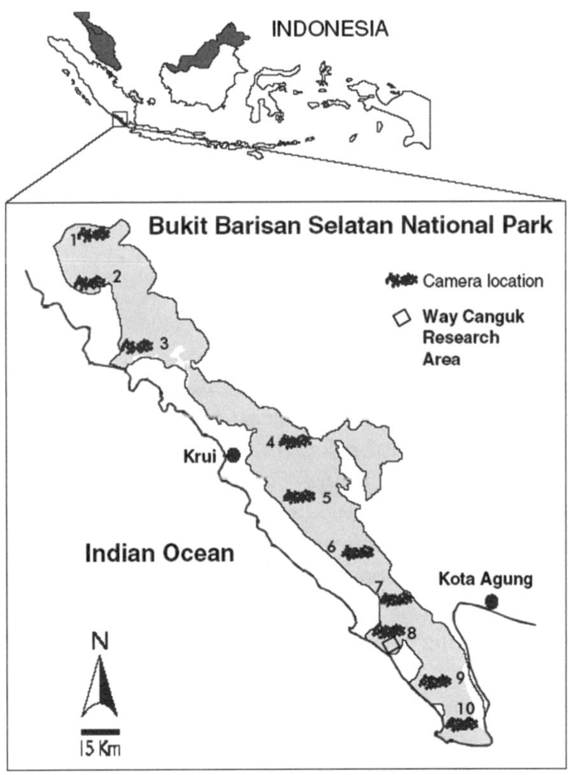

Fig. 13.1 Location of camera trap sampling blocks in the Bukit Barisan Selatan National Park, Sumatra, Indonesia

This sampling design has some useful features for general biodiversity surveys. First, the randomized placement of traps is appropriate for sampling a range of species because it breaks the tendency to find the "best" place to trap a particular species, a practice that may lead to subjective compromising of trap location or convenience sampling. The clustered sample within spatially separated blocks allows partitioning of diversity into its components. If we wished to treat the park as a landscape, we could decompose γ-diversity for the entire park into α-diversity at the level of sampling blocks and β-diversity between sample blocks. For the first example, however, we treat the southern half of Sumatra as the region and the Bukit Barisan Selatan National Park as a single site. The secondary samples are the 10 sampling blocks surveyed in each of five temporal replications. We will consider three measures of species richness: observed species richness based on the number of species actually photographed during a sampling interval, a detection-corrected estimate of species richness (based on model M_h: Burnham and Overton 1979; Nichols et al. 1998a, b) and relative species richness (Cam et al. 2000), another detection-corrected estimate of species richness.

13.3.1 Observed and Estimated Species Richness

To determine the observed and estimated species richness, we restricted the analysis to medium- and large-sized, terrestrial and semi-terrestrial mammals, and to the four largest species of terrestrial birds: the Red junglefowl *Gallus gallus*, Great argus pheasant *Argusianus argus*, Sumatran peacock-pheasant *Polyplectron sumatranus* and Salvadori's pheasant *Lophura inornata*. Given the height at which the cameras were positioned, we assumed that the cameras did not capture small mammals and small ground birds reliably or systematically. Thus, ground squirrels, treeshrews, rodents and arboreal primates were excluded from this analysis. The analysis used the software COMDYN (Nichols et al. 1998a, b; Hines et al. 1999) which incorporates Pollock's (1982) robust design approach to estimate species richness, rate of change in species richness, local extinction probability (rate), turnover rates and the number of colonizing species between two survey samples separated in space or in time. We follow Nichols et al. (1998a, b) definition of local extinction probability during the time period (i, j) as the likelihood that a species present in the community at time i is not present in the community at time j, for $j > i$. Turnover rate is defined as the probability that a species selected at random from the community at time j is a species that was not present in the community at time i. The number of colonizing species is the number of species not present at time i that entered the community between times i and j and are still present at time j.

Temporal surveys were compared sequentially such that the 1998 survey was compared to the 2000 survey, the 2000 survey was compared to the 2002 survey and so on. Data were summarized as follows. In each of 10 sampling blocks, we combined the species detected for all camera traps within the block over the month of sampling. The sampling blocks are analogous to survey stops within a single survey

Table 13.2 Species detection histories for Bukit Barisan Selatan National Park camera trap surveys between 1998 and 2005

Species data	# species detected	# species detected in i blocks ($i=1$ to 10)									
		f_1	f_2	f_3	f_4	f_5	f_6	f_7	f_8	f_9	f_{10}
Number of species detected (1998)	29	5	5	1	3	3	1	4	1	3	3
Number of species detected (2000)	27	2	8	2	0	1	2	2	4	3	3
Number of species detected (2002)	26	4	8	1	1	2	2	0	3	2	3
Number of species detected (2003)	28	6	2	5	4	1	2	3	2	1	2
Number of species detected (2005)	23	5	2	3	2	4	1	2	2	2	0
Number of species in 1998 detected in 2000	25	3	3	1	3	3	1	4	1	3	3
Number of species in 2000 detected in 1998	25	0	8	2	0	1	2	2	4	3	3
Number of species in 2000 detected in 2002	24	1	7	1	0	1	2	2	4	3	3
Number of species in 2002 detected in 2000	24	2	8	1	1	2	2	0	3	2	3
Number of species in 2002 detected in 2003	25	4	7	1	1	2	2	0	3	2	3
Number of species in 2003 detected in 2002	25	4	1	5	4	1	2	3	2	1	2
Number of species in 2003 detected in 2005	23	3	2	3	4	1	2	3	2	1	2
Number of species in 2005 detected in 2003	23	5	2	3	2	4	1	2	2	2	0

route (Nichols et al. 1998a, b) and are the secondary samples of the robust design. The species richness estimates and associated statistics are based on two sets of samples: f_i, $i=1, 2, \ldots, 10$ is the number of species detected in exactly i sampling blocks, and n_i is the number of species detected in sampling block i. We calculated f_i and n_i for each sample year and for a subset of the species detected each sample year composed of the f_i and n_i at the second time period (year) for the subset of species detected in the first time period (year) and f_i and n_i at the first time period for the subset of species detected in the second time period (Hines et al. 1999). Estimates associated with change in species richness over time are based on the subset of species that were detected in year 1 that were also detected in year 2, and the subset of species detected in year 2 that were also detected in year 1 (Table 13.2).

The rate of change in species richness (λ_{ij}) is a useful metric for biodiversity monitoring. It can be estimated as the ratio of the estimated species richness for times i and j ($\hat{S}_j / \hat{S}_i$). If the average detection probabilities are equal for the two time periods then λ_{ij} may be calculated directly from the number of species observed during each time period. Using the observed counts of species should result in smaller variances of $\hat{\lambda}_{ij}$ than using the estimated values, and when $\overline{p}_i = \overline{p}_j$ the estimate should be relatively unbiased.

The rate of local extinction is calculated in COMDYN as the complement of the likelihood that a species present at time i is still present at time j. If $S_{\mathrm{obs},i}$ is the number of species observed at time i and M_j^i, is the number of these species still present in period j, then the extinction rate (E) can be estimated as

$$\hat{E} = 1 - \frac{\hat{M}^{i}_{j}}{S_{obs,i}}. \tag{13.3}$$

The proportion of extinct species is one minus the proportion of species that persisted.

The turnover rate (T) is calculated in a similar manner as the extinction rate but in the reverse time order. If $S_{obs,j}$ is the number of species observed at time *j*, and M^{j}_{i} of the species were present at time *i*, then the turnover rate is calculated as the proportion of new species or species that were not present at time period *i* and is denoted by

$$\hat{T} = 1 - \frac{\hat{M}^{j}_{i}}{S_{obs,j}}. \tag{13.4}$$

The COMDYN software outputs a parameter PHI that is an estimate of the proportion of species present at time 2 among those present at time 1 that is used to calculate the complement, $\hat{E}$. Similarly, COMDYN outputs a parameter GAMMA that is an estimate of the of the proportion of species present at time 2 that were also present at time 1 and uses this to calculate the complement $\hat{T}$.

We considered a total of 32 species observed during the five replications in the analysis (28 of the 39 observed mammals [rodents and arboreal mammals were excluded] plus 4 terrestrial birds). The observed number of species tended to decline over time as did the estimated species richness although there is considerable overlap among confidence intervals in all five estimates of $\hat{S}_i$ (Table 13.3; Fig. 13.2). S_{obs} falls within the 95% confidence interval of $\hat{S}_i$ in every sample except 2000, a result consistent with the high average detection probabilities. The standard errors and confidence intervals of $\hat{S}_i$ increase substantially in the last two samples reflecting an increase in the proportion of rare species and decline in proportion of common species in the sample. The average detection probability did not differ significantly over time ($P > 0.1$, Table 13.3).

The estimated rates of local extinction during the four time intervals ranged from 0 to 0.138, while the turnover rate ranged from 0 to 0.084 (Table 13.3). Between 1998 and 2000, the estimated extinction probability exceeded turnover, and the number of colonizing species was moderately low (<5 species). Between 2000 and 2002, both local extinction and turnover rates were low and the estimate of colonizing species was 0. Between 2002 and 2003 we see a jump in turnover rate and estimated number of colonizing species rises to five species. Finally, between 2003 and 2005 we see an increase in the rate of local extinction, a drop in turnover and no new colonizing species. The overall conclusion for BBSNP is that there has been a net loss of species over time since 1998, with the number of local extinctions exceeding the number of colonizing species. We note that the rates of extinction and turnover are dependent on the time interval separating the sampling occasions. In this analysis, we have compared three sampling intervals of two years and one sampling interval of one year. Ideally, we would like to compare estimates of similar intervals, or make adjustments such that all estimates reflect a similar time interval.

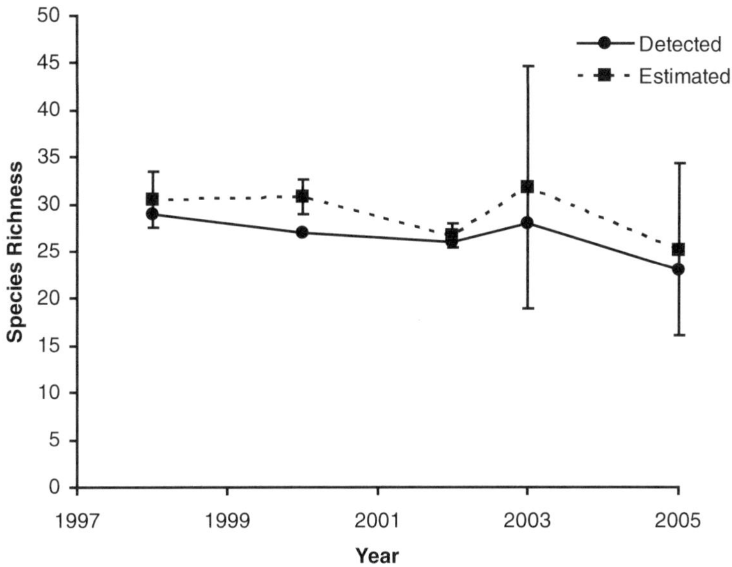

Fig. 13.2 Number of species observed in camera traps and estimated species richness (with standard *error bars*)

An important question regarding these data is the rate of change of species richness. The lack of significant difference among the estimated $\bar{P}$'s indicates that the $\hat{\lambda}_{ij}$ estimated directly from the observed number of species should provide a more precise estimate than the estimated values. In Fig. 13.3, the difference between the observed and estimated rate of change based on observed and estimated richness is very small at all time intervals indicating that the $\hat{\lambda}_{ij}$ based on observed species numbers are relatively unbiased. The standard errors and confidence intervals for the estimated $\hat{\lambda}_{ij}$ are similar to those based on observed number of species for the first two intervals but much larger for the last two time intervals. In this case, the use of either ratio makes little difference to the overall trend. Using the geometric mean of the ratio between 1998 and 2005 for estimated λ_{ij}, and λ_{ij} based on observed number of species, we find the trends average 2.9% and 2.4% decline in overall observed and estimated species per year, respectively. Both estimates suggest that species richness has declined over time in BBSNP.

13.3.2 *Relative Species Richness*

A second method to estimate and monitor species richness relies on the use of the regional species pool list to estimate the relative species richness, $\hat{\varphi}$, or the proportion of the regional species list present at a site. Estimation of $\hat{\varphi}$ may proceed using M_h estimates of species richness (Boulinier et al. 1998) adjusted by the regional

Table 13.3 Species richness and community dynamics estimates for camera trap samples in Bukit Barisan Selatan National Park calculated using COMDYN software. Each replicate includes the observed number of species (S_{obs}), estimated number of species ($\hat{S}$), average detection probability ($\bar{p}$), extinction rate ($\hat{E}$), turnover rate ($\hat{T}$) and number of colonizing species ($\hat{C}$), bootstrap estimates of standard error (SE), and bootstrap estimates of the 95% confidence intervals (95%CI). Extinction, turnover and number of colonizing species are calculated for the interval *t* to *t*+1

Year	S_{obs}	$\hat{S}$	SE	95%CI ($\hat{S}$)	$\bar{p}$	SE	95%CI ($\hat{S}$)	$\hat{E}$	SE	$\hat{T}$	SE	$\hat{C}$	SE
1998	29	30.5	2.95	[29.0, 39.5]	0.951	0.070	[0.726, 1.00]	0.138	0.0069	0	0.056	4.5	2.52
2000	27	30.8	1.94	[27.0, 32.8]	0.876	0.059	[0.799, 1.00]	0	0.070	0.042	0.054	0	1.78
2002	26	26.7	1.22	[26.6, 30.5]	0.974	0.038	[0.848, 1.00]	0	0.012	0.084	0.062	5.1	12.83
2003	28	31.8	12.60	[28.0, 68.9]	0.880	0.188	[0.389, 1.00]	0.101	0.088	0	0.061	0	6.31
2005	23	25.2	8.88	[23.0, 54.7]	0.914	0.167	[0.410, 1.00]	–		–			–

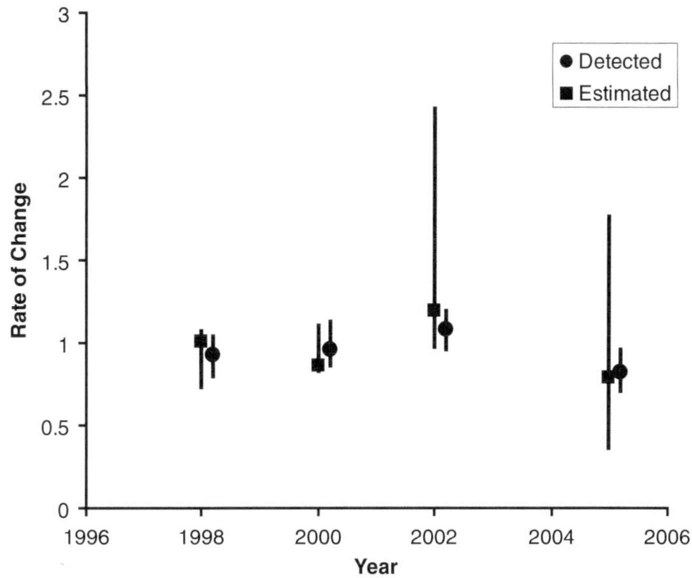

Fig. 13.3 Rates of change in species richness over time based on the number of species detected and on estimated species richness with bootstrap 95% confidence intervals. *Points* are offset for ease of reading

species pool (Cam et al. 2000) or by using occupancy analysis (MacKenzie et al. 2006). Species that have a zero probability of being detected should not be included in the regional species pool. In practice, however, it is often difficult to say unambiguously that a species has zero chance of detection unless we know for certain that it does not occur at the site, or we know for certain that our sampling method has no chance of detecting the species if it is present. For this example, we restrict the regional species pool to mammals (excluding small rodents) and retain mammals that are unlikely to be sampled because of specialized habitat needs but still may have detection probabilities > 0. Specifically, we include three species of otters and the otter civet, as well as giant squirrels, gibbons and langurs. The regional species pool therefore includes 37 terrestrial species, 4 semi-aquatic species and 13 arboreal species for a total of 54 species.

We followed MacKenzie et al. (2006) using occupancy analysis to estimate relative species richness. We treat the park as a single site, pooling the data from all camera sites into four observation periods of eight days each. We used a robust design (Pollock 1982) in which the five surveys (1998–2005) are the primary sampling periods and the four observation periods represent the secondary sampling periods within five primary sampling periods. In order to explore the possible effects of body size and preferred habitat strata we included body length (transformed to mean 0 and variance 1) and stratum (terrestrial, other to indicate arboreal and semi-aquatic) as covariates. We also included year as a covariate. We considered three sets of hypotheses related to the covariates. First, we predicted that the detection probability, extinction rate and colonization rate would vary over time. Second, we predicted that detection probability and extinction probability would increase with increasing body size. We believed that since there was ongoing poaching of large mammals for commercial markets and for subsistence, that body size might affect the likelihood of local extinction but not local colonization. Finally, we predicted that terrestrial mammals were more likely to be detected than arboreal or semi-aquatic mammals. So we only considered models in which colonization rate was constant or varied with time, extinction rate that was constant or varied with time and/or body size, and in which detection probability was modeled as a function of time, body size and substrate.

We used a multi-season occupancy model that treated the different species in the regional species pool as the "sites," and each species was known to be present during an observation period if it appeared at least once in a camera trap throughout the park during an 8-day observation period. For each of five primary sampling periods, we assume that a species is present or absent from the park and that system state does not change (community closure) within a sampling period. Changes in species presence may occur between primary sampling periods so that at primary time t:

$$\varphi_t = \varphi_{t-1}\left(1-\varepsilon_{t-1}\right)+\left(1-\varphi_{t-1}\right)\gamma_{t-1}, \tag{13.5}$$

where ε_{t-1} represents local extinction rate and γ_{t-1} represents the local colonization rate between time $t-1$ and t. The ratio of φ_t/φ_{t-1} can be used as a measure of the rate of change in species richness in a community.

We initially fit 64 relative species richness models in PRESENCE 2.0. We ranked models using the Akaike Information Criterion (AIC) to select the most appropriate model for describing the data well using the minimum number of parameters for adequate description (Burnham and Anderson 2002). The difference between the AIC for a model and the lowest value AIC model is ΔAIC which can be used to compute model likelihoods, measures of the relative strength of evidence for a model. AIC model weights are measures reflecting the evidence that a particular model is the "best" model for the data among the models being compared. Sometimes, the weight of evidence may support several alternative models, indicating that the "best" model might vary from data set to data set. Multi-model inference, based on model averaging, may be used to improve the stability of parameter estimates (Burnham and Anderson 2002). Model averaging uses the AIC weights to develop weighted parameter estimates.

The analysis resulted in 23 models with AIC weights greater than 0. Considering only the models with AIC weights ≥ 0.05 resulted in the selection of six possible models (Table 13.4). The model with low AIC [$\varphi\gamma(.)\varepsilon(\text{year})p(\text{year, size})$], included extinction probabilities that varied over time, and detection probabilities that varied by time and by the size of the species. The low AIC model received a model weight of 0.3657, indicating only moderate support for this model while the second model [$\varphi\gamma(.)\ \varepsilon(\text{year, size})\ p(\text{year})$] that included size as a covariate of extinction, received a weight of 0.2548. All models included an effect of time on extinction probabilities and three of six models included an effect of body size (sum of model weights equals 0.4209). All models included an effect of time and body size on detection probability, and two models included effects of stratum (sum of model weights equals 0.2678). There was only weak support for covariate effects on colonization; the last two models included an effect of year (sum of model weights equals 0.1126).

Because there are several models that might be supported given the same data set, model uncertainty is high. We used model averaging (Burnham and Anderson 2002) to develop weighted estimates of relative species richness, colonization, extinction and detection probabilities (Table 13.5). In this example, the relative richness based on detected species tracks $\hat{\varphi}$ reasonably well (Table 13.5) and observed φ's fall within the 95% confidence interval for $\hat{\varphi}$. This is an expected result when detection probabilities are high; in this example average detection probability exceeded 0.5 in every survey. Average detection probabilities varied by time and body size, as we predicted. For species with body length under 100 cm, detection probabilities ranged from 0.01 to 0.64 whereas for larger species, detection probabilities ranged from 0.15 to 0.99. There was some support for an effect of preferred habitat stratum on detection: average detection probabilities for

Table 13.4 Multi-season occupancy model selection statistics for analysis of relative species richness using occupancy statistics. The models estimate relative species richness for 1998 (φ), local colonization (γ), and extinction (ε), and detection probability (p). Covariates for p, γ and ε include time (year), transformed body size (size), and preferred habitat stratum (stratum). Results are given for only the top six models with AIC weights greater than or equal to 5%, and AIC weights have been adjusted to the model set

Model	AIC	ΔAIC	Weight (%)	−2 log likelihood	Parameters
$\varphi,\gamma(.),\varepsilon(\text{year}),p(\text{year, size})$	856.24	0.00	36.57	823.24	12
$\varphi,\gamma(.),\varepsilon(\text{year, size}),p(\text{year, size})$	856.97	0.73	25.38	830.97	13
$\varphi,\gamma(.),\varepsilon(\text{year}),p(\text{year, size, stratum})$	857.98	1.74	15.32	831.98	13
$\varphi,\gamma(.),\varepsilon(\text{year, size}),p(\text{year, size, stratum})$	858.56	2.32	11.46	830.56	14
$\varphi,\gamma(\text{year}),\varepsilon(\text{year}),p(\text{year, size})$	859.85	3.61	6.01	829.85	15
$\varphi,\gamma(\text{year}),\varepsilon(\text{year, size}),p(\text{year, size})$	860.12	3.88	5.25	828.12	16

Table 13.5 Model-averaged relative species richness and community dynamics estimates for camera trap samples in Bukit Barisan Selatan National Park calculated from results based on PRESENCE software. Each replicate includes the proportion of species detected (φ_{obs}) from the regional species pool, estimated proportion of species ($\hat{\varphi}$) with average detection probability ($\hat{p}$), bootstrap estimates of standard error (SE), extinction rate ($\hat{\varepsilon}$), and colonization rate ($\hat{\gamma}$). Extinction and colonization rates are calculated for interval t to $t+1$

Year	ϕ_{obs}	$\hat{\varphi}$	SE	$\hat{p}$	SE	$\hat{\varepsilon}$	SE	$\hat{\gamma}$	SE
1998	0.555	0.606	0.159	0.598	0.264	0.1762	0.1908	0.0649	0.0706
2000	0.537	0.540	0.272	0.608	0.251	0.0050	0.0121	0.0622	0.0666
2002	0.518	0.551	0.210	0.528	0.279	0.0000	0.0001	0.0650	0.0715
2003	0.593	0.600	0.205	0.563	0.285	0.3414	0.3409	0.0531	0.0625
2005	0.389	0.405	0.381	0.581	0.262				

terrestrial mammals were twice as large as for arboreal and semi-aquatic species. Semi-aquatic species were not detected during the five surveys, although they are known from the study area and have appeared in camera trap photographs outside this survey.

Extinction rates (Table 13.5) for relative species richness follow the same general trend as the results for estimated species richness (Fig. 13.3), but are much higher in the last interval. Probability of extinction varied by time and was high in the first and last interval but low in the middle intervals. There was an effect of body size on extinction probabilities but the relationship was opposite to our prediction: larger-bodied mammals had lower extinction probabilities. Mammals < 100 cm in length had extinction probabilities averaging 78% higher than mammals greater than 100 cm between the first and second sampling periods, and 55% greater between the fourth and fifth sampling periods. Finally, there was weak support for an effect of time on colonization with colonization estimates higher between the first and second sampling period and the third and fourth sampling period, compared to the other two estimates. We caution placing too much emphasis on the interpretation of short term variation because it is composed of sampling variation as well meaningful biological variation.

If we examine the rate of change in relative species richness between 1998 and 2005 using the geometric mean, we find that the rate of decline is 5% per year. Note that this is higher than reported above for the estimates of absolute richness, reflecting the higher estimated extinction rate for relative species richness between the fourth and fifth sampling interval. In this example, a 5% decline per year in the relative species richness translates to a loss of 1.6 species per year from the local area compared to 0.7 species per year based on the estimated species richness. Despite this apparent disparity, the annual rates of change in species richness for the two methods are not significantly different and both suggest a consistent decline of species richness over time.

So what do these results tell us about sampling terrestrial communities of larger-bodied mammals and birds in tropical forests using camera traps? First, the results indicate that camera traps appear to be effective in detecting mammalian and bird species and estimating number of species for terrestrial mammals and large terrestrial birds. Species were detected with consistently high detection probabilities in both analyses, and a large proportion of the expected species were detected during the sampling periods. The fact that several of these species are nocturnal and most have never shown up except as chance observations in the park suggests that camera traps are a cost-effective method of monitoring this component of biodiversity in tropical forests.

Second, this analysis points out some of the pitfalls faced when a large number of rare species turn up in a sample. In the analysis of absolute species richness using COMDYN in which we considered 10 spatial replicates per primary sample, the confirmed species or species that were actually photographed occur, on average, in four of the five primary sample periods and are detected in 2.3 replicates per primary sample. Within a sample, seven species on average are considered rare because they were detected in only one replication per primary sampling period, and we estimate that 1–3 rare species were undetected in each sample period. During the 2003 sample, 10 rare species were detected, and each of these species was consistently rare or absent in other sample periods. The large number of rare species in the 2003 sample increased the species richness estimate somewhat and decreased precision. In the relative species richness analysis, the high uncertainty associated with the best model led to a model averaging solution. Model averaging introduces an additional component of variation to the estimate of SE, the difference between the model-specific and model-averaged parameter values, and this reduces the precision of final estimates. Although the trend data suggest that absolute and relative species richness are declining, the high rates of colonization and extinction indicate that species are rapidly moving in and out of the local sample area.

Our inference about declining species richness in Bukit Barisan Selatan National Park is also supported by an analysis of occupancy and associated relative abundance (T. G. O'Brien, unpublished manuscript). We analyzed the trends in occupancy and relative abundance for 25 species in the park that had sufficient data and found evidence of a widespread decline in both occupancy and relative abundance among species, suggesting shrinkage of occupied habitat and smaller population sizes. Commercially exploited species (Sumatran tiger, elephant and rhinoceros) exhibited the most dramatic declines in occupancy and relative abundance, suggesting an important role of hunting. Wibisono (2006) reported that tigers in BBSNP declined by nearly 50% between 1999 and 2003. Wibisono also reported high hunting pressure on ungulate species and documented the spatial distribution of hunting. The combined results of these analyses support the trends predicted earlier by Kinnaird et al. (2003) and O'Brien et al. (2003) of continued erosion of species populations due to habitat loss and hunting.

Nichols et al. (1998a) also discuss potential problems with the estimation of the extinction, colonization, and turnover probabilities induced by variation among species in population sizes (see also Alpizar-Jara et al. 2004). If heterogeneity in detection probability due to population size is significant, then the observed species upon which we base our estimates will tend to have more individuals than species present but not observed and thus likely higher detection probabilities. To explore the effect of abundance on detection and vital rates, one might divide the species into high and low detection groups based on abundance and calculate vital rates for sub-groups (Alpizar-Jara et al. 2004) or incorporate abundance estimates directly as covariates in a relative species abundance analysis.

We believe that camera trapping can serve an important function in biodiversity monitoring programs that include monitoring the terrestrial bird and mammal component of biodiversity. We expect that use of camera traps for species richness estimation of terrestrial birds and mammals will prove most useful in forest, woodland and shrub habitats where visual detection range is limited and most species are cryptic. The methods will also be useful where nocturnal mammals form an important component of the community of interest.

Because many terrestrial bird and mammal species appear to be rare, camera trap studies of species richness need to be conducted over a sufficient length of time to ensure that rare species have a reasonable expectation of detection. We recommend that total effort in a sample exceed 1,000 trapdays.

Camera trap studies of species richness, as in studies of single species (e.g., Karanth et al. 2006) need to be conducted at the landscape scale appropriate to the most wide-ranging members of the community of interest. If trap placements are too close together, the sampling effort may miss species that are in the landscape but outside the trap array. We believe it is better to space traps as widely as is practical to maximize the total area covered by camera traps.

Acknowledgements Our research is a collaborative effort by the Wildlife Conservation Society and the Indonesian Ministry of Forestry's Department for Protection and Conservation of Nature. Our research was funded by the Wildlife Conservation Society, the Save the Tiger Fund, a special project of the National Fish and Wildlife Foundation in partnership with the Exxon Mobil Corporation, the US Fish and Wildlife Service Rhinoceros and Tiger Conservation Fund and Princeton University Alumni Association. We would like to thank J. Ginsberg, M. Rao, L. Krueger, F. Bagley, and J. Seidensticker for their support and advice during this project. We also thank U. Wijayanto and I. Tanjung for assistance in data collection. C. Foley, S. Durant, and J Kingdon kindly provided information on Tanzania mammals. The manuscript benefited greatly from the comments of J. D. Nichols, A. F. O'Connell, K.U. Karanth, and an anonymous reviewer.

References

Alpizar-Jara, R., J. D. Nichols, J. E. Hines, J. R. Sauer, K. P. Pollock, and C. S. Rosenberry. 2004. The relationship between species detection probability and local extinction probability. Oecologia 141:652–660

Balmford, A., P. Crane, A. P. Dobson, R. E. Green, and G. M. Mace. 2005. The 2010 challenge: data availability, information needs and extraterrestrial linsights. Philosophical Transactions of the Royal Society of London, Series B 360:221–228

Boulinier, T., J. D. Nichols, J. R. Sauer, J. E. Hines, and K. P. Pollock. 1998. Estimating species richness: the importance of heterogeneity in species detectability. Ecology 79:1018–10288

Bunge, J. and M. Fitzpatrick. 1993. Estimating the number of species: a review. Journal of American Statistical Association 88:364–373

Burnham, K. P. and D. E. Anderson. 2002. Model selection and multimodal inference: a practical information-theoretic approach, Second edition. Springer, New York

Burnham, K. P. and W. S. Overton. 1978. Estimation of the size of a closed population when capture probabilities vary among animals. Biometrika 65:623–633

Burnham, K. P. and W. S. Overton. 1979. Robust estimation of population size when capture probabilities vary among animals. Ecology 60:927–936

Cam, E., J. D. Nichols, J. R. Sauer, J. E. Hines, and C. H. Flather. 2000. Relative species richness and community completeness: avian communities and urbanization in the mid-Atlantic states. Ecological Applications 10:1196–1210

Cam, E., J. D. Nichols, J. R. Sauer, and J. E. Hines. 2002a. On the estimation of species richness based on the accumulation of previously unrecorded species. Ecography 25:102–108

Cam, E., J. D. Nichols, J. E. Hines, J. R. Sauer, R. Alpizar-Jara, and C. H. Flather. 2002b. Disentangling sampling and ecological explanations underlying species-area relationships. Ecology 83:1118–1130

Chambers, E. G. and G. U. Yule. 1942. Theory and observation in the investigation of accident causation. Journal of Royal Statistical Society Supplementum 7:89–109

Chao, A., S. M. Lee, and S. L. Jeng. 1992. Estimation of population size for capture–recapture data when capture probabilities vary by time and individual animal. Biometrics 48:201–216

Colwell, R. K. and J. A. Coddington. 1994. Estimating terrestrial biodiversity through extrapolation. Philosophical Transactions of the Royal Society of London, Series B 345:101–118

Corbett, A. S., R. A. Fischer, and C. B. Williams. 1943. The relation between the number of species and the number of individuals in a random sample of an animal population. Journal of Animal Ecology 12:42–58

Danielsen, F., N. D. Burgess, and A. Balmford. 2005. Monitoring matters: examining the potential of locally-based approaches. Biodiversity and Conservation 14:2507–2542

Good, I. J. 1953. The population frequencies of species and the estimation of population parameters. Biometrika 40:237–264

Hines, J. E., T. Boulinier, J. D. Nichols, J. R. Sauer, and K. P. Pollock. 1999. COMDYN: software to study the dynamics of animal communities using a capture-recapture approach. Bird Study 46 (suppl.):S209–S217

Karanth, K. U., J. D. Nichols, N. S. Kumar, and J. E. Hines. 2006. Assessing tiger population dynamics using photographic capture–recapture sampling. Ecology 87:2925–2937

Kinnaird, M. F., E. W. Sanderson, T. G. O'Brien, H. T. Wibisono, and G. Woolmer. 2003. Deforestation trends in a tropical landscape and implications for forest mammals. Conservation Biology 17:245–257

Lande, R. 1996. Statistics and partitioning of species diversity, and similarity among multiple communities. Oikos 76:5–13

Lynam, A. J., P. D. Round, and W. Y. Brockelman. 2006. Status of birds and large mammals in Thailand's Dong Phayayen – Khao Yai forest complex. Biodiversity Research and Training Program and Wildlife Conservation Society, Bangkok

MacKenzie, D. I., J. D. Nichols, J. E. Hines, M. G. Knuson, and A. B. Franklin. 2003. Estimating site occupancy, colonization, and local extinction when a species is detected imperfectly. Ecology 84:2200–2207

MacKenzie, D. I., J. D. Nichols, J. A. Royle, K. P. Pollock, L. L. Bailey, and J. E. Hines. 2006. Occupancy estimation and modeling: inferring patterns and dynamics of species occurrence. Academic, New York

Magurran, A. E. 1988. Ecological diversity and its measurement. Princeton University Press, Princeton

Mao, C. X. and R. K. Colwell. 2005. Estimation of species richness: mixture models, the role of rare species, and inferential challenges. Ecology 86:1143–1153

May, R. M. 1988. How many species are there on earth? Science 241:1441–1449

Nichols, J. D. and K. P. Pollock. 1983. Estimating taxonomic diversity extinction rates, and speciation rates from fossil data using capture–recapture models. Paleobiology 64:253–260

Nichols, J. D., R. W. Morris, C. Brownie, and K. P. Pollock. 1986. Sources of variation in extinction rates, turnover, and diversity of marine invertebrate families during the Paleozoic. Paleobiology 12:421–432

Nichols, J. D., T. Boulinear, J. A. Hines, K. P. Pollock, and J. R. Sauer. 1998a. Estimating rates of local species extinction, colonization, and turnover in animal communities. Ecological Applications 8:1213–1225

Nichols, J. D., T. Boulinear, J. A. Hines, K. P. Pollock, and J. R. Sauer. 1998b. Inference methods for spatial variation in species richness and community composition when not all species are detected. Conservation Biology 12:1390–1398

O'Brien, T. G., M. F. Kinnaird, and H. T. Wibisono. 2003. Crouching tigers, hidden prey: Sumatran tiger and prey populations in a tropical forest landscape. Animal Conservation 6:131–139

Otis, D. L., K. P. Burnham, G. C. White, and D. R. Anderson. 1978. Statistical inference from capture data on closed animal populations. Wildlife Monographs 62:1–135

Pollock, K. P. 1982. A capture–recapture design robust to unequality of capture. Journal of Wildlife Management 46:757–760

Pollock, K. P. and M. C. Otto. 1983. Robust estimation of population size in closed animal populations from capture–recapture experiments. Biometrics 52:639–649

Preston, F. W. 1948. The commonness and rarity of species. Ecology 29:254–283

Schluter, D. and R. E. Ricklefs. 1993. Species diversity: an introduction to the problem. Pages 1–10 *in* R. E. Ricklefs and D. Schluter, editors. Species diversity in ecological communities. University of Chicago Press, Chicago

Trolle, M. 2003a. Mammal survey in the Rio Jauaperi region, Rio Negro Basin, the Amazon, Brazil. Mammalia 67:75–83

Trolle, M. 2003b. Mammal survey in the southeastern Pantanal, Brazil. Biodiversity and Conservation 12:823–836

United Nations Environment Programme. 2002. Report on the sixth meeting of the Conference of the Parties to the Convention on Biological Diversity (UNEP/CBD/COP/6/20/Part2) Strategic Plan Decision VI/26

Whittaker, R. H. 1972. Evolution and measurement of species diversity. Taxon 21:213–251

Wibisono, H. T. 2006. Population ecology of Sumatran tigers (*Panthera tigris sumatrae*) and their prey in Bukit Barisan Selatan National Park, Sumatra, Indonesia. Unpubl. MSc. thesis, Univ. Massachusetts, 91 pp

Williams, C. B. 1964. Patterns in the balance of nature. Academic, London

Williams, B. K., J. D. Nichols, and M. J. Conroy. 2002. Analysis and management of animal populations. Academic, San Diego, CA

Zielinski, W. J., R. L. Truex, F. V. Schlexer, L. A. Campbell, and C. Carroll. 2005. Historical and contemporary distributions of carnivores in forests of the Sierra Nevada, California, USA. Journal of Biogeography 32:1385–1407

Chapter 14
Camera Traps in Animal Ecology and Conservation: What's Next?

James D. Nichols, Allan F. O'Connell, and K. Ullas Karanth

14.1 Introduction

As documented in the preceding chapters, the use of camera traps in animal ecology has undergone an appropriate and substantive evolution. This evolution has included the general uses of camera traps and the resulting data, as well as more specific topics such as equipment and statistical inference methods. Collectively, the contributions of this volume should not be viewed as an endpoint summary, but as a milestone along this evolutionary path. The various authors have attempted to briefly summarize that evolution, to describe current methods and uses of camera trap data, and to provide some new methods that we expect to see increased use in the future. In this chapter, we use the preceding chapters to provide brief summaries of the current state of the art and science of camera trap use and then provide speculation and recommendations about changes that we anticipate and hope for in the next decade. In terms of organization, we first focus on the overall uses of camera traps and resulting data, as these uses provide the framework needed to evaluate all further methodological developments. We then discuss equipment and finish with a review of statistical inference methods.

14.2 Uses of Camera Trap Data

Initial uses of camera traps were primarily by nature photographers and natural historians to obtain snapshot observations of secretive animals (Kucera and Barrett, Chap. 2). Currently, scientists and managers employ camera traps to investigate

J.D. Nichols (✉)
U.S. Geological Survey, Patuxent Wildlife Research Center, Laurel, MD 20708, USA
e-mail: jnichols@usgs.gov

A.F. O'Connell
U.S. Geological Survey, Patuxent Wildlife Research Center, Beltsville, MD 20705, USA

K.U. Karanth
Wildlife Conservation Society – India Program, Centre for Wildlife Studies,
26-2, Aga Abbas Ali Road (Apt: 403), Bangalore, Karnataka 560042, India

A.F. O'Connell et al. (eds.), *Camera Traps in Animal Ecology: Methods and Analyses*, DOI 10.1007/978-4-431-99495-4_14, © Springer 2011

animal behavior (Bridges and Noss, Chap. 5) and to estimate state variables and rate parameters that characterize populations and communities of animals (e.g., Karanth et al., Chap. 7; Maffei et al., Chap. 8; Karanth et al., Chap. 9; O'Brien et al., Chap. 13). As noted by O'Brien (Chap. 6), efforts to use camera trap data to draw inferences about quantities such as population size effectively began in the early 1990s, as Karanth (1995) exploited the natural markings of tigers to estimate tiger abundance using formal capture–recapture (CR) methods. As documented in O'Brien's (Chap. 6) review, many of the investigations using camera traps over the last decade have focused on estimation methodology. We view this focus as appropriate and reasonable for an emergent methodology.

We expect methodological development to continue (see below), but we hope that subsequent efforts to estimate parameters of animal populations and communities are better integrated into overall programs of science or management. All too frequently, abundance estimates of animals have been viewed as endpoints, rather than as components of larger programs of inquiry or management. We certainly understand this initial focus on estimation, as it is natural to want to obtain the first real abundance estimates of secretive animals for which such inference has been previously impossible. However, it is difficult to argue for the importance of such estimates, without the context provided by a larger program of science or conservation/management. Because camera trap studies are relatively expensive of funds and effort, we believe that they should be focused on estimates that are useful either for discriminating among competing hypotheses (science) or for informing management decisions (Nichols et al., Chap. 4).

In various parts of the world, camera traps are being used to monitor animal populations. However, just as isolated estimates of abundance have low inherent value, series of such estimates across time and/or space are also of limited utility when viewed alone. Such estimates are most useful when the monitoring program is embedded within a larger program of science or management (Yoccoz et al. 2001; Nichols and Williams 2006). In these cases, it is clear exactly how the resulting estimates are to be used as parts of the larger endeavor (see Chap. 4).

As an example of the use of camera trap data in scientific inquiry, Karanth et al. (2004) tested an a priori hypothesis about a predicted positive relationship between tiger and prey densities. The nature of the hypothesis required estimates of density of tigers and principal prey species at a number of sites across India. Camera traps were used to estimate tiger densities, and distance sampling was used to estimate densities of prey. The approach to spatial monitoring was inherited from the tested hypothesis, as appropriate in the conduct of science. Unfortunately, we believe that this tailoring of sampling design and data needs to an a priori question or set of questions is relatively rare for those using camera traps. Our hope is that the future will bring fewer studies initiated simply to provide estimates of animal abundance or density and more studies designed to provide such estimates for the purpose of discriminating among competing ideas or hypotheses.

As outlined in Chap. 4, the components of a good program of informed management include: (1) objectives, (2) potential management actions, (3) model(s) for predicting system response to management actions and (4) a monitoring program.

In informed management, the monitoring program is needed to produce estimates of state variables such as population size to serve at least three distinct purposes: (1) making state-dependent management decisions, (2) assessing the degree to which objectives are being met, and (3) discriminating among competing hypotheses about how the system responds to management actions (the scientific component of informed management in the face of uncertainty). In the future, we hope to see camera traps integrated into conservation programs designed to bring about increases in animal numbers by such actions as establishment of corridors connecting previously isolated animal populations, and increased law enforcement to protect predators and their prey. As with other areas of conservation and wildlife management, biologists and managers have generally done a relatively poor job of incorporating information from monitoring and scientific investigations into their management decisions. Camera trap investigations have frequently focused on species of high conservation value, and we hope to see these investigations properly integrated into serious programs of management and conservation.

14.3 Camera Trap Equipment and Photographic Data

In Chap. 3, Swann et al. showed that a variety of camera trap designs and manufacturers is currently available, suggesting that systems can be found to meet the needs of almost any scientific study or survey. Nevertheless, prototypes of camera traps based on cell phone platforms are now in development. We expect that such devices will improve system efficiency and reduce size and cost (see Chap. 7). The worlds of mobile phone technology and video imaging have experienced major advances in both software applications and hardware innovations (Greene 2006). A number of these technological advances are certain to influence camera trap systems and how we use these systems in the future. Geo-tagging, the integration of positioning technology with photographic images, will allow camera traps (i.e., using the new cell phone platform mentioned earlier) to have Geographic Positioning System (GPS) capability, thereby allowing practitioners to generate geo-referenced data more efficiently. Research continues in the field of image and video searching that may affect how to evaluate or analyze photographic data. Image recognition software continues to be developed and, for camera traps, such programs can play an important role in identifying individual animals (see below for a further discussion). In keeping with the cell phone platform referred to earlier, incorporating projection systems into cell phone-sized camera traps could allow for remote viewing of traps, tracking animal movement over time and space, again insuring a more efficient and cost-effective sampling process.

Whatever the choice or need with respect to the type of camera trap (and we include videographic devices under this umbrella), miniaturization, digital recording, wireless, networked platforms and even animal-borne video and environmental data systems (AVEDs) are expanding the frontier of remote imaging (Moll et al. 2007). Wireless, networked camera systems are now being used to generate information ranging from presence–absence data, to more complex behavioral observations of breeding birds, to the amount of time animals spend in traps (i.e., in order to limit time in pitfall traps and reduce associated deaths) (Hamilton et al. 2007; Taggart et al. 2007). Related work on image processing and analysis (e.g., image rectification) improve the use of camera traps and other imaging systems. Admittedly, some systems, such as AVEDs, may not permit the population-wide inferences that we have come to expect from camera traps and that we have discussed throughout this volume, but these systems can be used to make advances in other areas such as animal behavior, habitat use, and species interactions, especially for elusive species in harsh, inaccessible environments (see Chap. 5). For example, miniaturized camera systems (Little Leonardo Co., Ltd, Tokyo, Japan) triggered by depth have been used to document group foraging behavior in Adélie *Pygoscelis adeliae* and chinstrap *Pygoscelis antarctica* penguins (Takahashi et al. 2004), feeding behavior in gentoo penguins *Pygoscelis papua* (Takahashi et al. 2008), and predatory behavior in European shags *Phalacrocorax aristotelis* (Watanuki et al. 2008).

14.4 Statistical Inference Methods

As is the case for equipment, statistical methods for drawing inferences from camera trap data are undergoing rapid evolution. Initial uses of camera trap data for statistical inference were focused on abundance of individually identifiable animals (Karanth 1995). Over the past 15 years there has been some development of new models for abundance estimation and a great deal of development in the companion problem of translating abundance estimates into inferences about density. Recent work has also moved from time-specific snapshots of abundance and/or density to dynamics of these state variables over time. The last decade has also brought recognition that camera trap data for animals that are not individually identifiable can be used for inference about two additional state variables, occupancy and species richness. Occupancy models permit inference about the occurrence of a species over space and habitat types, and dynamic models permit inference about changes in distribution over time. Species richness within a specified taxonomic group is a community-level state variable, and its dynamics can also be studied with camera trap data.

14.4.1 Abundance and Density

Closed population CR models are used to estimate abundance from camera trap data for individually identifiable animals (Karanth 1995, Chap. 6). The basic set of models available for use with such data has not changed much since the early 1990s. Exceptions to this general statement are models dealing with heterogeneous capture probabilities across individual animals. Mixture models (Norris and Pollock 1996; Pledger 2000) and a class of parametric models (Dorazio and Royle

2003) offer new and flexible approaches to this difficult estimation problem and have begun to see use with camera trapping data (e.g., Karanth et al. 2006).

In order to translate abundance estimates into estimates of density, it is necessary to draw inferences about the spatial distribution of animals that are exposed to capture efforts. For this purpose, Karanth and Nichols (1998, 2002) initially proposed the use of information on animal movement found in multiple captures of individual animals during a camera trap study. Their use of such data required two steps, was ad hoc, and simply represented the only approach that seemed practical at the time. The weakness of this approach is recognized (e.g., Williams et al. 2002) and has been the basis for arguments in the literature about which movement statistics to use (Soisalo and Cavalcanti 2006, Chap. 6). One of the most exciting new developments in the analysis of camera trap data has been the use of spatially explicit CR models (Efford 2004; Borchers and Efford 2008; Royle and Young 2008; Royle and Dorazio 2008; Royle et al. 2009a, b; Royle and Gardner, Chap. 10). As noted by Royle and Gardner (Chap. 10), this approach deals nicely with two basic problems of the historical two-step approach to density estimation. First, it deals explicitly with heterogeneity in detection probabilities caused by different spatial locations of animals with respect to the camera trap array. Second, it presents a formal, single-step approach to simultaneously estimate abundance and density in a manner that is readily defended. Likelihood-based inferences can be obtained with user-friendly software DENSITY (e.g., Borchers and Efford 2008), and the new software SPACECAP has been recently developed to implement flexible Bayesian approaches (Singh et al. 2010). We believe that spatially explicit CR modeling will eventually replace previous approaches to the analysis of camera trap data and become the method of choice for inference from such data.

Camera trap studies of individually identifiable animals carried out at the same locations across multiple years can make use of CR models for open populations (Karanth et al. 2006, Chap. 9). Such models permit inference about population dynamics and the vital rates (e.g., survival rates) responsible for such dynamics. For large animals that are difficult to capture (e.g., large felids), use of camera trap data for inference about survival should lead to larger sample sizes and stronger inferences than radio-telemetry, the other method used to estimate survival for such animals. There are now several camera trap studies that have been ongoing for several years at the same locations, and we anticipate much greater use of these data to study population dynamics and, hopefully, population responses to management actions. Some camera trap studies have been extended in space to cover multiple subpopulations (K. U. Karanth, unpubl. data). Data from such programs can be used with multistate CR models (e.g., Williams et al. 2002) and will permit inferences about dispersal and connectivity within metapopulation systems. One of the most exciting areas of methodological development will involve spatially explicit CR models for open populations (Chap. 10; Gardner et al. in press). Such models will permit all of the usual inferences obtained from open models, but will additionally permit inference about home range dynamics (e.g., relatively static versus dynamic across years), a topic for which formal inference methods have not been previously available.

We believe that an important area of research in the next decade will be the combination of camera trap data with other kinds of data on individual animals. For example, Soisalo and Cavalcanti (2006) used radio telemetry to obtain information on animal movement and spatial distribution that should be superior to the kinds of data provided by multiple recaptures of individuals in camera traps. As noted, such data are critical to the translation of abundance estimates into inferences about density. Approaches that integrate telemetry data into spatial CR modeling will be an exciting area of research. DNA extracted from scat samples can be used to identify individuals for CR analyses (Lukacs and Burnham 2005; Yoshizaki et al. 2009). K. U. Karanth's use of both camera traps (Karanth and Nichols 1998, 2000, 2002) and DNA (Mondol et al. 2009) to sample tigers on the same study areas leads to the obvious question of how best to combine these types of data to draw inferences about tiger population dynamics.

The methods discussed above for the estimation of abundance and density require detection history data for individual animals. The methods thus presuppose the ability to unambiguously identify individual animals. In some cases this is readily accomplished, but as studies become more extensive in time and space, the use of pattern recognition software will play a larger role in camera trap studies. L. Hiby has been a pioneer in development of such software for use with the difficult case of animal photographs that are not standardized by distance from camera or orientation of animals (e.g., Hiby and Lovell 1990, 2001; Hiby et al. 2009; Kelly 2001). Such software can select from digitized photographic libraries "most probable" animals for matches with new photos, thus greatly facilitating the task of the investigator in matching animals to develop detection histories. We foresee increased need for such software as camera trapping studies are extended in space and time and to species for which individual identification is possible but difficult (e.g., Sarmento et al. 2009).

A final topic in abundance and density estimation in which we expect substantial development involves the general topic of study design (e.g., see Kelly 2008). Karanth and Nichols (2002) addressed various aspects of this issue by discussing desirable features of camera trap study designs. Recognizing that many researchers were limited by camera equipment, Karanth and Nichols (2002) also described designs that would permit inferences based on a limited number of camera traps. More recent work has considered the issue of trap spacing in more detail (e.g., Dillon and Kelly 2007). Royle et al. (2009b) have recently developed recommendations for optimal design of studies that intend to use newly-developed spatial CR models for density estimation with camera trap data. We expect the general topic of study design to co-evolve with inference methods, and camera trap studies of the next decade should benefit from this co-evolutionary work.

14.4.2 Occupancy

The development of occupancy models for inferences about whether or not a local area is occupied by a species (MacKenzie et al. 2006) provides a potential use for camera trap data on animals for which individuals do not have natural marks that permit individual identification (e.g., Chap. 6 and O'Connell and Bailey, Chap. 11). Either spatial (multiple camera traps per sample unit) or temporal (multiple trap-nights of a single camera trap in each sample unit) replication provides the sort of species detection versus non-detection data needed to estimate both detection probability and occupancy. Recent multistate occupancy models permit characterization of occupied locations in terms of reproductive output, disease presence, and relative abundance (Nichols et al. 2007; MacKenzie et al. 2009), permitting more detailed inferences than simply presence of the species or not. We anticipate greater use of camera trap data in conjunction with occupancy modeling for investigating questions about animal range and habitat use (MacKenzie et al. 2005, Chaps. 6 and 11).

Some occupancy studies use camera traps together with other remote sampling devices at each sample unit (O'Connell et al. 2006). Such studies permit inference about the relative detection probabilities associated with the different devices, thus providing evidence about the effectiveness of camera traps relative to the other sampling approaches (O'Connell et al. 2006; Nichols et al. 2008). These multiple device studies also permit inference about species occupancy at two spatial scales, the scale of the sample units, and the scale of the actual location of sampling devices within the sample units, with occupancy at the latter scale potentially varying across sampling occasions (Nichols et al. 2008).

Conducting camera trap studies at multiple locations across time (e.g., for multiple years) permits inferences about the dynamics of occupancy processes (MacKenzie et al. 2006). Specifically, a robust design approach (geographic and temporal replication within each season/year for multiple seasons/years) permits inferences about probabilities of local extinction and colonization across seasons/years. The state variable of such studies can be viewed as the proportion of area or sample units occupied by the focal species. Such studies of occupancy dynamics offer the potential to draw inferences about the efficacy of management actions applied to different spatial units or years.

Under some sampling circumstances, species-level detection data can be used to estimate abundance (Royle and Nichols 2003, Chap. 6). This ability arises from the relationship between the probability of detecting a species at a sample unit and the number of individuals of that species in the sample unit. This approach can be used to estimate abundance using camera trap data and, if assumptions are reasonably met, offers some potential for inference about abundance for species that cannot be individually identified from photographs.

In summary, the use of occupancy modeling with camera trap detection data at the species level permits inferences about animal range and habitat use even for species for which individuals cannot be identified from photographs. These methods can be extended to study the processes underlying occupancy dynamics. In some sampling situations, occupancy modeling can even be used to draw inferences about the distribution of animal abundance across the landscape. We anticipate increased use of occupancy models with camera trap data for species that cannot be individually identified from photographs. We also expect accompanying work on camera trap deployment and other design issues for studies that are based on the occupancy modeling framework.

14.4.3 Species Richness

Another use of camera trap data that appears to hold great potential and can be used even for species that are not individually identifiable is the estimation of species richness (Kéry, Chap. 12). Inference is based on species detection–nondetection data from camera traps for multiple species within some group of interest. Kéry (Chap. 12) describes two general approaches that use camera trap data for inference about species richness. The first approach substitutes species identity for individual identity and utilizes closed population CR modeling for community level inferences (e.g., Burnham and Overton 1979, Williams et al. 2002). The second approach simply extends occupancy modeling to multiple species (Dorazio and Royle 2005, MacKenzie et al. 2006). Both approaches are reasonable, and the occupancy based approach permits use of species-level covariates, which will make it the clear choice in some situations.

Deployment of camera traps at one or more locations with replicate (temporal or spatial) sampling within seasons/years and extending across multiple seasons/years permits inference about changes in species richness over time. Such changes are brought about by local rates of species extinction and colonization, and these vital rates can be estimated as well using models patterned after those used for CR robust design analyses (e.g., Nichols et al. 1998). Occupancy modeling similarly permits inferences about species-specific rates of local colonization and extinction (MacKenzie et al. 2006) for the members of a community. Occupancy modeling (e.g., MacKenzie et al. 2009) also permits inferences about the effects of presence of one species at a location on the rates of local extinction or colonization for another species at that location, thus providing mechanistic modeling of possible competitive interactions that may be determinants of community dynamics.

As with the occupancy state variable, we expect to see increased use of camera trap data for estimation of species richness and community dynamics. For example, we could envision simultaneously estimating the richness of predator and prey communities for the purpose of drawing inferences about the relationship between the two quantities. Similarly, we can envisage assessment of the effectiveness of conservation actions using species richness and related metrics. For example, we might expect establishment of corridors to produce increases in colonization rates and species richness. Camera traps deployed across the landscape and operated before and after corridor establishment would yield data that could be used to develop relevant inferences. As with the state variables of abundance, density and occupancy, we expect new work on the design of camera trap studies focused on a variety of animal community attributes.

14.5 Conclusions

The contributions of this volume have documented the evolution of camera trap use from photography of secretive animals to investigations of animal behavior to sophisticated analyses of population and community dynamics. As rapid and impressive as this evolution has been, we expect it to continue at a similar pace into the next decade. We expect, and hope for, a broadening of the emphasis of camera trap studies from primarily methodological development to investigation of serious scientific and conservation questions. Clearly, data acquisition capabilities of camera traps and other remote photographic hardware used in ecological field studies have benefited from technological advances in electrical and computer engineering. In addition, we anticipate further development of statistical inference methods used to study state variables ranging from abundance to occupancy to species richness. This methodological development will likely be accompanied by research on design of camera trap studies that are focused on one or more of these state variables. Our basic conclusion is that a book written on scientific and conservation uses of camera trapping 10 years from now will look substantially different than the present volume. We welcome such an outcome and hope that the present volume will contribute to the changes that we anticipate.

References

Borchers, D. L. and M. G. Efford. 2008. Spatially explicit maximum likelihood methods for capture–recapture studies. Biometrics 64:377–385

Burnham, K. P. and W. S. Overton. 1979. Robust estimation of population size when capture probabilities vary among animals. Ecology 60:927–936

Dillon, A. G. and M. J. Kelly. 2007. Ocelot activity, trap success, and density in Belize: the impact of trap spacing and animal movement on density estimates. Oryx 41:469–477

Dorazio, R. M. and J. A. Royle. 2003. Mixture models for estimating the size of a closed population when capture rates vary among individuals. Biometrics 59:351–364

Dorazio, R. M. and J. A. Royle. 2005. Estimating size and composition of biological communities by modeling the occurrence of species. Journal of the American Statistical Association 100:389–398

Efford, M. G. 2004. Density estimation in live-trapping studies. Oikos 106:598–610

Gardner, B., J. Reppucci, M. Lucherini, and J. A. Royle. 2010. Spatially-explicit inference for open populations: estimating demographic parameters from camera-trap studies. Ecology (in press)

Greene, K. 2006. Technology review. Published by MIT. (http://www.technologyreview.com/infotech/17937/). (Accessed October 2009)

Hamilton, M. P., E. A. Graham, P. W. Rundel, M. F. Allen, W. Kaiser, M. H. Hansen, and D. L. Estirn. 2007. New approaches in embedded networked sensing for terrestrial ecological observatories. Environmental Engineering Science 24:192–204

Hiby, L. and P. Lovell. 1990. Computer aided matching of natural markings: a prototype system for grey seals. Reports of the International Whaling Commission, Special Issue 12:57–61

Hiby, L. and P. Lovell. 2001. An automated system for matching the callosity patterns in aerial photographs of southern right whales. Journal of Cetacean Research and Management Special Issue 2

Hiby, L., P. Lovell, N. Patil, N. S. Kumar, A. M. Gopalaswamy, and K. U. Karanth. 2009. A tiger cannot change its stripes: using a three-dimensional model to match images of living tigers and tiger skins. Biology Letters 5:383–386

Karanth, K. U. 1995. Estimating tiger *Panthera tigris* populations from camera-trap data using capture-recapture models. Biological Conservation 71:333–338

Karanth, K. U. and J. D. Nichols. 1998. Estimation of tiger densities in India using photographic captures and recaptures. Ecology 79:2852–2862

Karanth, K. U. and J. D. Nichols. 2000. Ecological status and conservation of tigers in India. Final Technical Report to the Division of International Conservation, U.S. Fish and Wildlife Service, Washington, D.C., and Wildlife Conservation Society, New York. Centre for Wildlife Studies, Bangalore, India. 124 pp

Karanth, K. U. and J. D. Nichols, editors. 2002. Monitoring tigers and their prey. A manual for wildlife managers, researchers, and conservationists. Centre for Wildlife Studies, Bangalore, India. 193 pp

Karanth, K. U., J. D. Nichols, N. S. Kumar, W. A. Link, and J. E. Hines. 2004. Tigers and their prey: predicting carnivore densities from prey abundance. Proceedings of the National Academy of Sciences USA 101:4854–4858

Karanth, K. U., J. D. Nichols, N. S. Kumar, and J. E. Hines. 2006. Assessing tiger population dynamics using photographic capture–recapture sampling. Ecology 87:2925–2937

Kelly, M. J. 2001. Computer-aided photograph matching in studies employing individual identification: an example from Serengeti cheetahs. Journal of Mammalogy 82:440–449

Kelly, M. J. 2008. Design, evaluate, refine: camera trap studies for elusive species. Animal Conservation 11:182–184

Lukacs, P. M. and K. P. Burnham. 2005. Estimating population size from DNA-based closed capture-recapture data incorporating genotyping error. Journal of Wildlife Management 69:396–403

MacKenzie, D. I., J. D. Nichols, N. Sutton, K. Kawanishi, and L. L. Bailey. 2005. Suggestions for dealing with detection probability in population studies of rare species. Ecology 86:1101–1113

MacKenzie, D. I., J. D. Nichols, J. A. Royle, K. H. Pollock, L. A. Bailey, and J. E. Hines. 2006. Occupancy modeling and estimation. Academic, San Diego, CA. 324 pp

MacKenzie, D. I., J. D. Nichols, M. E. Seamans, and R. J. Gutierrez. 2009. Dynamic models for problems of species occurrence with multiple states. Ecology 90:823–835

Moll, R. J., J. J. Millspaugh, J. Beringer, J. Sartwell, and Z. He. 2007. A new 'view' of ecology and conservation through animal-borne video systems. Trends in Ecology and Evolution 22:660–668

Mondol, S., K. U. Karanth, N. S. Kumar, A. M. Gopalaswamy, A. Andheria, and U. Ramakrishnan. 2009. Evaluation of non-invasive genetic sampling methods for estimating tiger population size. Biological Conservation 142:2350–2360

Nichols, J. D. and B. K. Williams. 2006. Monitoring for conservation. Trends in Ecology and Evolution 21:668–673

Nichols, J. D., T. Boulinier, J. E. Hines, K. H. Pollock, and J. R. Sauer. 1998. Estimating rates of local extinction, colonization and turnover in animal communities. Ecological Applications 8:1213–1225

Nichols, J. D., J. E. Hines, D. I. MacKenzie, M. E. Seamans, and R. J. Gutierrez. 2007. Occupancy estimation with multiple states and state uncertainty. Ecology 88:1395–1400

Nichols, J. D., L. L. Bailey, A. F. O'Connell, Jr., N. W. Talancy, E. H. Campbell, E. H. C. Grant, A. T. Gilbert, E. M. Annand, T. P. Husband, and J. E. Hines. 2008. Multi-scale occupancy estimation and modeling using multiple detection methods. Journal of Applied Ecology 45:1321–1329

Norris, J. L. and K. H. Pollock. 1996. Nonparametric MLE under two closed capture–recapture models with heterogeneity. Biometrics 52:639–649

O'Connell, A. F. Jr., N. W. Talancy, L. L. Bailey, J. R. Sauer, R. Cook, and A. T. Gilbert. 2006. Estimating site occupancy and detection probability parameters for meso- and large mammals in a coastal ecosystem. Journal of Wildlife Management 70:1625–1633

Pledger, S. 2000. Unified maximum likelihood estimates for closed capture–recapture models using mixtures. Biometrics 56:434–442

Royle, J. A. and R. M. Dorazio. 2008. Hierarchical modeling and inference in ecology: the analysis of data from populations and communities. Academic, San Diego, CA

Royle, J. A. and J. D. Nichols. 2003. Estimating abundance from repeated presence absence data or point counts. Ecology 84:777–790

Royle, J. A. and K. Young. 2008. A hierarchical model for spatial capture–recapture data. Ecology 89:2281–2289

Royle, J. A., J. D. Nichols, K. U. Karanth, and A. M. Gopalaswamy. 2009a. A hierarchical model for estimating density in camera trap studies. Journal of Applied Ecology 46:118–127

Royle, J. A., K. U. Karanth, A. M. Gopalaswamy, and N. S. Kumar. 2009b. Bayesian inference in camera trapping studies for a class of spatial capture–recapture models. Ecology 90:3233–3244

Sarmento, P., J. Cruz, C. Eira, and C. Fonseca. 2009. Evaluation of camera trapping for estimating red fox abundance. Journal of Wildlife Management 73:1207–1213

Singh, P., A. M. Gopalaswamy, J. A. Royle, N. S. Kumar, and K. U. Karanth. 2010. SPACECAP: a program to estimate animal abundance and density using Bayesian spatially-explicit capture-recapture models. Wildlife Conservation Society-India Program, Centre for Wildlife Studies, Bangalore, India. Version 1.0

Soisalo, M. K. and S. M. C. Cavalcanti. 2006 Estimating the density of a jaguar population in the Brazilian Pantanal using camera-traps and capture–recapture sampling in combination with GPS radio-telemetry. Biological Conservation 129:487–496

Taggart, M., E. Graham, M. Hamilton, S. Ahmadian, M. Rahimi, J. Sharon, C. Hicks, and J. King. 2007. Deployment of wireless, networked camera systems and sensors for observation of avian and reptile behavior. Page 19 *in* R. W. Kays and M. Wikelski, editors. Review of National Science Foundation sponsored animal tracking and physiological monitoring workshop. Report published online at http://www.movebank.org/assets/ATPM_whitepaper.pdf

Takahashi, A., K. Sato, Y. Naito, M. J. Dunn, P. N. Trathan, and J. P. Croxall. 2004. Penguin-mounted cameras glimpse underwater group behaviour. Proceedings of the Royal Society London B Biological Sciences 271:S281–S282

Takahashi, A., N. Kokubun, Y. Mori, and H. Shin. 2008. Krill-feeding behaviour of gentoo penguins as shown by animal-borne camera loggers. Polar Biology 31:1291–1294

Watanuki, Y., F. Daunt, A. Takahashi, M. Newell, S. Wanless, K. Sato, and N. Miyazaki. 2008. Microhabitat use and prey capture of a bottom-feeding top predator, the European shag, shown by camera loggers. Marine Ecology Progress Series 356:283–293

Williams, B. K., J. D. Nichols, and M. J. Conroy. 2002. Analysis and management of animal populations. Academic, San Diego. 817 pp

Yoccoz, N. G., J. D. Nichols, and T. Boulinier. 2001. Monitoring of biological diversity in space and time. Trends in Ecology and Evolution 16:446–453

Yoshizaki, J., K. H. Pollock, C. Brownie, and R. A. Webster. 2009. Modeling misidentification errors in capture–recapture studies using photographic identification of evolving marks. Ecology 90:3–9

Index

Made in the USA
Monee, IL
14 September 2026

63193269R00160